PAS PROTEINS: REGULATORS AND SENSORS OF DEVELOPMENT AND PHYSIOLOGY

PAS PROTEINS:
Regulators and Sensors of Development and Physiology

edited by

Stephen T. Crews
The University of North Carolina at Chapel Hill
Chapel Hill, NC

KLUWER ACADEMIC PUBLISHERS
Boston / New York / Dordrecht / London

Distributors for North, Central and South America:
Kluwer Academic Publishers
101 Philip Drive
Assinippi Park
Norwell, Massachusetts 02061 USA
Telephone (781) 871-6600
Fax (781) 681-9045
E-Mail: kluwer@wkap.com

Distributors for all other countries:
Kluwer Academic Publishers Group
Post Office Box 322
3300 AH Dordrecht, THE NETHERLANDS
Telephone 31 786 576 000
Fax 31 786 576 254
E-Mail: services@wkap.nl

Electronic Services <http://www.wkap.nl>

Library of Congress Cataloging-in-Publication Data

PAS PROTEINS: Regulators and Sensors of Development and Physiology / edited
by Stephen T. Crews.
    p. cm.
  Includes bibliographical references and index.
  ISBN 1-4020-7586-3
    1. Transcription factors. 2. Cellular signal transduction. I. Crews, Stephen T.
QP552.T68P375 2003
572'.69--dc22   2003060110

The Publisher offers discounts on this book for course use and bulk purchases.
For further information, send email to <Laura.Walsh@wkap.com>.

Contents

Contributing Authors

Christopher A. Bradfield

The McArdle Laboratory for Cancer Research, University of Wisconsin Medical School, Madison, WI 53706

Michael K. Chan

Department of Biochemistry, The Ohio State University, The Ohio State University, Columbus, OH 43210

Stephen T. Crews

Department of Biochemistry and Biophysics, Program in Molecular Biology and Biotechnology, The University of North Carolina at Chapel Hill, Chapel Hill, NC 27599

Chen-Ming Fan

Department of Embryology, Carnegie Institution of Washington, Baltimore, MD 21211

Bing Hao

Department of Biochemistry, The Ohio State University, The Ohio State University, Columbus, OH 43210

Eric B. Harstad

The McArdle Laboratory for Cancer Research, University of Wisconsin Medical School, Madison, WI 53706

John B. Hogenesch

The Genomics Institute of the Novartis Research Foundation, San Diego, CA, 92121

Mark S. Johnson

Department of Microbiology and Molecular Genetics, School of Medicine, Loma Linda University, Loma Linda, CA 92350

Steve A. Kay

Department of Cell Biology, The Scripps Research Institute, San Diego, CA, 92037; The Genomics Institute of the Novartis Research Foundation, San Diego, CA, 92121

Denise J. Montell

Department of Biological Chemistry, The Johns Hopkins University School of Medicine, Baltimore, MD 21205

Jo Anne Powell-Coffman

Department of Zoology and Genetics, Iowa State University, Ames, IA 50011

Gregg L. Semenza

McKusick-Nathans Institute of Genetic Medicine, The Johns Hopkins University School of Medicine, Baltimore, MD 21287

Barry L. Taylor

Department of Microbiology and Molecular Genetics, School of Medicine, Loma Linda University, Loma Linda, CA 92350

Kylie J. Watts

Department of Microbiology and Molecular Genetics, School of Medicine, Loma Linda University, Loma Linda, CA 92350

Thomas G. Wilson

Department of Entomology, The Ohio State University, Columbus, OH 43210

Guang Yao

The McArdle Laboratory for Cancer Research, University of Wisconsin Medical School, Madison, WI 53706

Xuejun Zhong

Department of Biochemistry, The Ohio State University, The Ohio State University, Columbus, OH 43210

Preface

see that ye be not troubled: for all these things must come to PAS, but the end is not yet. from: Matthew 24:6-8.

Thomas Hunt Morgan and Calvin Bridges, two of the pioneers of modern genetic research, reported in 1923 the discovery of a large number of *Drosophila* mutants (1). Included in this collection was the *spineless* mutant, which likely represents the first identified PAS gene. In the subsequent 80 years, hundreds of PAS genes have been identified. Their investigators have rarely gone unrewarded, since these genes participate in a wide variety of biological and biochemical processes. While most animal PAS proteins are transcription factors, prokaryotic and plant PAS proteins are biochemically diverse and frequently act as environmental sensors. The subject of this volume is a comprehensive examination of the biochemistry and biology of all classes of known PAS proteins, with the intention that readers will find insight into their own work and ideas from study of the multitude of PAS protein functions.

The PAS domain itself was first recognized by sequence comparison of two proteins, *Drosophila* Single-minded (Sim) and Period (Per) (2). The sequence identity between Sim and Per was a long (>250 aa), poorly-conserved stretch containing two short repeats, PAS-A and PAS-B, each 51 aa in length. With the discovery of a third PAS protein, the human Aryl hydrocarbon nuclear receptor translocator (Arnt), the PAS domain got its name (Per-Arnt-Sim) (3). Additional bHLH-PAS proteins continued to be isolated from animals revealing their high degree of evolutionary conservation and interesting developmental and physiological functions.

Biochemical experiments initially with Per (4) and soon after the aryl hydrocarbon receptor (Ahr) (5-7) indicated that the PAS domain could mediate protein-protein interactions both between other PAS proteins and non-PAS proteins. One observation drove much of the thinking about bHLH-PAS proteins: the function of the Ahr bHLH-PAS protein was dependent on binding to ligands, such as dioxin, suggesting that other PAS proteins might also be dependent on small molecule signaling for function. While this analogy may not be generalizable, the function of some bHLH-PAS proteins can respond to physiological conditions, such as oxygen and carbon monoxide concentrations, and the modulation of bHLH-PAS protein function by signaling molecules remains a subject of interest.

Bioinformatic and experimental approaches provided an important advance to the field when it was discovered that a large number of plant and prokaryotic proteins shared sequence homology to the PAS domain (8-10). Many of these proteins are physiological sensors that combine a PAS domain with a variety of biochemical activities and ligand-binding properties. In some cases, the PAS domain includes ligand-binding residues, and in other cases, ligand binding occurs outside of the PAS domain. Just as important, comparative and structural analyses brought a new definition of the PAS domain. The structures of multiple PAS proteins revealed a common domain >100 aa in length that encompassed the previously noted PAS repeat. The newly-defined, shorter PAS domain is an authentic protein domain, and bHLH-PAS proteins possess two PAS domains (designated PAS-1 and PAS-2) separated by a spacer (11). Some chapters use the newer designation and others, for historical consistency, refer to the PAS-A and PAS-B repeats. The following chapters describe in great detail the diversity of PAS protein function. Besides the wealth of information regarding these genes, proteins, and their biochemistry and biology, the reader may find it useful to consider several issues. What do these proteins tell us about the evolution of PAS genes and the cell types, genetic pathways, and biochemical processes they govern? Are there common features to these diverse processes? What is special about PAS domains, and their arrangements and functions in the various PAS proteins? Are there also unique features to the PAS domain that impart exceptional properties to the proteins that possess it?

REFERENCES

1. Bridges, C., and T. H. Morgan. 1923. The third-chromosome group of mutant characters of Drosophila melanogaster. *Carnegie Inst. Wash. Publ.*:1-251.

2. Crews, S. T., J. B. Thomas, and C. S. Goodman. 1988. The Drosophila *single-minded* gene encodes a nuclear protein with sequence similarity to the *per* gene product. *Cell* 52:143-151.

3. Nambu, J. R., J. L. Lewis, K. A. Wharton, and S. T. Crews. 1991. The Drosophila *single-minded* gene encodes a helix-loop-helix protein which acts as a master regulator of CNS midline development. *Cell* 67:1157-1167.

4. Huang, Z. J., I. Edery, and M. Rosbash. 1993. PAS is a dimerization domain common to *Drosophila* period and several transcription factors. *Nature* 364:259-262.

5. Whitelaw, M., I. Pongratz, A. Wilhelmsson, J. A. Gustafsson, and L. Poellinger. 1993. Ligand-dependent recruitment of the Arnt coregulator determines DNA recognition by the dioxin receptor. *Mol. Cell. Biol.* 13:2504-14.

6. Reyes, H., S. Reisz-Porszasz, and O. Hankinson. 1992. Identification of the Ah receptor nuclear translocator protein (Arnt) as a component of the DNA binding form of the Ah receptor. *Science* 256:1193-5.

7. Chan, W. K., R. Chu, S. Jain, J. K. Reddy, and C. A. Bradfield. 1994. Baculovirus Expression of the Ah Receptor and Ah Receptor Nuclear Translocater. *J. Biol. Chem.* 269:26464-26471.

8. Lagarias, D. M., S.-H. Wu, and J. C. Lagarias. 1995. Atypical phytochrome gene structure in the green alga Mesotaenium caldariorum. *Plant Molec. Biol.* 29:1127-1142.

9. Ponting, C. P., and L. Aravind. 1997. PAS: a multifunctional domain family comes to light. *Curr. Biol.* 7:674-677.

10. Zhulin, I. B., B. L. Taylor, and R. Dixon. 1997. PAS domain S-boxes in Archae, Bacteria and sensors for oxygen and redox. *Trends Bioch. Sci.* 22:331-333.

11. Taylor, B. L., and I. B. Zhulin. 1999. PAS domains: internal sensors of oxygen, redox potential, and light. *Microbiol. Mol. Biol. Rev.* 63:479-506.

Stephen Crews

The University of North Carolina at Chapel Hill

Chapter 1

STRUCTURE OF THE PAS FOLD AND SIGNAL TRANSDUCTION MECHANISMS

Xuejun Zhong, Bing Hao & Michael K. Chan
The Ohio State University, Columbus, OH 43210

Despite their diverse origin and function, the structures of PAS domains from four distinct classes of PAS proteins support the notion of a common fold. The structures that help to define this fold are (i) the *Ectothiorhodospira halophila* photoactive yellow protein (PYP) (1-3), a self-contained, bacterial blue-light receptor, with a covalently attached 4-hydroxycinnamoyl chromophore, (ii) the oxygen-sensing heme domains of *Bradyrhizobium japonicum* FixL (BjFixLH) (4-6) and *Rhizobium meliloti* FixL (RmFixLH) (7), (iii) the FMN-containing LOV2 domain from the plant blue-light receptor phy3 (8), and (iv) the N-terminus domain of the HERG (human *eag*-related gene) voltage-dependent potassium channel (9). This chapter describes the structural features that comprise a PAS domain, and reviews the current understanding into the mechanism of signal transduction utilized by each of the four structurally characterized classes of PAS proteins.

1.   THE PAS DOMAIN FOLD

PAS domains adopt a glove like-fold dominated by a central antiparallel β sheet (Figure 1). Two schemes have been developed for describing their structural features. One is based on the original structure of PYP and consists of an (i) N-terminal cap, (ii) a PAS core, made up of the two N-terminal β-sheets and three helices, (iii) a helical connector, composed of the long α helix that diagonally crosses the β-sheets, and (iv) a β-scaffold, composed of the three long β-strands that comprise the second half of the central β-sheet (Figure 1A) (3). Since the fold of the N-terminal CAP is not conserved between the structures of PYP and RmFixLH, for the purposes of this review, this region is not included as part of the PAS domain proper.

The second scheme (4), that omits this N-terminal CAP region, is based on the conserved sequential arrangement of the secondary structural elements in the PAS fold (Figure 2-1A). Here, the α-helices and β-sheets are designated sequentially by alphabet, A_β, B_β, C_α, D_α E_α, F_α, G_β, H_β, I_β, while the loops are named according to the secondary structures that flank them. For example, the region between the F_α-helix and G_β-strand is known as the FG loop. This system facilitates the discussion of the conserved residues found in various PAS domains. Using this nomenclature the PAS domain can be described as a left-handed glove with the G_β and H_β strands forming the thumb, the C_α, D_α, E_α, and F_α helices, the palm, and the H_β, I_β, A_β, and B_β strands, the fingers.

The structures of each class of PAS domain are provided in Figure 1. An overlap of these structures is depicted in Figure 2A. This superposition illustrates the general conservation of the PAS fold with the main differences being the location of the F helix. Like the palm of a hand, this helix shifts to accommodate various cofactors within the domain. In BjFixLH and RmFixLH, the F-helix (palm) adopts a position that sandwiches the heme between itself and the GH strands (the thumb). In the phy3 LOV2 domain, this helix shifts away from the C_α, D_α, E_α helices, to form a channel for the FMN cofactor.

One possible mechanism for regulation of the F-helix position is suggested by the sequence and structural alignments of the five structurally characterized PAS domains (Figure 2). In all cases, the regions comprising the EF loop and F-helix have the same number of residues (Figure 2B) and start at the same point and end at the same point. What is different is how this region is folded. This appears to be linked to length of the F-helix (Figure 2A). BjFixLH and RmFixLH have the greatest number of residues wound into the F-helix (longest) and therefore have the shortest EF loop. The small size of the EF loop possibly restricts the movement of the F helix thus leaving a large pocket suitable for the heme cofactor. Conversely, in LOV2 and HERG, fewer residues are wound into the F-helix, and this results in a longer EF loop enabling the F-helix to move in and fill the cofactor-binding pocket. This ability to regulate the size of the central pocket by this simple mechanism could help to explain how PAS domains are able to bind so many different types of cofactors and may account for the ubiquitous nature of PAS domains over a wide variety of kingdoms and sensory systems.

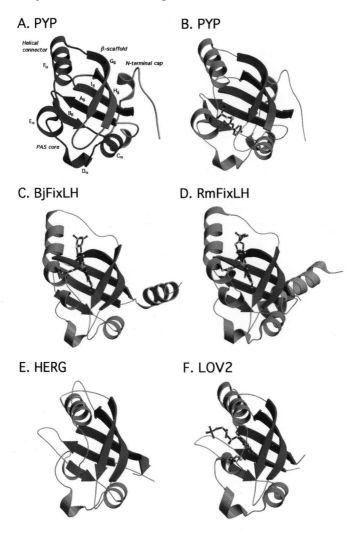

Figure 1. PAS domain structures: (A) E. halophila PYP colored according to the original PYP nomenclature (3); N-terminal cap, tan; PAS core, violet, helical connector, green; and β-scaffold, red. The remaining structures (B-D) are colored by secondary structure elements: α-helix in cyan, β-strand in red, and random coil in green. The regions that are not part of the PAS domain are colored in tan. (B) Structure of PYP (3); (C) the heme-containing PAS domain of B. japonicum FixL (4, 5), (D) one subunit from structure of the R. meliloti FixL heme domain (7), (E) the PAS domain from the human HERG protein (9), (F) the LOV2 domain from the phototropin module of *Adiantum* phy3 (8).

A.

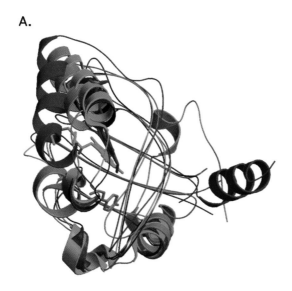

B.

PAS core

		A$_\beta$	B$_\beta$	C$_\alpha$	D$_\alpha$	E$_\alpha$
Bj FixL	155-	A M I V I **D G** - - - H G I	I Q L F **S** T A A E R L F G W S E L E A	I G Q N - - V N	I L M P E
Rm FixL	149-	A T V V **S A T** - - - D G T	I V S F N A A A V R Q F G Y A E E E V	I G Q N - - L R	I L M P E
Eh PYP	29-	G A I Q L **D G** - - - D G N	I L Q Y N A A E G D I T G R D P K Q V	I G K N - F F	K D V A P C
Hs HERG	28-	K F **I I A N A R** V E N **C A V** I Y	C N D G F C E L C G Y S R A E V **M Q R P**	C T C D F	L H G P
Ad LOV2	930-	S F **V I T D P R** L P D N P I	I F A **S** D R F L E L T **E** Y T R E E V L G N N - -	C R F	L Q G R

Helical connector | Beta-scaffold

		F$_\alpha$		G$_\beta$
Bj FixL	195-	P D R S R H D S Y **I** S R Y R T T S D P H I	I G I G R I **V** T G K R R **D G** T
Rm FixL	189-	P Y R H E H D G Y **L** Q R Y M A T G E K R I	I G I D R V **V** S G Q R K **D G** S
Eh PYP	70-	T D S P E F Y G K **F** K E G V A S G N - - -	L N T M F E **Y** T F D - Y Q M T
Hs HERG	73-	C T Q R R A A A Q I A Q A L L G A E - - - -	E R K V E I A **F** Y R K **D G** S
Ad LOV2	973-	G T D R K A V Q L **I** R D A V K E Q R - - - -	D V T V Q **V** L N Y T K G **G R**

Beta-scaffold

		H$_\beta$	I$_\beta$
Bj FixL	231-	T F **P** M H L S I G E M Q S - - **G G** E P Y **F** T G F V R D L
Rm FixL	225-	T F **P** M K L A V G E M R S - - **G G** E R F **F** T G F I R D L
Eh PYP	102-	**P** T K V K **V H** M K K A L S - - **G** - - D S Y W V F V K R V
Hs HERG	105-	C F L C L **V** D V V **P** V K N E D **G** A V I M **F** I L N F E V V
Ad LOV2	1005-	A F W N L **F H L** Q V M R D E N **G** D V Q Y **F** I G V Q Q E M

Figure 2. Structure and structure-based sequence alignment of PAS domains. (A) Overlap of PAS domain regions of PYP (yellow), HERG (blue), BjFixLH (red), and LOV2 (green) (B) Structure-based sequence alignment of BjFixL, RmFixL, PYP, HERG and LOV2 PAS domains. Helices are represented by rectangular boxes, strands by arrows and loops by lines. The striped region at the beginning of F-helix varies depending on the length of the helix. The amino acids labeled in red are active site residues that interact with cofactors. Conserved hydrophobic residues, Gly/Pro residues, and conserved residues whose side chains form hydrogen bonds to main chain atoms are highlighted in green, yellow and blue, respectively.

One other element that varies in PAS domains is the E-helix. In LOV2 and HERG, the E helix is a 3_{10} helix turn. In FixL, it is an α-helix, In PYP, it is a π-helix plus an extremely short β-strand preceding the π-helix. While these differences lead to only small changes in the cofactor-binding pocket, they still may play a role. In the case of PYP, additional β-strand allows the E-helix to shift outward and opens up the pocket for PYP cofactor binding.

2. STRUCTURE-BASED SEQUENCE ALIGNMENT OF PAS DOMAINS

While PAS domains can be aligned solely based on sequence, the quality of this fit can be improved by taking into account their structural similarities. One advantage of such structure-based alignment (Figure 2B) is that it assists in the identification of the conserved residues that stabilize the PAS fold.

Conserved Gly and Pro residues. In the five structures of PAS domains, several conserved Gly/Pro residues are found in the loops between secondary structural elements (Figure 2B yellow colored). For the several conserved Gly residues, this is presumably due to its greater flexibility required to form kinks and turns. In some cases, the ability of Gly to adopt the left-handed backbone conformation may enable the formation of critical hydrogen-bonding interactions. Conversely, the conserved proline at the end of the E_{α}-helix presumably helps to terminate this helix due to its rigidity. Proline also lacks main chain amide hydrogen, and thus cannot form a hydrogen bond to the carbonyl of the (n-3) residue. This hydrogen-bonding interaction is critical for stabilizing the continuation of the α-helical chain.

Conserved residues whose side chains form hydrogen bonds to main chain atoms of other residues. The side chains of residues Asp34, Gln43, and Gln61 were observed to hydrogen bond to the main-chain amide nitrogens of neighboring residues in the structure of PYP. These interactions were suggested to be critical for stabilizing specific turns associated with the overall fold (3). While there are exceptions, the high conservation of these residues (Figure 2B, highlighted in blue) for the five structurally characterized PAS domains, appears to support this assertion.

Conserved hydrophobic residues. The structure-based sequence alignment reveals several highly conserved hydrophobic residues (Figure 2B, highlighted in green). These residues are usually buried, and presumably help to stabilize the PAS fold by forming the hydrophobic core. (3).

3.   PYP - STRUCTURE AND MECHANISM

The photoactive yellow protein (PYP) is a blue-light photoreceptor (10-12) found in the phototrophic bacterium *Ectothiorhodospira halophila* that is proposed to regulate negative phototaxis (13). The light response is mediated by a 4-hydroxycinnamoyl chromophore whose light-induced excited state structures differ from that of the ground state (1, 2). This chromophore is covalently attached to the protein via a thioester linkage to the side chain of Cys69 that places it within the palm of the PAS fold within the EF loop. Conformational changes of the cofactor induce localized changes within the protein that are presumably associated with signal propagation. Currently, the downstream partners associated with PYP have not been identified.

Light absorption by the 4-hydroxycinnamoyl chromophore of PYP initiates a self-contained light cycle (14-17). After photon absorption, the yellow-colored PYP ground state (P) converts rapidly ($<< 10$ ns) to a red-shifted intermediate (I_1), and then quickly to a bleached, long-lived intermediate (I_2), which is presumed to be the signaling state. The I_2 intermediate then reverts back to the initial ground state via protein-mediated thermal relaxation.

The ground state structure of PYP (Figure 3A) determined by Getzoff and coworkers (18) reveals that the 4-hydroxycinnamoyl chromophore adopts a trans conformation and that its carbonyl group forms a hydrogen bond with the main-chain backbone amide of Cys 69. The phenolic oxygen of the chromophore forms hydrogen-bonded interactions with the side chains of Tyr42 and Glu46, while the guanidinium side chain of Arg52 forms hydrogen bonds to the carbonyl oxygens of Tyr98 and Thr50.

The structure of the I_1 early intermediate (Figure 3B) was determined at better than 1 Å resolution by trapping this intermediate at cryogenic temperatures (2). The structure of the I_1 intermediate suggests that light absorption by the 4-hydroxycinnamoyl chromophore leads to isomerization of its thioester linkage without large-scale movement of its aromatic ring. This isomerization causes the thioester carbonyl oxygen O2 to rotate from its hydrogen bond to the main chain backbone amide of Cys 69 in P towards a hydrophobic cleft in I_1. As the O2 - backbone amide hydrogen bond is present in the structures of both the ground state and long lived I_2 intermediate, reformation of this hydrogen bond may help to drive the rotation of the cis-isomerized chromophore to its observed orientation in the I_2 state (see below). While this mechanism appears reasonable, it should be noted that this structure has been assigned to an earlier intermediate (I_0^+) based on a series of 1.8 Å resolution structures determined at ambient temperature by time-resolved x-ray crystallography (19).

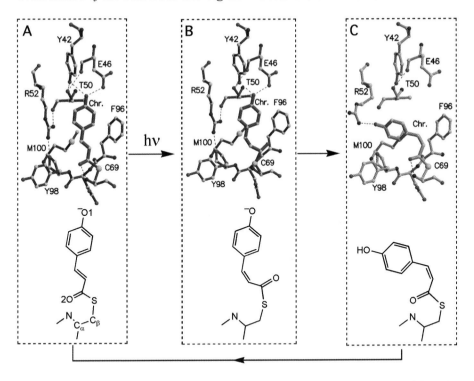

Figure 3. Active-site structures (top) and the chromophore conformations (bottom) of the (A) ground state P (18), (B) early intermediate I1 (2), and (C) t he bleached i ntermediate I2 o f PYP (1).

   The atomic structure of the long-lived I_2 intermediate of PYP (Figure 3C) has been studied by millisecond time-resolved multiwavelength Laue crystallography and optical spectroscopy (1). This structure reveals that in the I_2 state, the conjugated C-C double bond between the phenolic ring and the thioester linkage of the 4-hydroxycinnamoyl chromophore retains the cis-isomerization generated in the I_1 state. A major change, however, is the rotation of the chromophore around the double bond thereby reforming the O2 - backbone amide hydrogen bond present in the ground state structure. The aromatic ring of the chromophore has moved toward the protein surface, and the phenolic oxygen becomes solvent exposed and protonated in the bleached state. This different I_2 state configuration of the chromophore leads to subtle protein conformational changes within the regions between and including the D_α and E_α helices. The most important of these changes may be those involving Arg52. In the ground state, the deprotonated phenolic oxygen of trans-isomerized 4-hydroxycinnamoyl chromophore forms hydrogen-bonding interactions with the side chains of Tyr42 and Glu46, while the guanidinium side chain of Arg52 forms hydrogen bonds to the carbonyl oxygens of Tyr98 and Thr50. In the I_2 state, however, Arg52 shifts to hydrogen bond with the phenolic oxygen of the rearranged chromophore. As a result of this shift, Arg52 becomes more solvent exposed, and together with the chromophore protonation forms a positive electrostatic patch on the protein surface. These changes in surface potential and shape are thought to affect the interaction between PYP and its yet to be identified downstream partner, thereby regulating the negative phototactic response.

4.   FIXL HEME DOMAIN - STRUCTURE AND MECHANISM

FixL is a heme-based O_2-sensing protein that regulates the expression of the nitrogen fixation genes in *Rhizobia* (20-24). Its N-terminal region consists of a heme-binding PAS domain that regulates the activity of a C-terminal histidine kinase domain as a function of the O_2 level (22-26). In the absence of O_2, the kinase is active ('on' state) and FixL phosphorylates its two-component partner, FixJ. Binding of phospho-FixJ to the nifK gene promoter initiates transcription of the nitrogen fixation genes. In the presence of O_2, however, binding of O_2 to FixL's heme results in a conformational change that represses its histidine kinase activity ('off' state). Understanding how O_2 mediates these conformational changes is one of the fundamental questions in regards to this protein.

Structures of various unliganded and ligand-bound forms of *Bradyrhizobium japonicum* FixL heme domain (BjFixLH) have been determined (4-6). These structures reveal a characteristic PAS fold, with the heme lying at the center of the glove covalently attached to the protein via His200. While the overall structures of these are nearly the same, one difference is the shift of a loop between the F_α-helix and G_β-strand (FG loop) upon binding of certain ligands (O_2, CN$^-$, imidazole) to the heme. This heme-mediated conformational change is presumably linked to the regulation of the histidine kinase activity in the full-length FixL protein.

In addition to these BjFixLH structures, the structures of the ferric and ferrous forms of the unliganded *Rhizobium meliloti* FixL heme domain (RmFixLH) have recently been determined (7). This RmFixLH protein contained an additional N-terminal region not present in the BjFixLH structure. Importantly, addition of this N-terminal region leads to a different quaternary structure than that for BjFixLH, with the additional region serving to promote homodimerization of the RmFixLH protein. While it is likely that this region plays a similar role in the full-length protein, further study will be required to confirm this. Homodimerization, however, does not appear to affect the fold of the PAS domain. The PAS region of RmFixLH is virtually identical to that of BjFixLH. Presumably its mechanism of signal transduction will be similar to BjFixLH, though no ligand-bound forms of RmFixLH are available to confirm this.

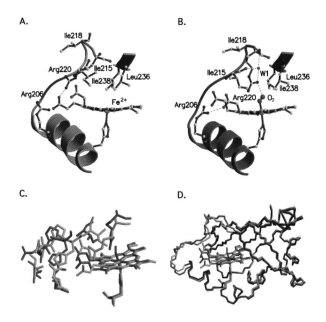

Figure 4. Ribbons and ball-and-stick model diagrams of the BjFixL heme-binding pocket for (A) deoxy-BjFixL and (B) oxy-BjFixL. Conformation difference in the heme pocket (C) and in the FG loop (D) between deoxy-BjFixLH (cyan and blue) and oxy-BjFixLH (red and tan) (4-6).

Comparison of the deoxy-BjFixLH (unliganded ferrous BjFixLH) and the oxy-BjFixLH (O_2-bound ferrous BjFixLH) structures reveals a heme-mediated conformational change that is distinct from hemoglobin (Figure 4). Instead of the movement of the proximal histidine and its associated helix as in hemoglobin, in BjFixLH binding of O_2 to the heme leads to a flattening of the heme plane. This releases Arg220 from a salt bridge with heme-propionate 7, and allows it to move into the heme pocket to hydrogen bond to the bound dioxygen ligand. When in the heme pocket, Arg220 serves as a steric barrier, promoting the shift of the FG loop (residues 211 to 215) to a position farther away from the heme pocket.

A detailed analysis of the various structures of BjFixLH, suggests that Arg220 may also play a central role in O_2 recognition and discrimination. While the binding of O_2 to the heme iron results in the shift of the FG loop, the bindings of CO and NO do not. Presumably the movement of Arg220 into the heme pocket is required to drive the conformational change for these smaller ligands, and the greater ability of Arg220 to hydrogen bond O_2

ligand as compared to NO and CO leads to the different conformational response. In support of this role, binding of CN⁻ to the heme iron leads to movement of Arg220 into the heme pocket and as in oxy-BjFixLH, the structure of cyanomet-BjFixLH exhibits a shift of its FG loop. This use of ligand dependent allostery for ligand recognition and discrimination appears to be a common feature of heme-sensory proteins (27).

5.   PHY3 LOV2 DOMAIN- STRUCTURE AND MECHANISM

Phototropin is a major blue-light receptor that regulates phototropism in seed plants (28, 29). It contains a C-terminal serine/threonine kinase and two upstream flavin-bound PAS domains (30). These PAS domains undergo a fully reversible photocycle characterized by a photo-induced blue shift of three major absorbance bands that absorb in the blue, with spontaneous recovery of these bands in the dark (31). The spectrum of the blue-shifted state suggests the formation of a covalent C(4a)-cysteinyl adduct as an intermediate (32). The three-dimensional structure of the second PAS domain (LOV2 domain) from the phototropin module of *Adiantum* phy3 in the dark state is solved by Crosson and Moffat at 2.7Å resolution (8). This section describes the characteristics of the flavin-binding pocket and the mechanism proposed for its response to light.

The phy3 LOV2 domain contains a noncovalently bound FMN molecule. Eleven residues (Figure 2B, red letters) conserved in phototropin LOV domains interact with FMN molecule through a network of hydrogen bonding, van der Waals, and electrostatic interactions (8). Many interactions come from N965, R967 and Q970 of the consensus sequence NCRFLQ, which constitutes the 3_{10} helix (E_α) in phy3 LOV2. The phosphate of FMN interacts with the guanidinium groups of R967 and R983 (from F_α), forming salt bridges. Side chain carbonyl oxygen atoms of N965 and Q970 form hydrogen bonds with hydroxy groups of the ribityl chain of FMN. The isoalloxazine ring of FMN form hydrogen bonds with the side chains of N998, N1008 and Q1029. F1010 is stacked on the *re* face of the isoalloxazine ring. Residues V986, L1012, F1025, T934, and V932 define the hydrophobic pocket around the dimethylbenzene moiety of the isoalloxazine ring. There is no hydrogen bond to the N5 of the isoalloxazine ring.

Dark-minus-light difference absorption spectra suggest the formation of a C(4a)-cysteinyl adduct after light absorption (31). A cysteine residue C966 was indeed found near isoalloxazine ring of FMN (8). The S_γ of C966 is 4.2 Å from C(4a) of FMN in the LOV2 structure. Facile rotation around the C_α-

C_β bond would bring S_γ within 2.6 Å of C(4a) and could promote the flavin-cysteinyl adduct formation.

A model for flavin-cysteinyl adduct formation has been proposed by Crosson and Moffat (8) (Figure 5). In the dark state, the pKa of the buried thiol of C966 is sufficiently high to prevent adduct formation. The photon absorption by the isoalloxazine ring causes charge redistribution on the ring, decreases the basicity of N1 and increases that of N5. Thus photoexcitation of FMN could promote base abstraction of the thiol proton by N5 and nucleophilic attack of the thiolate anion on C(4a) to form the flavin-cysteinyl adduct. Dark-state recovery of this adduct could be driven thermally or could be protein assisted. Structural studies of the light-activated phototropin will provide further insight into the details of flavin-mediated signal transduction by the phy3 LOV2 domain.

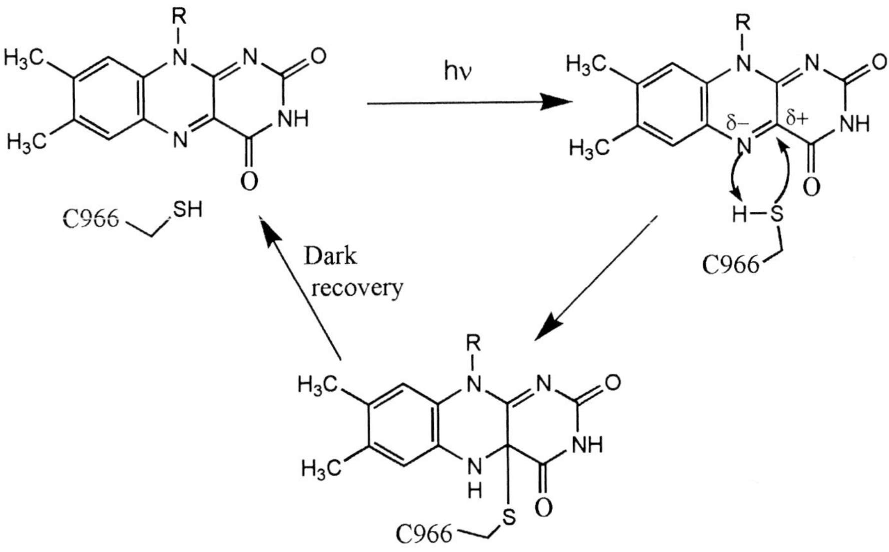

Figure 5. Proposed model for cystein-C(4a) covalent adduct formation in response to light absorption by LOV domain (8).

6. HERG EAG DOMAIN- STRUCTURE

HERG (human eag-related gene product) is a member of the *eag* (ether-a-go-go) K^+ channel family (33). It plays a central role in cardiac electrical excitability, and when defective, underlies one form of the long QT syndrome, a chromosome 7-linked inherited disorder causing sudden death from ventricular tachyarrhythmia (34). The functional channel is a tetramer with a central ion conduction pathway (35). Each subunit consists of six membrane-spanning stretches, a large cytosolic C-terminus and a cytosolic N-terminal EAG domain. The structure of the cytosolic N-terminal EAG domain reveals that it is a PAS domain (9). Its features are most similar to LOV2, though unlike LOV2, no cofactor was present in the structure. It is thought that this N-terminal PAS domain regulates the activity of HERG via its association with the body of the K+ channel – perhaps through a hydrophobic patch (Phe 29, Ile 31, Ile 42, Tyr 43, Met 60, Val 113, Val 115, Ile 123, Met 124, Ile 126) on its surface (9). In the crystal, however, this hydrophobic surface is involved in homodimerization.

7.   HOMODIMERIZATION IN PAS DOMAIN STRUCTURES

PAS domains not only sense signals from the environment but also transduce the signals to receiver domains or proteins via protein-protein interactions (36, 37). PAS domains sometimes form homodimers, but more often are involved in heterodimer formation. The N-terminal PAS domain of HERG can form tetramers in solution (38). RmFixL forms a 2:1 complex with RmFixJ in solution (39). In higher eukaryotes, the PAS proteins usually contain an N-terminal basic-helix-loop-helix (bHLH) motif that binds the target DNA element, two PAS domains (PAS-A, PAS-B repeats), and a C-terminal transcriptional active domain. PAS domains in these proteins function as surface for both homotypic interactions with other PAS proteins and heterotypic interactions with cellular chaperons (37). The aryl hydrocarbon receptor nuclear translocator (ARNT) protein forms heterodimers with aryl hydrocarbon receptor (AHR) and mammalian hypoxia-inducible factor (HIF-1α) in addition to forming homodimers (36). One fundamental issue is how these PAS domains interact

The best model for homodimerization of PAS proteins comes from the RmFixLH structures (Figure 6A) (7). Here, RmFixLH is homodimerized through a 2-fold symmetry interaction of its additional N-terminal helix I-loop-helix II (HLH) region that precedes its PAS domain. The short N-terminal helix (helix II) of one subunit interacts with the same helix of its partner through a series of leucine-zipper like hydrophobic interactions. This involves Leu 139, Ile 142, Leu 143, Val 146, Pro 147 and Phe 246, which are located on one side of the amphipathic helix II. Interactions are also observed between the N-terminal helix I and certain residues of the PAS HI loop of the alternative molecule. These appear to form a fairly stable interaction that buries 2788.6 Å^2 of hydrophobic surface area.

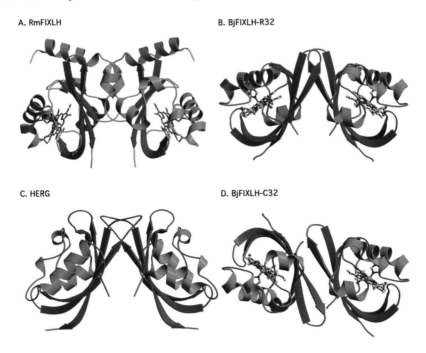

Figure 6. Crystal symmetry related homodimerization interfaces observed in the crystal structures of (A) RmFixLH (7), (B) BjFixLH (R32 crystal form), (C) HERG (9), (D) BjFixLH (C2 form).

It should be noted that a different mode of homodimerization has been observed in structures of BjFixLH and N-HERG structures (Figure 6B-D). Here, the crystal symmetry related dimerization interface is formed by a hydrophobic patch located on one site of the primary β-sheet of each subunit. While the common use of this hydrophobic region for dimerization is intriguing, the interaction of the N-terminal regions of PYP and RmFixLH with this region suggests that this dimerization is likely coincidental. This is supported in part by the finding of an alternative hydrophobic packing interaction of this same surface in a second crystal form of BjFixLH, and the lower buried surface area for this mode of dimerization, 1051.8 Å^2. Presumably, as in RmFixLH, other hydrophobic regions of the protein interact with this surface in full-length forms of these proteins.

8.   CONCLUSION

Despite the low sequence homology between various PAS domains in the PAS superfamily, the structures of the PAS domains from PYP, BjFixL, RmFixL, Phototropin and HERG reveal a common fold. The flexibility of the F_α helix may help to explain how PAS domains can bind to so many different types of cofactors, and could account for their ubiquitous occurrence over a wide variety of kingdoms and sensory systems. The mechanisms of signal transduction elucidated from the structures of PYP and BjFixLH may suggest general strategies for how structural changes take place within the PAS domain. Structures of these PAS domains with their partner domains and proteins, however, are required to provide further information into the protein-protein interactions underlying these signaling mechanisms. Additionally, structures of bHLH-PAS proteins from eukaryotes will be important to understand their roles in hypoxia, angiogenesis, vasculogenesis, circadian rhythms, neural development, respiratory system development, carcinogen metabolism, hormone responsiveness, and cell motility, as will be discussed in detail in later chapters.

REFERENCES

1. Genick, U. K., G. E. O. Borgstahl, N. K., Z. Ren, C. Pradervand, P. M. Burke, V. Srajer, T.-Y. Teng, W. Schildkamp, D. E. McRee, et al. 1997. Structure of a protein photocycle intermediate by millisecond time-resolved crystallography. *Science* 275:1471-1475.

2. Genick, U. K., S. M. Soltis, P. Kuhn, I. L. Canestrelli, and E. D. Getzoff. 1998. Structure at 0.85 angstrom resolution of an early protein photocycle intermediate. *Nature* 392:206-209.

3. Pellequer, J.-L., K. A. Wager-Smith, S. A. Kay, and E. D. Getzoff. 1998. Photoactive yellow protein: a structural prototype for the three-dimensional fold of the PAS domain superfamily. *Proc. Natl. Acad. Sci. USA* 95:5884-5890.

4. Gong, W., B. Hao, S. S. Mansy, G. Gonzalez, M. A. Gilles-Gonzalez, and M. K. Chan. 1998. Structure of a biological oxygen sensor: a new mechanism for heme-driven signal transduction. *Proc. Natl. Acad. Sci. USA* 95:15177-15182.

5. Gong, W., B. Hao, and M. K. Chan. 2000. New mechanistic insights from structural studies of the oxygen-sensing domain of Bradyrhizobium japonicum FixL. *Biochemistry* 39:3955-3962.

6. Hao, B., C. Isaza, J. Arndt, M. Soltis, and M. K. Chan. 2002. Structure-based mechanism of oxygen sensing and ligand discrimination by FixL heme domain of Bradyrhizobium japonicum. *Biochemistry* 41:12952-12958.

7. Miyatake, H., M. Mukai, S.-Y. Park, S.-I. Adachi, K. Tamura, H. Nakamura, K. Nakamura, T. Tsuchiya, T. Iizuka, and Y. Shiro. 2000. Sensory mechanism of oxygen sensor FixL from Rhizobium meliloti: crystallographic, mutagenesis and resonance raman spectroscopic studies. *J. Mol. Biol.* 301:415-431.

8. Crosson, S., and K. Moffat. 2001. Structure of a flavin-binding plant photoreceptor domain: Insights into light-mediated signal transduction. *Proc. Natl. Acad. Sci. USA* 98:2995-3000.

9. Cabral, J. H. M., A. Lee, S. L. Cohen, B. T. Chait, M. Li, and R. Mackinnon. 1998. Crystal structure and functional analysis of the HERG potassium channel N terminus: A eukaryotic PAS domain. *Cell* 95:649-655.

10. Meyer, T. E. 1985. Isolation and characterization of soluble cytochromes, ferredoxins and other chromophoric proteins from the halophilic phototrophic bacterium Ectothiorhodospira Halophila. *Biochim. Biophys. Acta* 806:175-183.

11. Meyer, T. E., G. Tollin, J. H. Hazzard, and M. A. Cusanovich. 1989. Photoactive yellow protein from the purple phototrophic bacterium, Ectothiorhodospira Halophila - quantum yield of photobleaching and effects of temperature, alcohols, glycerol, and sucrose on kinetics of photobleaching and recover. *Biophys. J.* 56:559-564.

12. Meyer, T. E., E. Yakali, M. A. Cusanovich, and G. Tollin. 1987. Properties of a water-soluble, yellow protein isolated from a halophilic phototrophic bacterium that has photochemical activity analogous to sensory rhodopsin. *Biochemistry* 26:418-423.

13. Sprenger, W. W., W. D. Hoff, J. P. Armitage, and K. J. Hellingwerf. 1993. The Eubacterium Ectothiorhodospira halophila is negatively phototactic, with a wavelength dependence that fits the absorption-spectrum of the photoactive yellow protein. *J. Bacteriology* 175:3096-3104.

14. Ng, K., E. D. Getzoff, and K. Moffat. 1995. Optical studies of a bacterial photoreceptor protein, photoactive yellow protein, in single crystals. *Biochemistry* 34:879-890.

15. Meyer, T. E., M. A. Cusanovich, and G. Tollin. 1993. Transient proton uptake and release is associated with the photocycle of the photoactive yellow protein from the purple phototrophic bacterium Ectothiorhodospira Halophila. *Arch. Biochem. Biophys.* 306:515-517.

16. Meyer, T. E., G. Tollin, T. P. Causgrove, P. Cheng, and R. E. Blankenship. 1991. Picosecond decay kinetics and quantum yield of fluorescence of the photoactive yellow protein from the halophilic purple phototrophic bacterium, Ectothiorhodospira Halophila. *Biophys. J.* 59:988-991.

17. McRee, D. E., T. E. Meyer, M. A. Cusanovich, H. E. Parge, and E. D. Getzoff. 1986. Crystallographic characterization of a photoactive yellow protein with photochemistry similar to sensory rhodopsin. *J. Biol. Chem.* 261:3850-3851.

18. Borgstahl, G. E. O., D. R. Williams, and E. D. Getzoff. 1995. 1.4 angstrom structure of photoactive yellow protein, a cytosolic photoreceptor - unusual fold, active-site, and chromophore. *Biochemistry* 34:6278-6287.

19. Ren, Z., B. Perman, V. Srajer, T. Y. Teng, C. Pradervand, D. Bourgeois, F. Schotte, T. Ursby, R. Kort, M. Wulff, et al. 2001. A molecular movie at 1.8 angstrom resolution displays the photocycle of photoactive yellow protein, a eubacterial blue- light receptor, from nanoseconds to seconds. *Biochemistry* 40:13788-13801.

20. Fisher, R. F., and S. R. Long. 1992. Rhizobium - plant signal exchange. Nature 357:655-660.

21. Stock, J. B., A. J. Ninfa, and A. M. Stock. 1989. Protein-phosphorylation and regulation of adaptive responses in bacteria. *Microbiol. Rev.* 53:450-490.

22. Gilles-Gonzalez, M. A., G. Ditta, and D. R. Helinski. 1991. A haemoprotein with kinase activity encoded by the oxygen sensor of Rhizobium meliloti. *Nature* 350:170-172.

23. Gilles-Gonzalez, M. A., and G. Gonzalez. 1993. Regulation of the kinase activity of heme protein FixL from the two-component system FixL/FixJ of Rhizobium meliloti. *J. Biol. Chem.* 268:16293-7.

24.   Gilles-Gonzalez, M. A., G. Gonzalez, M. F. Perutz, L. Kiger, M. C. Marden, and C. Poyart. 1994. Heme-based sensors, exemplified by the kinase FixL, are a new class of heme protein with distinctive ligand binding and autoxidation. *Biochemistry* 33:8067-73.

25.   Gilles-Gonzalez, M. A. 2001. Oxygen signal transduction. *IUBMB Life* 51:165-173.

26.   Gilles-Gonzalez, M. A., G. Gonzalez, and M. F. Perutz. 1995. Kinase activity of oxygen sensor FixL depends on the spin state of its heme iron. *Biochemistry* 34:232-236.

27.   Jain, R., and M. K. Chan. 2003. Mechanisms of ligand discrimination by heme proteins. *J. Biol. Inorg. Chem.* 8:1-11.

28.   Lasceve, G., J. Leymarie, M. A. Olney, E. Liscum, J. M. Christie, A. Vavasseur, and W. R. Briggs. 1999. Arabidopsis contains at least four independent blue-light-activated signal transduction pathways. *Plant Physiol.* 120:605-614.

29.   Christie, J. M., P. Reymond, G. K. Powell, P. Bernasconi, A. A. Raibekas, E. Liscum, and W. R. Briggs. 1998. Arabidopsis NPH1: A flavoprotein with the properties of a photoreceptor for phototropism. *Science* 282:1698-1701.

30.   Christie, J. M., M. Salomon, K. Nozue, M. Wada, and W. R. Briggs. 1999. LOV (light, oxygen, or voltage) domains of the blue-light photoreceptor phototropin (NPH1): binding sites for the chromophore flavin mononucleotide. *Proc. Natl. Acad. Sci. USA* 96:8779-8783.

31.   Salomon, M., J. M. Christie, E. Knieb, U. Lempert, and W. R. Briggs. 2000. Photochemical and mutational analysis of the FMN-binding domains of the plant blue light receptor, phototropin. *Biochemistry* 39:9401-9410.

32.   Miller, S. M., V. Massey, D. Ballou, C. H. Williams, M. D. Distefano, M. J. Moore, and C. T. Walsh. 1990. Use of a site-directed triple mutant to trap intermediates - demonstration that the flavin-C(4a)-thiol adduct and reduced flavin are kinetically competent intermediates in mercuric ion reductase. *Biochemistry* 29:2831-2841.

33.   Warmke, J. W., and B. Ganetzky. 1994. A family of potassium channel genes related to eag in drosophila and mammals. *Proc. Natl. Acad. Sci. USA* 91:3438-3442.

34.   Curran, M. E., I. Splawski, K. W. Timothy, G. M. Vincent, E. D. Green, and M. T. Keating. 1995. A molecular-basis for cardiac-arrhythmia - HERG mutations cause long QT syndrome. *Cell* 80:795-803.

35.   Sigworth, F. J. 1994. Voltage gating of ion channels. Q. Rev. Biophys. 27:1-40.

36.   Taylor, B. L., and I. B. Zhulin. 1999. PAS domains: internal sensors of oxygen, redox, and light. Microbiol. *Mol. Biol. Rev.* 63:479-506.

37.   Gu, Y. Z., J. B. Hogenesch, and C. A. Bradfield. 2000. The PAS superfamily: sensors of environmental and developmental signals. *Annu. Rev. Pharmacol. Toxicol.* 40:519-561.

38.   Li, X. D., J. Xu, and M. Li. 1997. The human Delta 1261 mutation of the HERG potassium channel results in a truncated protein that contains a subunit interaction domain and decreases the channel expression. *J. Biol. Chem.* 272:705-708.

39.   Tuckerman, J. R., G. Gonzalez, and M. A. Gilles-Gonzalez. 2001. Complexation precedes phosphorylation for two-component regulatory system FixL/FixJ of Sinorhizobium meliloti. *J. Mol. Biol.* 308:449-455.

Chapter 2

SIGNAL TRANSDUCTION IN PROKARYOTIC PAS DOMAINS

Barry L. Taylor, Mark S. Johnson, and Kylie J. Watts
Loma Linda University, Loma Linda, CA 92350

1. INTRODUCTION

With the progress in genome sequencing and *in silico* analysis, there has been a rapid increase in identified PAS domains in the SMART database (Simple Modular Architecture Research Tool <http://smart.embl-heidelberg.de/>). At the time of writing (September, 2002) more than 2000 non-redundant PAS domains are listed in SMART, making the PAS domain one of the most common domains found in sensory proteins. Of the known PAS domains, sixty percent are in prokaryotes where they are ubiquitous in both eubacteria and archaea.

Readers should be aware of variant usages of 'PAS' (1, 2). In this volume the PAS domain is defined to include the conserved elements of the three-dimensional crystal structure (3). This is equivalent to combined PAS and PAC domains in the bioinformatic analysis used for SMART [Compare reference (4)]. However, SMART identifies ~1,000 more PAS domains than PAC domains. Since, with a few exceptions, PAS and PAC domains do not occur without each other, the apparent difference in PAS and PAC numbers reflects a limitation of the SMART algorithm used to identify PAS domains rather than a true disparity between the numbers of these subdomains (I. Zhulin, personal communication).

PAS domains are associated with sensory proteins and, in prokaryotes, are predominantly associated with sensing oxygen, light, redox and energy levels (1). Of particular interest is the finding that, although PAS domains are sensory input elements, they are located in the cytoplasm and not on the external surface of the cell (2). The specificity of a PAS domain for sensory input signals is determined, in part, by the cofactor associated with the PAS domain. Identified cofactors include a 4-hydroxycinnamyl chromophore in photoactivated yellow protein (5), heme in the FixL protein (6), flavin adenine dinucleotide (FAD) in NifL (7, 8) and Aer (9-12), and flavin mononucleotide (FMN) in phototropin in plants (13, 14). Some PAS domains appear to function without a cofactor. Thus the PAS domain appears to be a versatile domain that has been adapted to many different roles in the cell. It is likely that further investigations will continue to reveal novel roles for PAS domains.

Many signals detected by prokaryotic PAS domains are signals that report the cellular energy level. For example, in phototrophic organisms the presence or absence of light profoundly affects the generation of energy. A PAS-containing photoreceptor that detects a loss of light may initiate an aversion response that returns the cell to an illuminated environment where the cell can maintain optimal energy production. Similarly, oxidative energy production decreases when a heterotrophic organism becomes anaerobic. An oxygen-binding PAS domain may initiate an aversion response which returns the cell to an aerobic environment.

Such monitoring of a single component from which energy is derived can be beneficial to bacteria but does not protect the cell from all circumstances that deplete cellular energy. An oxygen sensor does not detect an energy loss in an aerobic environment if starvation for carbon source lowers respiration and the proton motive force (PMF). However, many bacteria have adapted PAS domains for more versatile sensing of energy depletion by sensing a central component of energy production, such as respiratory redox potential or PMF. Figure 1 summarizes some of the immediate events that follow a shift from aerobic to anaerobic conditions. Under aerobic conditions, the components of the electron transport system are in a relatively oxidized steady state, and the PMF is high. As the oxygen supply becomes limiting, the components of the electron transport system are reduced and the PMF decreases; reducing equivalents (e.g., NADH) build up, ATP levels drop, and fermentation products are produced as a means to reoxidize NADH so that glycolysis can continue. At pH 7.5, the weak acid products of fermentation dissociate, releasing protons that tax the buffering capacity of the cytosol. By sensing energy in the form of redox potential or PMF, prokaryotes not only have a more versatile sensing mechanism, they can have a faster response time (15, 16). In a switch from aerobic to anaerobic conditions, changes in PMF and redox potential occur in milliseconds (17, 18). This is roughly four orders of magnitude faster than an ATP drop, which we previously measured at 23% of wild-type levels after 10 minutes in the presence of arsenate and the non-metabolizable, PEP-depleting glucose analog, alpha-methylglucoside (19).

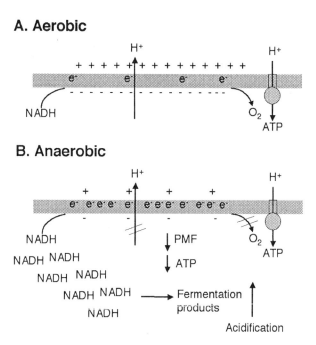

Figure 1. Cellular changes in response to a sudden move from (A) an aerobic to (B) an anaerobic environment (see text for details).

Energy taxis as we define it includes behavior in response to a limitation in respiratory electron transport by depletion of an electron acceptor, decreased electron-donating substrates (usually carbon sources), or diversion of electrons from the electron transport system (20). This includes direct oxygen taxis, aerotaxis, electron acceptor taxis, and aerotaxis-like responses to alternative electron acceptors in anaerobic cells (16, 21, 22), redox taxis to quinone (23) and metabolism-dependent taxis to a carbon source in *E. coli* (24) and other bacteria (25-27). The energy taxis responses are comparable to the 'fight or flight' responses mediated by epinephrine in mammals.

2. STRUCTURE FUNCTION

Presently, the crystal structures of five PAS structures have been resolved: photoactive yellow protein (PYP) from *Ectothiorhodospira halophila* (3, 5), FixL from *Bradyrhizobium japonicum* (6) and *Rhizobium meliloti* (28), Human ether-a-go-go related gene (HERG) (29), and the LOV2 domain from the chimeric plant photoreceptor, phy3 (14). In addition, an NMR structure for the PAS domain from human PAS kinase (hPASK) has been resolved (30). Of these, structure function relationships have most extensively been studied in PYP.

The general mechanism by which the unique structure of a PAS domain determines functionality remains to be elucidated. The overall architecture of the PAS domain has been described as a left-handed glove that can enclose a cofactor [see (6) and Chapter 1 of this volume], and as a hand is able to hold a variety of objects, the PAS domain superfamily can bind chemically diverse cofactors. Although the scope of cofactors is still unknown, and the mechanism of cofactor binding has been studied in only a few proteins, certain common themes are emerging. The EF loop, the F-helix and the FG loop are important regions for cofactor binding [(5, 6, 12, 14) and Chapter 2 of this volume]. Cofactor sensing involves a "state" change in the cofactor that alters local interactions in the binding cleft. These changes are transmitted and amplified to other regions of the domain. It is too early to predict whether these conformational changes will be transmitted in similar ways within the PAS superfamily.

Only recently has there been structural data that yields insights into how a PAS regulatory domain might complex with its downstream, kinase domain (30). Human PAS kinase (hPASK) is a PAS-containing serine/threonine kinase that has a PAS domain and a PAS kinase domain. The PAS domain of PAS kinase inhibits the kinase activity when added in trans (31). The orthologous PAS kinase in yeast has recently been shown to block the production of glycogen by phosphorylating and inhibiting glycogen synthase (32). It may also be involved in the control of protein synthesis. NMR structural studies on hPASK showed that it can bind specific small molecules to its hydrophobic core, and that the kinase domain of hPASK binds to a flexible loop (within the FG loop) on the PAS domain surface (30). Several mutations linked structural and functional segments for ligand binding and inhibition of kinase activity. The predominant site of interaction between the PAS and PASK domains were within the FG loop of the PAS domain, but interactions were also present within the F-helix and the HI loop.

3.   TWO-COMPONENT SYSTEMS

Many prokaryotic PAS domains are in, or associated with, histidine kinase sensor proteins (1, 4, 33) in two-component regulatory systems. Prototypical two-component pathways consist of a histidine kinase sensor protein and a cognate response regulator (34). An input module in the N-terminus of the histidine kinase senses stimuli directly or indirectly in association with an upstream receptor. A C-terminal transmitter module includes a conserved histidine that is autophosphorylated in response to sensory stimuli. The phosphoryl moiety is transferred from the conserved histidine to an aspartate residue in a receiver module on the cognate response regulator (Figure 2). As a result of phosphorylation, the output domain of the response regulator is activated so that it is able to bind to DNA or another signaling protein. Most response regulator proteins are transcriptional activators. However, in *Escherichia coli* aerotaxis, the CheY protein is a free-standing receiver domain that when phosphorylated binds directly to the FliM switch protein, which controls the directions of rotation of the flagellar motors (35, 36).

The individual modules of the two-component system have been variously combined in nature to construct more complex phosphorelay circuits. The four-step His-Asp-His-Asp phosphorelay that controls the initiation of sporulation in *Bacillus subtilis* illustrates the role of PAS domains in a complex circuit (37-39) (Figure 2). The KinA histidine kinase of *B. subtilis* has three PAS domains in the N-terminal segment, two of which are known sensory input sites. KinA autophosphorylates a His residue, and then transfers the phosphoryl group from His to Asp to His to Asp, through a series of three separate proteins including Spo0F, Spo0B and ultimately the transcriptional regulator Spo0A. In contrast, the virulence gene regulator BvgS from *Bordatella spp.* employs a modified phosphorelay where the first three phosphotransfer events are entirely within the same protein (40, 41). The ArcB/ArcA phosphorelay that regulates the aerobic metabolism modulon in *E. coli* has a His-Asp-His-Asp relay with several alternative pathways (Figure 2) (42). In another variation, the NifL/NifA signaling pathway has a histidine kinase-like NifL protein with a FAD-PAS domain (Figure 2), but there is no autophosphorylation of NifL or phosphorelay to the NifA transcriptional regulator.

4. PAS DOMAINS IN MICROBIAL GENOMES

A broader picture of the role of PAS domains in prokaryote species has followed the increase in fully sequenced microbial genomes that are

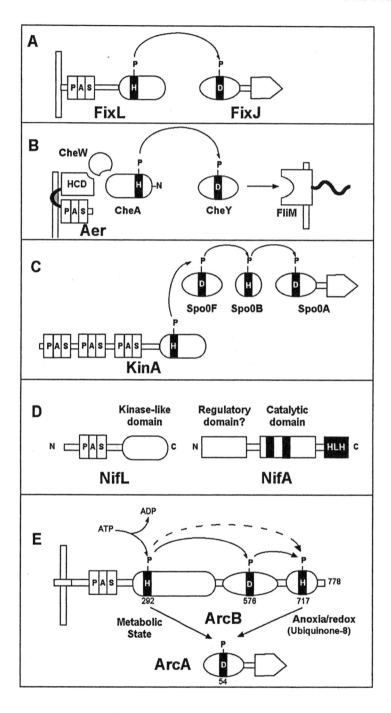

Figure 2. Phosphorelay control from A) FixL, *S. meliloti*, B) Aer, *E. coli* C) KinA, *B. subtilis* D) NifL, *A. vinelandii*, and E) ArcB, *E. coli* [modified from (2) with permission].

accessible. The number of PAS domain proteins reported ranges from zero in some species, to 61 in the filamentous *Cyanobacterium anabaena*. The PAS-proteins have been assigned diverse functions, such as those listed in Table 1. However, many of the putative assignments are based solely on *in silico* analysis and are of varying reliability. For example, the number of assigned aerotaxis proteins is 8, whereas more than 30 are now known. (Q. Ma and I. Zhulin, unpublished data). PAS proteins may have single, dual or up to six PAS domains in one protein (2). One analysis of eleven prokaryotic genomes found a correlation between the total number of PAS domains in a genome and the number of components of the respiratory and photosynthetic electron transport-associated proteins (33).

Table 1. Proposed annotations assigned to bacterial PAS-domain-proteins in SMART database.

Annotation	Number of Assigned Proteins
Aerotaxis	8
ArcB	5
Bacteriophytochrome	4
Chemotaxis protein	21
Diguanylate cyclase	5
FixL	12
Histidine kinase	142
NifL	5
NtrB.NirB.NR(II)	16/1/1=18
NtrC	5
NtrY	4
PhoR	15
Phototropin	3
PYP	5
Serine/Threonine kinase	1
Sigma factor	8

Some of the multiple PAS domains in proteins probably originated from a duplication of one domain (33). Multiple copies of a similar domain may provide a selective advantage by amplifying the sensory signal. Other PAS domains in one protein are diverse and have different origins. This probably represents a diverse function for the individual PAS domains, since individual PAS sequences are well maintained over long phylogenetic distances. The fidelity with which differentiated PAS sequences have been maintained throughout the phylogenetic tree is best explained by a differentiated function for individual branches of the PAS domain lineage (2). Where different types of PAS domains are present, one sensor protein may respond to multiple input signals, each activating a specialized PAS domain.

5.   GLOBAL REGULATION OF CELL METABOLISM AND BEHAVIOR

The remainder of this chapter looks at the sensory transduction function of selected PAS domain-containing proteins in prokaryotes. The coverage is not exhaustive and is concentrated on progress reported since 1999. For earlier findings the reader is referred to cited references and to previous reviews (2, 15).

5.1   Bacterial Behavior: Aerotaxis

Aerotaxis is the behavior in which motile bacteria sense a gradient of oxygen and navigate in the gradient to a niche that has an oxygen concentration optimal for growth (15, 20). This results in veils of bacteria in bodies of water, where each species collects at the depth where the oxygen concentration is optimal for that species (43-45). Aerotaxis is also essential in plant-microbe interactions in the rhizosphere. Aerotaxis apparently guides *Azospirillum* to regions of the root where the oxygen concentration is low enough (0.4%) to avoid oxygen inactivation of the nitrogen fixation complex (25, 46).

5.1.1   Aerotaxis Transducer Aer

Aerotaxis has been most extensively studied in *E. coli* where the transducer for the response is the 55 kDa Aer protein [for a review see refs. (15, 16)]. Aer has an N-terminal PAS domain that contains noncovalently bound FAD, a membrane anchoring domain, and a C-terminal signaling domain which is similar to the signaling domain in chemotaxis receptors (9, 11, 16) (Figure 3). The C-terminal signaling domain is linked to the membrane by a HAMP domain which is a recently identified signal transduction domain (47, 48). The predicted topology for the Aer protein has the protein anchored in the membrane by its central hydrophobic sequence, and the N- and C-terminal domains localized in the cytoplasm (Figure 3) (10, 12). The membrane binding topology of Aer was recently confirmed in our laboratory using disulfide crosslinking between substituted cysteine residues in the dimeric Aer protein (M. Brandon, D. Amin, and M. Johnson, unpublished observation). The cytoplasmic location of the Aer PAS domain also supports the role of Aer in sensing the intracellular environment (16). As in other membrane-bound PAS proteins, the Aer PAS domain is located adjacent to the transmembrane region, leaving open the possibility that the PAS domain can interact with domains of other membrane proteins. A possible target for the Aer PAS domain is a component of the electron transport system. The Aer protein does not sense oxygen directly, but senses changes in the respiratory electron transport system that result from depletion or resupply of oxygen (15, 16, 21). As a result, Aer can detect any change in cellular energy levels that modulate respiration.

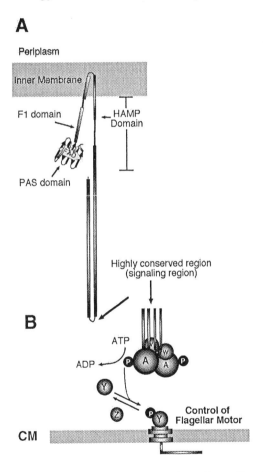

Figure 3. A) Overall topology of Aer. B) Signaling region of homodimeric Aer and the two-component phosphorelay that controls flagellar rotation.

5.1.2   Aer PAS Domain and Signaling Pathway

It is likely that the oxidation/reduction of FAD triggers a conformational change that changes the Aer PAS domain from an inactive to an active form (16). In order to determine which Aer PAS residues are critical for signaling, we serially replaced more than 40 residues in the PAS domain with cysteine (12). Residues were mutated that are both conserved and non-conserved in the PAS superfamily. Conserved residues will include residues that are critical for the scaffolding that gives PAS domains their characteristic alpha/beta fold. Residues that are essential for FAD binding and signal transduction are more likely to be non-conserved in other PAS domains.

Mutations at seven residues (Gly42, Arg57, His58, Asp60, Asp68, Trp79 and Gly90) produced a null phenotype and a Glu47 mutation eliminated more than ninety percent of the response to an oxygen increase (12) (Figure 4). Five of these mutations are in conserved residues. Four of the null mutations, replacement of Arg57, His58, Asp60, Asp68, also abolished FAD binding. Three of these are variable residues, in and near the putative EF loop (Figure 4). Further mutagenesis studies determined that the charge on Arg57 and Asp60 is critical for FAD binding.

Replacement of Asn34, Phe66 and Asn85 generated a signal-on (tumbling) bias (12). Asn34 and Phe66 are conserved residues and Asn85 is a variable residue. Replacement of Tyr111 inverted signaling by the Aer protein so that a positive signal elicited negative behavioral responses and *vice versa* (Figure 4). These findings indicate that FAD binding and the active site for signaling are centered around the EF loop and the PAS core region (Figure 4). Most of the residues that are involved in FAD binding and signal transduction are predicted to project into an internal pocket that corresponds to the FMN binding site in the crystal structure of the PAS domain of phototropin (14). The EF loop forms one boundary of this pocket (Figure 4). Our working hypothesis is that residues that interact with the isoalloxazine ring of the bound FAD transduce redox changes in FAD into a conformational change in the PAS domain. The latter propagates the aerotaxis signal to the C-terminal domain of Aer.

The signal that is generated by a conformational change in the FAD-PAS domain could be transmitted linearly to the signaling domain through the membrane segment of Aer, or transversely through protein-protein interaction between the PAS and HAMP domains (compare Figure 3). There is now persuasive evidence for direct interaction of the PAS and HAMP domains. Both the transmembrane and HAMP domains are essential for correct folding of the PAS domain and FAD binding (S. Herrmann, Q. Ma, M. Johnson, and B. Taylor, manuscript in preparation). Point mutations in the HAMP domain disrupt FAD binding in the PAS domain (10, 48). An allele-specific second site suppressor for a C253R mutation in the HAMP domain was identified in the PAS domain, indicating that the primary and suppressor mutations are in close proximity (K. Watts, unpublished observation).

In summary, the available evidence supports a model for signal transduction for Aer in which a redox change in the FAD bound to the Aer PAS domain generates a conformational change that is propagated through the PAS domain and across a PAS/HAMP contact domain. It is proposed that the signal induces a conformational change in the HAMP domain that is transmitted through the C-terminal domain to the highly conserved signaling domain. Bacterial chemoreceptors are clustered at the cell pole and there is cross communication between the clustered receptors (49-51). In *E. coli,* the unit in the cluster is proposed to be a trimer of dimeric receptors (52). Signal propagation from the HAMP domain to the signaling domain may involve protein-protein interactions. The signaling domain feeds into the *E. coli* chemotaxis pathway and regulates the rate of autophosphorylation of the CheA protein and phosphorylation of the protein. Phosphoryl-CheY binds to the FliM protein in the flagellar motor complex and reverses the direction of flagella rotation [see (36, 53, 54) and Figure 3B].

5.2 Energy Metabolism

5.2.1 Arc B Regulation of Energy Metabolism

The ArcB/ArcA two-component system of *E. coli* controls the global anoxic redox modulon, which coordinates the responses of more than 30 operons to anaerobiosis (55-57). Genes controlled include those encoding citric acid cycle enzymes, glyoxylate cycle enzymes, many dehydrogenases for aerobic growth, fatty acid oxidation enzymes, cytochrome *o* oxidase, cytochrome *d* oxidase, pyruvate formate-lyase and enzymes involved in cobalamin synthesis (57). Microaerobic control often functions in concert with FNR. The sensor for this system is ArcB, which has an N-terminal membrane binding region and a C-terminal cytoplasmic region that includes a PAS domain, a transmitter module, a receiver module, and secondary transmitter domain (HPt) (Figure 3). Autokinase activation occurs under reducing growth conditions, where phospho-transfer occurs serially through conserved His292, Asp576 and His717 residues, and ultimately on to Asp54 in the receiver domain of ArcA. Soluble constructs of ArcB missing the membrane binding N-terminus are constitutively active (58), suggesting that sensor control requires the membrane milieu for its redox control or for native conformation. However, residues in the hydrophobic domain are not conserved, and function is retained with hydrophobic replacements. This fact, along with more recent data, suggests that membrane-bound redox components are required for function.

In a recent study, aerobic-activated phosphorylation of purified, soluble (missing the N-terminal membrane-binding region) ArcB (residues 78-778) was inhibited by oxidized soluble ubiquinone-0 and menadione (59). The half maximal inhibition was 5 µM for ubiquinone-0 and 50 µM for menadione. The reduced forms of these quinones did not inhibit phosphorylation suggesting that the oxidized form of the electron carrier quinones are the inhibitory signal that is sensed *in vivo*. The expression of the reporter gene, F(*cydA'-lacZ*), increased 3-fold in a *ubiCA* mutant that was unable to synthesize the quinone precursor 3-octylprenyl-4-hydroxybenzoate, but was unaffected in a *ubiA arcB* double mutant, consistent with aerobic respiratory control being directed by ArcB under the control of the oxidative state of ubiquinone-8.

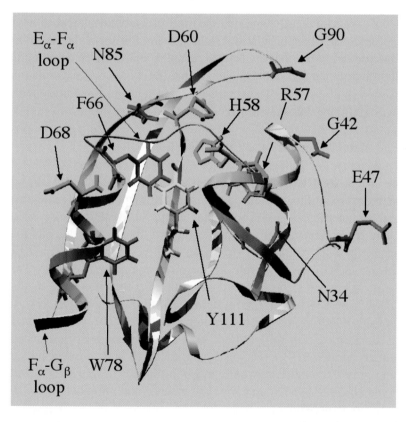

Figure 4. Putative model of the Aer PAS domain highlighting residues important for function. Cysteine replacements of specific residues altered swimming bias (purple), inverted the response aerotactic response (yellow), abolished behavior but not FAD binding (red) or abolished behavior and FAD binding (green).

The PAS domain is presumed to be involved in sensing and control. Amino acid replacements at the PAS residues C242G (60), C242A (42), or conserved N181A (42), abolished ArcB regulation. The conserved Asp181 residue in the PAS domain of ArcB corresponds to Asp34 of the Aer PAS domain which appears to be a critical residue in signaling between the PAS and HAMP domains in Aer [(12), K. Watts, unpublished observation]. In ArcB, the N181A and C242A replacements did not prevent phospho-transfer between transmitter and receiver domains *in vitro*, suggesting that the role of the PAS domain in ArcB is exclusively relegated to that of sensor. However, the PAS domain is missing in ArcB from *Hemophilus influenzae*, and this ArcB is functional when expressed in *E. coli* (59), indicating that there is not an absolute requirement for a PAS domain in the class of ArcB receptors. To date, the ArcB PAS domain has not been shown to bind a cofactor. FAD added to everted vesicles did not stimulate activity (61), nor did extracts containing overexpressed ArcB show absorption spectra characteristic of flavin (59).

5.2.2   Aer Regulation of Anaerobic Respiration and the Entner-Doudoroff Pathway

Recent studies using DNA microarrays and phenotype microarrays found that the Aer protein is a global regulator of five operons for anaerobic respiration and the operon expressing the enzymes of the Entner-Doudoroff pathway (62). An *aer* deletion decreased expression of several anaerobic pathways under aerobic growth conditions. However, no anaerobic studies were performed. Expression of Aer is induced by the transcriptional activators FlhD and FlhC. On-going studies in our laboratory indicate that there is a decrease in Aer levels in *E. coli* grown anaerobically compared to those grown aerobically (S-H. Huang, A.C-W. Lin, H. Wang, M. Johnson, R. Gunsalus, B. Taylor, unpublished observation). Therefore the FAD-PAS-containing Aer protein has dual roles in regulating *E. coli* behavior (aerotaxis) and metabolism.

The mechanism by which Aer regulates anaerobic respiration is unknown, but regulation does not require the CheA-CheY phosphorelay of aerotaxis and bacterial chemotaxis (62). The absence of a putative DNA binding domain in Aer suggests that Aer may interact with a transcriptional activator in regulating anaerobic respiration and the Entner-Doudoroff pathways. The PAS domain is a possible site for heterodimerization with a transcriptional regulator but no candidate transcriptional activators were identified. It is likely that Aer functions in concert with ArcAB and Fnr in the global regulation of anaerobic respiration.

5.2.3   Regulation of Photosystem II by PpsR and CrtJ

Anoxygenic phototrophic bacteria such as *Rhodobacter* can derive energy from oxygen under aerobic conditions and from light when dissolved oxygen concentrations fall below 1%. Under aerobic conditions, the synthesis of the integral membrane photosystem is repressed; under low oxygen conditions, synthesis is derepressed, and a new light-driven, energy-generating photosystem is packaged into the membrane. Genetic control of this system is by DNA binding proteins, PpsR (for "photopigment suppression") in *Rhodobacter sphaeroides* and CrtJ in *R. capsulatus*. They share 53% homology (63) but their mechanisms may not be identical. Under aerobic conditions, both of these regulatory proteins repress genes coding for assembly and structural proteins of photosystem II. The genes include the *puc* operon which encodes light harvesting complex II proteins, and the genes encoding bacteriochlorophyl (*bch*) and carotenoids (*crt*). Null mutations in *ppsR* or *crtJ* cause constitutive overexpression of the photosystem under aerobic conditions (64, 65).

The PpsR and CrtJ proteins have a carboxyl helix loop helix (bHLH) domain that is involved in DNA binding, two PAS domains that may be involved in the tetramerization and possibly signal sensing (although mutations in the PAS domain still allowed tetramerization) and an N-terminal region that is required for structure. To isolate mutations in *ppsR*, Gomelsky et al. (63) took advantage of earlier work showing that second-site suppressors to another redox regulator, AppA, which has antagonistic roles to PpsR, mapped in the *ppsR* gene (66). Using this selection method, three classes of mutations were found, with different degrees of suppression. Five suppressors mapped in the PAS domains: N176I, R242C, L248P and L249P in the first PAS domain and G288E in the second PAS domain. These results suggest that the PpsR and AppA proteins interact physically. Interestingly, we recently found intragenic second-site suppressors of HAMP domain mutations in the Aer PAS domain at positions similar to two of these critical residues in the *ppsR* PAS domain. They include the conserved N176 (N34D in Aer) and G288 (S28G in Aer) (K. Watts, unpublished observation). Signaling via physical interaction between PAS domains and a signaling protein/domain may be a recurring paradigm in PAS domain signaling.

Although the control mechanism is not completely understood, recent data suggest that the PpsR protein binds DNA after disulfide bond formation between PAS domain residue C251 (M112 Aer homologue) and amino acid C424 in the HTH motif, which is the domain known to be directly responsible for DNA binding (64). The oxidized conformation binds to palindromic sequences of the promoters that it regulates. Since the redox potential of *R. sphaeroides* cytosol remains at −200 mV, the mechanism by which a reversible reduction/oxidation of PpsR could occur has been questioned. A model recently introduced by Masuda and Bauer proposes that the flavoprotein AppA binds to PpsR and reduces the PpsR disulfide bond (67). A complex between PpsR and AppA is then formed and PpsR dissociates from the DNA, relieving the repression. The ultimate source of electrons to reduce AppA could be the quinone pool (68). A unique situation occurs in the presence of blue light, whereby the repression continues even under low concentrations of dissolved oxygen. Recently, it has been proposed that a blue light-dependent conformational change occurs in AppA that is directed by the FAD cofactor (69). This altered conformation prevents PpsR-AppA complexation, allowing a PpsR/DNA complex despite the reduced state. AppA is the first example of dual sensing in a protein, where both redox and light signals are integrated. The *R. capsulatus* PpsR homolog, CrtJ, lacks AppA and there is no effect of light on photosystem synthesis.

5.2.4   Direct Oxygen Sensors

There are two types of PAS domains now known to contain hemes: one is represented by FixL, which signals through phosphorylation of the regulatory protein, FixJ, and another is represented by the direct oxygen sensor protein (Dos) of *E. coli* and the PDEA1 protein of *Acetobacter xylinum* (70, 71). The Dos and PDEA1 proteins have a heme-binding PAS domain that binds oxygen directly and regulates the phosphodiesterase catalytic domain. The phosphodiesterase linearizes cyclic bis (3'-5') diguanylic acid which is a key regulator of bacterial cellulose synthesis. This is the first example of oxygen regulation of a second messenger (70, 71).

Hemes bind CO much more tightly than they bind oxygen. The Dos heme-PAS domain, however, exhibits similar dissociation constants for the two ligands, effectively showing a large discrimination between CO and oxygen. Dos has a unique hexacoordinate heme-iron, which is coordinated to histidine and methionine residues at the axial positions. Replacement of methionine with isoleucine prevented hexacoordination as well as the high degree of ligand discrimination between oxygen and CO (71). The unusual coordination is likely responsible for the slow rate of ligand association, enablement of a large conformation change between hexa- and pentacoordination for signal transduction, and the ability to discriminate between oxygen and CO.

5.2.5   Bacteriophytochromes

Phytochromes are red and far red light sensors that were previously thought to be unique to plants and algae (72). Recently, proteins similar to phytochromes were reported in the cyanobacteria *Synechocystis* and *Fremyella diplosiphon* (73-76), in non-photosynthetic eubacteria *Deinococcus radiodurans* and *Pseudomonas aeruginosa* (77) and in several gram negative bacteria and members of the alpha-proteobacterium family. These phytochromes have homology to the two-component histidine kinase family (78). Also, unlike plant phytochromes, which have a phytochromobilin cofactor, these bacteriophytochromes bind the heme-degradation product, biliverdin. Several of the bacteriophytochrome operons have a gene for a heme oxygenase that can synthesize biliverdin. Also unlike plant phytochromes, bacteriophytochromes are missing the cysteine responsible for the thioether bond to the cofactor, and bind the biliverdin through a schiff linkage to a histidine that is adjacent to the conserved cysteine in plant phytochromes.

Jiang et al. (79) reported a unique protein in *Rhodospirillum centenum* that they named Ppr (for PYP-phytochrome related). Ppr has an N-terminal, cis-hydroxycinnamic-acid binding PYP domain, a central phytochrome-type domain and a C-terminal histidine kinase domain. The spectrum of the photocycle was typical of PYP, but the kinetics were protracted (79). The Ppr protein presented a novel opportunity to determine a physiological function for a PYP domain, as no role had yet been demonstrated despite the large body of knowledge on the protein itself. In a *ppr* strain, phototaxis was normal, but light-dependent regulation of chalcone synthase, an enzyme involved in flavenoid synthesis in plants, was altered (79).

Recently, two bacteriophytochromes that have a PAS domain replacing the histidine kinase domain and that control photosystem synthesis were found in the anoxygenic bacteria *Bradyrhizobium ORS278* and *Rhodopseudomonas palustris* but not the closely related *Rhodobacter, Rubrivivax or Rhodospirillum* (80). The sensors switch from an inactive red light absorbing state, with an absorbance maximum of 676 nm, to an active far red light absorbing state (752 nm). Furthermore, these two bacteria contain a *ppsR* gene (see section 5.1.3 on PpsR) upstream of the bacteriophytochrome that may belong to the same light regulatory pathway (80).

5.2.6   Bacterial Phototropins

Phototropins are the predominant photoreceptor for phototropism in plants. They are membrane-bound kinases that autophosphorylate in response to blue light. The architecture of phototropins includes two N-terminal FMN-binding PAS domains (originally called "LOV', for light, oxygen and voltage sensor) and a C-terminal serine/threonine domain. In response to blue-light, the FMN forms a reversible FMN-cysteine C4a-thiol adduct whereby FMN is reduced and the thiol is oxidized (13, 81, 82). The crystal structure of the LOV2 domain from the Adiantum phytochrome/phototropin chimeric photoreceptor phy3 was recently solved at 2.73-Å resolution (14).

The discovery of three bacterial phototropins was recently reported by Losi et al. (83). They include YtvA in *B. subtilis*, Q9ABE3 in *Caulobacter crescentus*, and Q55576 in *Synechocystis* sp. strain PCC 6803. YtvA bound FMN and exhibited similar photochemistry to the plant phototropins, forming a Thio383 adduct with FMN C4a (83). Although a role for bacterial phototropins remains to be determined, they are certain to comprise a large family. A recent cursory BLAST search yielded nearly 30 likely members of this domain (M. Johnson, unpublished data).

5.2.7 Photoactive Yellow Protein (PYP)

The first member of the PAS domain superfamily for which the crystal structure was resolved is the light sensor PYP from *E. halophila* (5, 84, 85). The protein is a 14-kDa cytosolic protein that contains a covalently bound p-hydroxycinnamoyl chromophore that undergoes a photocycle (86, 87). This chromophore exhibits a negative charge in the ground state that is stabilized by hydrogen bonding to the protonated carboxyl group of Glu46 as well as to the hydroxyl group of Tyr42 (5, 88). Light triggers a transition from trans to cis between double-bonded first and second aliphatic carbons of the chromophore, leading to protonation of the phenolate anion from Glu46 and ultimately to a conformational change in the protein. Computational theory predicts that the protonation, not the cis to trans conversion of the chromophore, leads to a structure change in the protein that propagates signaling (89, 90). Although structural, kinetic and mechanistic data for the PYP protein far exceeds similar data for other PAS domains, little is known about the components with which PYP interacts or about the biological function of PYP. As earlier mentioned, an opportunity to study PYP interactions with another domain was found in *Rhodospirillum centenum*, which has a phytochrome with an N-terminal PYP domain and a C-terminal histidine kinase (79). Named Ppr, for PYP-phytochrome related protein, the protein regulates chalcone synthase in response to light (see section 5.2.5 on bacteriophytochromes).

5.3   Sporulation

When the environment surrounding *B. subtilis* is adverse, the bacterium responds by forming dormant, heat-resistant endospores. More than 125 genes are involved in the sporulation process, and the trigger to sporulate involves integrating diverse environmental and physiological signals within a signal integration circuit (91). To accomplish this, an expanded, two-component phosphorelay system is employed to integrate the multiple signals. In *B. subtilis*, there are five closely related histidine kinases identified by bioinformatic studies that may phosphorylate the response regulator SpoOF: KinA, KinB, KinC, KinD, and KinE (92). Both KinA and Kin C contain PAS domains. KinA has been identified as the major source of phosphoryl groups for SpoOF (93), and KinB the secondary source. KinA is therefore the major sensor kinase controlling the initiation of development in *B. subtilis*.

KinA autophosphorylates a histidine residue in the histidine kinase domain. The phosphoryl moiety is passed through the His-Asp-His-Asp phosphorelay (94). That is, the phosphoryl group is transferred from KinA to an aspartate residue in the response regulator, SpoOF, then to a histidine in the phosphotransferase, SpoOB, and lastly to an aspartate in the response regulator SpoOA. An increase in SpoOA-P activates the transcription of genes that control entry into sporulation and transition to a two compartment sporangium in which gene transcription is regulated differentially (91). Phosphatases, specific for SpoOF-P or SpoOA-P may end transduction of the signal, however, if the conditions for sporulation are not favorable (95).

The KinA protein is a soluble cytoplasmic kinase that senses the internal environment of the cell and functions as a homodimer. KinA has a 400 amino acid N-terminal sensing region and a 200 amino acid C-terminal autokinase domain (96) (Figure 2C). The N-terminus is divided into three PAS domains, PAS-A, PAS-B, and PAS-C (4). The KinB protein does not contain a PAS domain and possibly responds to different signals than KinA (97). KinC protein is able to phosphorylate mutant forms of SpoOA and allows bypass of the phosphorelay (98, 99). It has a PAS domain (4), but it does not appear to be involved in the initiation of sporulation.

The PAS domains in KinA may contribute to the sporulation signal by sensing unfavorable conditions for cell growth. Dimerization of the KinA N-terminus is accomplished through the PAS-B/PAS-C region, while PAS-A is not dimerized (96). PAS-A however is important for KinA enzymatic activity. Wang et al. reported that deleting PAS-A caused more than 90% loss of the initial rate of autophosphorylation in the isolated protein (96). Amino acid changes in PAS-A (F77S) and in PAS-C (I280T) affected the *in vivo* activity of KinA, suggesting that both PAS domains are required for signal sensing. The N-terminal region with its PAS domains therefore appears to be essential for both signal sensing and for maintaining the correct conformation of the autokinase domain. This may be similar to the situation with the *E. coli* Aer protein where the C-terminus is required for maintaining the conformation of the N-terminal PAS domain (48). Wang et al. also noted that purified KinA has no absorption spectrum for any of the ligands known to bind to PAS domains (FAD, FMN, or heme), or any other chromophore (96).

What the PAS domains bind and how they regulate KinA activity remains a mystery. It was recently demonstrated, however, that isolated PAS-A fragment binds ATP and catalyzes exchange of phosphate between ATP and nucleoside diphosphates (38). A C75A mutation in PAS-A increased the affinity for ATP five-fold and stimulated KinA-dependent sporulation. It is counterintuitive to correlate high ATP concentration with the energy-depleting conditions that initiate sporulation (38). Stephenson and Hoch proposed that hydrolysis of ATP may energize conformational signaling in response to the environmentally important stimuli that are detected by KinA (38). PAS domains are widely documented as sensory input domains without reported catalytic activity, therefore the finding of intrinsic nucleoside diphosphate kinase activity is a newly proposed role for PAS domains. If it is not an artifact, this observed activity of the KinA PAS-A domain is noteworthy.

5.4   Nitrogen Fixation and Nitrogen Metabolism

As an essential component of the global nitrogen cycle, symbiotic and free-living bacteria are responsible for biological nitrogen fixation. This process is very sensitive to oxygen due to an oxygen-labile nitrogenase. To avoid inactivation to the nitrogenase, expression of nitrogen fixation genes in diazotrophs is regulated by oxygen. All known oxygen sensors that regulate nitrogen fixation, and other putative sensors that regulate nitrogen fixation and nitrogen metabolism contain PAS domains.

5.4.1   FixL: A Direct Oxygen Sensor

FixL is the second of two known direct oxygen sensors (see section 5.2.4 on Dos). The crystal structures of FixL from *B. japonicum* and *R. meliloti* have been characterized, and a detailed mechanistic description can be found in Chapter 1 of this volume. The FixL protein is a paradigm for oxygen sensors. It senses changes in the oxygen tension while symbiotically associated with plant roots, and signals to the response regulator, FixJ, to regulate the expression of genes involved in nitrogen fixation (100). Oxygen binds to a heme cofactor bound to the PAS domain. Phosphorylation is stimulated in the deoxy- state, whereas in the oxy- state, autophosphorylation is inhibited.

The P50 of oxygen for FixL is in the range of 17 to 70 mm Hg (101). Concerted changes occur due to steric hindrance after oxygen binds. The sixth coordination site is surrounded by a hydrophobic triad of non-polar residues and oxygen binding causes a shift in this triad that relieves the stress of these hydrophobic residues with oxygen (102). Their movement is transmitted to the FG loop which is thought to communicate with the kinase domain.

5.4.2   NifL

The sensor protein NifL and the cognate transcriptional activator NifA constitute an atypical two-component regulatory system that regulates the expression of genes involved in nitrogen fixation in free-living bacteria such as *Azoarcus spp.*, *Enterobacter agglomerans*, *Klebsiella pneumoniae*, *Klebsiella oxytoca* and *Azotobacter vinelandii* [(103), and SMART database]. The system is atypical because signal transduction occurs via complex formation between NifL and NifA, rather than by the classical phosphotransfer mechanism (Figure 2D). NifA activates the transcription of σ^N-dependent *nif* promoters involved in the synthesis of nitrogenase (103). The expression of NifA in *A. vinelandii* is constitutive, and regulation of NifA activity by NifL is the primary level of nitrogen control (104). In response to high external oxygen or fixed nitrogen, NifL inhibits the response regulator NifA to prevent the synthesis of nitrogenase in physiologically unfavorable conditions. Stoichiometric levels of NifL and NifA are required for regulation, supporting the hypothesis that signal transduction involves protein: protein interaction (and conformational changes) rather than covalent modification.

The NifL protein is composed of two domains joined by a Q linker (105). The N-terminal domain contains a PAS domain (1) with significant homology to FixL proteins and the *bat* gene product of *Halobacterium salinarum* (106), indicating that this domain may be responsive to oxygen signals (105). The C-terminal domain shows significant homology to histidine kinase transmitter domains including the region containing the conserved histidine residue (105). There is however, no evidence for phosphotransfer from NifL to NifA, and the conserved histidine that is phosphorylated in other kinases, is not necessary for the inhibitory activity of NifL (107). In addition, no autophosphorylation of NifL or phosphotransfer to NifA has been detected (8, 103). Instead, the C-terminal region of NifL is involved in ADP-dependent stimulation of NifL-NifA complex formation, and a fragment containing the last 127 residues is sufficient to bind and inhibit NifA (108, 109) (Figure 2D).

The N-terminal PAS domain of NifL is predicted to act primarily as a sensor that transmits a redox signal to the C-terminus, because the PAS domains of *A. vinelandii* and *K. pneumoniae* NifL both bind FAD noncovalently [(7) and (7, 8) respectively]. In *A. vinelandii* NifL, residues 1-147 are essential for FAD binding (110) and the truncated protein, which lacks the PAS domain, does not respond to redox changes. The PAS domain is distinct from the C-terminal nitrogen-responsive and nucleotide-binding domains as the removal of the PAS domain from NifL does not prevent ADP-dependent stimulation of NifL binding to NifA (110). The *A. vinelandii* NifL truncated protein (residues 147-519) purifies as a dimer, whilst the PAS domain (residues 1-284) purifies as a tetramer, suggesting that the PAS domain may also contain oligomerization determinants (110).

The redox-sensitive PAS domain acts as a switch for the regulatory activity of NifL. The oxidized form of *A. vinelandii* NifL inhibits NifA activity and when FAD is reduced (under anaerobic conditions), NifA activity is unaffected (7). The redox potential of NifL FAD is -226mV at pH8.0 and it can readily be reduced *in vitro* by various electron donors and NAD(P)H-dependent enzymes (111). Recently, *K. pneumoniae* NifL was shown to be membrane-associated if the bacteria were grown under oxygen- and nitrogen-limited conditions, suggesting that reduction of NifL FAD may be carried out by a membrane-associated electron donor (112). The authors suggested that membrane-association might create a spatial gap between NifL and NifA that keeps NifL from interacting with NifA in the reduced form. The inhibitory activity of NifL is also effected by adenosine nucleotides *in vitro*, which suggests an ability to sense cellular energy (109). The response to nucleotides overrides the redox switch and the possibility exists that NifL complexing with NifA is controlled by the ATP/ADP ratio (109).

The response to nitrogen status in several species of bacteria containing NifA is believed to be transduced by PII proteins such as PII-like GlnK protein. This protein has been implicated in the regulation of nitrogen fixation in response to NH_4^+ by relieving or stimulating NifL inhibition of NifA (104). Unmodified GlnK stimulates NifL inhibition and uridylylation of GlnK in response to nitrogen limitation prevents this function in *A. vinelandii* (104). Rudnick and colleagues indicated that this is different from the proposed situation in *K. pneumoniae*, in which unmodified GlnK relieves NifL inhibition instead of stimulating it. Another difference in NifL regulation is found in *K. pneumoniae* where the primary oxygen receptor appears to be Fnr (113). Fnr is thought to signal a lack of oxygen to NifL by activating the transcription of genes whose products relieve NifL inhibition through reduction of FAD. In contrast, the reduction of *A. vinelandii* NifL appears to occur unspecifically in response to the availability of reducing equivalents in the cell (113).

5.4.3   NtrB

The nitrogen regulatory proteins, NtrB and NtrC, comprise a two-component system that has been identified in 18 species of bacteria [(114), SMART database]. NtrB contains an N-terminal PAS domain. Phosphorylation of NtrC by NtrB under nitrogen-limiting conditions results in the binding of NtrC-P to promoters and transcriptional activation of nitrogen-regulated genes (114). Earlier it was proposed that NtrC-P might activate the transcription of a gene(s) whose product(s) functions to relieve NifL inhibition under nitrogen-limitation (115). More recently, microarray studies were used to detect all operons under the control of NtrC in *Escherichia coli*, and many of the operons were found to encode transport systems for nitrogen-containing compounds (116).

The NtrB protein is a histidine kinase. It has been divided into a sensor domain (S region) in the N-terminus, and a transmitter domain (H, N and G regions) in the central region and C-terminus. The transmitter domain contains the phosphotransferase, phosphatase, and kinase functions. The H region contains the conserved histidine residue and the glycine-rich G region is thought to be important for ATP-binding. The sensor domain contains the PAS domain (2) that is important for intramolecular signal transduction (117) and is joined to the transmitter module by a Q linker. NtrB is a bifunctional histidine kinase; it can autophosphorylate the conserved histidine and donate the phosphate to the response regulator, NtrC, or it can act as a phosphatase, promoting the rapid hydrolysis of NtrC-P. NtrB senses the nitrogen status of the cell via the PII protein and PII is responsible for inhibiting the kinase activity of NtrB and activating its phosphatase activity. PII is uridylylated under nitrogen limiting conditions, and deuridylylated under conditions of nitrogen excess (118). It is the nonuridylylated form of PII that interacts with NtrB to determine the balance of the phosphorylation and dephosphorylation reactions (119, 120), which in turn affects NtrC phosphorylation. NtrB functions as a dimer (121), and the sensor domain, particularly residues 111-115 at the C-terminal end of the PAS domain, provide the major dimerization determinants for *K. pneumoniae* NtrB (122). In contrast, interaction between NtrB and NtrC involves the HN region of the NtrB transmitter domain and the NtrC receiver domain (122).

One of the functions of the sensor PAS domain is to control the interaction of the H and G regions (123); compare reference (2). When these domains interact, NtrB acts as a positive regulator, otherwise, NtrB acts as a negative regulator. Kramer and Weiss suggested that the function of the sensor PAS domain may be to change the orientation of the transmitter domains within the NtrB dimer. The G domain of one subunit reorients so that it is able to interact with the H domain of the other subunit, leading to phosphorylation of the histidine. Jiang et al. (117) similarly suggested that all domains of NtrB are involved in regulating the kinase and phosphatase activities and showed that the N-terminus controls access to the active site for an isolated dimeric central domain added in trans. They also showed that PII interacts with the isolated NtrB transmitter module and not the sensor domain (117). However the phosphatase activity of the isolated transmitter module was low, even in the presence of PII, again suggesting that the sensor domain was necessary for the central region of NtrB to assume the correct conformation. Although these functions have been attributed to the sensor domain in the N-terminus of NtrB, direct involvement of the PAS domain has yet to be elucidated.

5.4.4   NtrY

The NtrY and NtrX proteins are probably involved in nitrogen fixation and metabolism and comprise a two-component system that has been identified in 10 species of bacteria [SMART database, (124) and references within]. NtrY is likely to be located in the membrane and may be a sensor of extracellular nitrogen concentration (125). A PAS domain has been identified in the cytoplasmic region of NtrY upstream of the histidine kinase-like domain in the C-terminus (1, 4). The NtrX protein, which shares a high degree of homology with NtrC proteins, is probably a response regulator (125). In *Azorhizobium caulinodans*, expression of the *ntrYX* operon was repressed in an *ntrC* mutant grown in the presence of nitrate, suggesting interaction between the *ntrYX/ntrBC* systems (125). In the same study NtrY/NtrX were also found to modulate *NifA* expression and *ntrYX* transcription was shown to be partially under the control of NtrC (125). The exact role of the PAS domain in NtrY function has not yet been shown, however it may have a role in oxygen (redox) sensing in the regulation of nitrogen metabolism (2, 125).

5.5   Virulence

The BvgS and BvgA proteins comprise a two-component system responsible for regulating virulence gene expression in the human pathogens *Bordetella   pertussis*,   *Bordetella   parapertussis*,   and   *Bordetella bronchiseptica* [(126) and SMART database]. B. pertussis is the etiological agent   for   the   contagious   childhood   disease,   whooping   cough;   B.

parapertussis causes a pertussis-like syndrome, while B. bronchieptica is associated with human infections in severely compromised individuals. The 135 kDa BvgS protein, which contains a PAS domain, is a transmembrane sensor kinase that detects environmental signals (Figure 5).

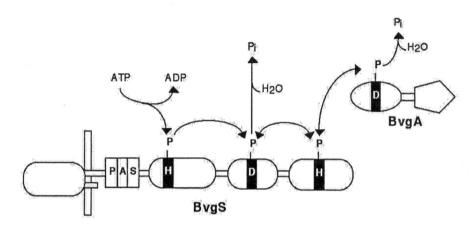

Figure 5. Phosphorelay system in BvgS [modified from (2)].

BvgS has an N-terminal periplasmic domain linked by a transmembrane region to a cytoplasmic PAS domain, transmitter, receiver, and histidine phosphotransfer domains (2, 40, 41) The BvgA protein is a 23-kDa cytoplasmic response regulator which has a receiver domain and a C-terminal DNA-binding domain (127). Transfer of the phosphoryl group from BvgS to BvgA increases the affinity of BvgA for promoters that activate the transcription of genes involved in virulence (128, 129).

In response to environmental signals, BvgS and BvgA mediate the transition between different phases (130). During the virulent phase, all three *Bordetella* bacteria produce nearly identical virulence factors that include adhesins such as filamentous hemagglutinin, fimbriae and pertactin, as well as toxins such as a bifunctional adenylate cyclase/hemolysin, dermonecrotic toxin, tracheal cytotoxin, a *B. pertussis* specific pertussis toxin, and *B. bronchiseptica* specific type III secreted proteins [(130), and references within]. Pathogenicity requires that several of the *Bordetella* virulence genes be differentially regulated (131).

The BvgAS system also employs a His-Asp-His-Asp phosphorelay, however unlike the Kin phosphorelay (see section 5.3 on KinA), BvgS contains the first three phosphotransfer domains within the same protein (40, 41). BvgA is a typical response regulator and contains the fourth phosphotransfer domain (40, 41, 132)

Although earlier studies of the BvgS protein identified functional mutations in the "linker region", our analysis shows that these mutations are located in the PAS domain (K. Watts, unpublished observation). A putative ATP binding site in the PAS domain of BvgS has been identified (133) and mutations within the site (G624V with K625L) inactivated BvgS function. ATP binding to the PAS-A domain of KinA was demonstrated recently [see section 5.3 on KinA and (38)], however this has not been confirmed for the BvgS PAS domain. Transcription of the *bvgA*-activated genes can nearly be eliminated by the presence of sulfate anion or nicotinic acid and growth at low temperatures (134, 135). Various mutations shown to confer a constitutive phenotype, and desensitize BvgS to these environmental signals, are located in the PAS domain. These include S682P (133), G680S (136) and G688S (135). The G688S mutation also eliminates the requirement for an intact periplasmic domain for functional BvgS. G688S is equivalent to the mutation G110S in the Aer PAS domain that confers a null phenotype (10). Certain deletion or insertion mutations that inactivate BvgS (135) are also located in the PAS domain.

Bock and Gross recently suggested that BvgS is connected to the oxidation status of the cell via a link to the ubiquinone pool (137). Oxidized ubiquinone-0 was found to be a strong inhibitor of BvgS kinase activity *in vitro* with half maximal inhibition occurring at 11μM (137). Kinase activity can be turned off and on by changing the oxidation status of the quinone (137). Ubiquinone is also reported to inhibit the kinase activity of ArcB [see section 5.2.1 on ArcB and (59)]. The PAS domains of BvgS and ArcB are suggested as likely candidates for interacting with quinones, although proof is lacking. Furthermore, mutations introduced into the PAS domain caused only a weak decrease in quinone-dependent inhibition of autophosphorylation.

6. CONCLUSION

The rapid expansion of the prokaryotic PAS domain superfamily presents an excellent opportunity to study the general mechanisms of signal transduction as well as the common themes by which PAS domains sense and signal. Many of these domains, such as those found in FixL, KinA, ArcB, Ppr, NtrB and BvgS, are found in histidine kinase proteins, and are involved in two-component or multi-component phosphorelay cascades. Other diverse roles are being compiled. The unique direct oxygen sensor, Dos, is the first example of second messenger regulation by oxygen. A new catalytic role for a PAS domain has been proposed for KinA. The isolated PAS-A domain of KinA binds ATP and catalyzes phosphate exchange between ATP and nucleoside diphosphates. PpsR, which regulates photosystem II in *Rhodobacter spp.*, forms a complex with AppA, the first reported dual sensing protein, where both redox and light signals are integrated. Association between AppA and PpsR likely occurs through the PAS domain of PpsR. The discovery of the bacteriophytochrome PprR which has both PYP and histidine kinase domains, presents a unique opportunity to study PYP interactions with a histidine kinase domain.

Whereas early studies limited the function of PAS domains to one of a dimerization domain, more recent work demonstrates a more versatile role for PAS domains that includes sensing, homodimerization, heterodimerization, domain/domain association, and signaling. Studies on Aer and PpsR suggest that signaling via physical interaction between PAS domains and a signaling protein/domain may be a recurring paradigm. This model is supported by recent NMR studies on human PAS kinase, which demonstrated specific contact points between the PAS and kinase domains when added in trans. As similar studies are performed on other PAS proteins, additional mechanisms are likely to emerge.

REFERENCES

1. Zhulin, I. B., B. L. Taylor, and R. Dixon. 1997. PAS domain S-boxes in Archaea, Bacteria and sensors for oxygen and redox. *Trends Biochem. Sci.* 22:331-3.

2. Taylor, B. L., and I. B. Zhulin. 1999. PAS domains: internal sensors of oxygen, redox potential, and light. *Microbiol. Mol. Biol. Rev.* 63:479-506.

3. Pellequer, J. L., K. A. Wager-Smith, S. A. Kay, and E. D. Getzoff. 1998. Photoactive yellow protein: a structural prototype for the three-dimensional fold of the PAS domain superfamily. *Proc. Natl. Acad. Sci. USA* 95:5884-90.

4. Ponting, C. P., and L. Aravind. 1997. PAS: a multifunctional domain family comes to light. *Curr. Biol.* 7:R674-7.

5. Borgstahl, G. E., D. R. Williams, and E. D. Getzoff. 1995. 1.4 A structure of photoactive yellow protein, a cytosolic photoreceptor: unusual fold, active site, and chromophore. *Biochemistry* 34:6278-87.

6. Gong, W., B. Hao, S. S. Mansy, G. Gonzalez, M. A. Gilles-Gonzalez, and M. K. Chan. 1998. Structure of a biological oxygen sensor: a new mechanism for heme-driven signal transduction. *Proc. Natl. Acad. Sci. USA* 95:15177-82.

7. Hill, S., S. Austin, T. Eydmann, T. Jones, and R. Dixon. 1996. *Azotobacter vinelandii* NIFL is a flavoprotein that modulates transcriptional activation of nitrogen-fixation genes via a redox-sensitive switch. *Proc. Natl. Acad. Sci. USA* 93:2143-8.

8. Schmitz, R. A. 1997. NifL of *Klebsiella pneumoniae* carries an N-terminally bound FAD cofactor, which is not directly required for the inhibitory function of NifL. *FEMS Microbiol. Lett.* 157:313-8.

9. Bibikov, S. I., R. Biran, K. E. Rudd, and J. S. Parkinson. 1997. A signal transducer for aerotaxis in *Escherichia coli. J. Bacteriol.* 179:4075-9.

10. Bibikov, S. I., L. A. Barnes, Y. Gitin, and J. S. Parkinson. 2000. Domain organization and flavin adenine dinucleotide-binding determinants in the aerotaxis signal transducer Aer of *Escherichia coli. Proc. Natl. Acad. Sci. USA* 97:5830-5.

11. Rebbapragada, A., M. S. Johnson, G. P. Harding, A. J. Zuccarelli, H. M. Fletcher, I. B. Zhulin, and B. L. Taylor. 1997. The Aer protein and the serine chemoreceptor Tsr independently sense intracellular energy levels and transduce oxygen, redox, and energy signals for *Escherichia coli* behavior. *Proc. Natl. Acad. Sci. USA* 94:10541-6.

12. Repik, A., A. Rebbapragada, M. S. Johnson, J. O. Haznedar, I. B. Zhulin, and B. L. Taylor. 2000. PAS domain residues involved in signal transduction by the Aer redox sensor of *Escherichia coli. Mol. Microbiol.* 36:806-16.

13. Christie, J. M., M. Salomon, K. Nozue, M. Wada, and W. R. Briggs. 1999. LOV (light, oxygen, or voltage) domains of the blue-light photoreceptor phototropin (nph1): binding sites for the chromophore flavin mononucleotide. *Proc. Natl. Acad. Sci. USA* 96:8779-83.

14. Crosson, S., and K. Moffat. 2001. Structure of a flavin-binding plant photoreceptor domain: insights into light-mediated signal transduction. *Proc. Natl. Acad. Sci. USA* 98:2995-3000.

15. Taylor, B. L., I. B. Zhulin, and M. S. Johnson. 1999. Aerotaxis and other energy-sensing behavior in bacteria. *Annu. Rev. Microbiol.* 53:103-28.

16. Taylor, B. L., A. Rebbapragada, and M. S. Johnson. 2001. The FAD-PAS domain as a sensor for behavioral responses in *Escherichia coli. Antioxid. Redox Signal* 3:867-79.

17. Harold, F. M., and P. C. Maloney. 1996. Energy transduction by ion currents, p. 283-306. In F. C. Neidhardt (ed.), *Escherichia coli* and *Salmonella:* Cellular and Molecular Biology, second ed, vol. 1. ASM Press, Washington, D.C.

18. Hyung, S., J. Saw, S. Hou, R. W. Larsen, K. J. Watts, M. S. Johnson, M. A. Zimmer, G. W. Ordal , B. L. Taylor, and M. Alam. 2002. Aerotactic responses in bacteria to photoreleased oxygen. *FEMS Microbiol. Lett.* 217:237-242.

19. Johnson, M. S., and B. L. Taylor. 1993. Comparison of methods for specific depletion of ATP in *Salmonella typhimurium. Appl. Environ. Microbiol.* 59:3509-12.

20. Taylor, B. L., and I. B. Zhulin. 1998. In search of higher energy: metabolism-dependent behaviour in bacteria. *Mol. Microbiol.* 28:683-90.

21. Taylor, B. L. 1983. Role of proton motive force in sensory transduction in bacteria. *Annu. Rev. Microbiol.* 37:551-73.

22. Taylor, B. L., J. B. Miller, H. M. Warrick, and D. E. Koshland, Jr. 1979. Electron acceptor taxis and blue light effect on bacterial chemotaxis. *J. Bacteriol.* 140:567-73.

23. Bespalov, V. A., I. B. Zhulin, and B. L. Taylor. 1996. Behavioral responses of *Escherichia coli* to changes in redox potential. *Proc. Natl. Acad. Sci. USA* 93:10084-9.

24. Zhulin, I. B., E. H. Rowsell, M. S. Johnson, and B. L. Taylor. 1997. Glycerol elicits energy taxis of *Escherichia coli* and *Salmonella typhimurium. J. Bacteriol.* 179:3196-201.

25. Alexandre, G., S. E. Greer, and I. B. Zhulin. 2000. Energy taxis is the dominant behavior in *Azospirillum brasilense. J. Bacteriol.* 182:6042-8.

26. Zhulin, I. B., V. A. Bespalov, M. S. Johnson, and B. L. Taylor. 1996. Oxygen taxis and proton motive force in *Azospirillum brasilense. J. Bacteriol.* 178:5199-204.

27. Zhulin, I. B., M. S. Johnson, and B. L. Taylor. 1997. How do bacteria avoid high oxygen concentrations? *Biosci. Rep.* 17:335-42.

28. Miyatake, H., M. Mukai, S. Y. Park, S. Adachi, K. Tamura, H. Nakamura, K. Nakamura, T. Tsuchiya, T. Iizuka, and Y. Shiro. 2000. Sensory mechanism of oxygen

sensor FixL from *Rhizobium meliloti:* crystallographic, mutagenesis and resonance Raman spectroscopic studies. *J. Mol. Biol.* 301:415-31.

29.   Morais Cabral, J. H., A. Lee, S. L. Cohen, B. T. Chait, M. Li, and R. Mackinnon. 1998. Crystal structure and functional analysis of the HERG potassium channel N terminus: a eukaryotic PAS domain. *Cell* 95:649-55.

30.   Amezcua, C., S. Harper, J. Rutter, and K. Gardner. 2002. Structure and interactions of PAS Kinase N-terminal PAS domain. Model for intramolecular kinase regulation. *Structure (Camb)* 10:1349.

31.   Rutter, J., C. H. Michnoff, S. M. Harper, K. H. Gardner, and S. L. McKnight. 2001. PAS kinase: an evolutionarily conserved PAS domain-regulated serine/threonine kinase. *Proc. Natl. Acad. Sci. USA* 98:8991-6.

32.   Rutter, J., B. Probst, and S. McKnight. 2002. Coordinate Regulation of Sugar Flux and Translation by PAS Kinase. *Cell* 111:17.

33.   Zhulin, I. B., and B. L. Taylor. 1998. Correlation of PAS domains with electron transport-associated proteins in completely sequenced microbial genomes. *Mol. Microbiol.* 29:1522-3.

34.   Stock, J. B., A. J. Ninfa, and A. M. Stock. 1989. Protein phosphorylation and regulation of adaptive responses in bacteria. *Microbiol. Rev.* 53:450-90.

35.   Eisenbach, M. 1996. Control of bacterial chemotaxis. *Mol. Microbiol.* 20:903-10.

36.   Barak, R., and M. Eisenbach. 1996. Regulation of interaction between signaling protein CheY and flagellar motor during bacterial chemotaxis. *Curr. Top. Cell. Regul.* 34:137-58.

37.   Grimshaw, C. E., S. Huang, C. G. Hanstein, M. A. Strauch, D. Burbulys, L. Wang, J. A. Hoch, and J. M. Whiteley. 1998. Synergistic kinetic interactions between components of the phosphorelay controlling sporulation in *Bacillus subtilis.* *Biochemistry* 37:1365-75.

38.   Stephenson, K., and J. A. Hoch. 2001. PAS-A domain of phosphorelay sensor kinase A: a catalytic ATP-binding domain involved in the initiation of development in *Bacillus subtilis. Proc. Natl. Acad. Sci. USA* 98:15251-6.

39.   Appleby, J. L., J. S. Parkinson, and R. B. Bourret. 1996. Signal transduction via the multi-step phosphorelay: not necessarily a road less traveled. *Cell* 86:845-8.

40.   Uhl, M. A., and J. F. Miller. 1996. Central role of the BvgS receiver as a phosphorylated intermediate in a complex two-component phosphorelay. *J. Biol. Chem.* 271:33176-80.

41.   Uhl, M. A., and J. F. Miller. 1996. Integration of multiple domains in a two-component sensor protein: the *Bordetella pertussis* BvgAS phosphorelay. *EMBO J.* 15:1028-36.

42.   Matsushika, A., and T. Mizuno. 2000. Characterization of three putative sub-domains in the signal-input domain of the ArcB hybrid sensor in *Escherichia coli. J. Biochem. (Tokyo)* 127:855-60.

43.   Jorgensen, B. B. 1982. Ecology of the bacteria of the sulphur cycle with special reference to anoxic-oxic interface environments. *Philos. Trans. R. Soc. Lond. B Biol. Sci.* 298:543-61.

44.   Canfield, D. E., and D. J. Des Marais. 1991. Aerobic sulfate reduction in microbial mats. *Science* 251:1471-3.

45.   Donaghay, P. L., Rimes, H. M., Sieburth, J. McN. 1992. Simultaneous sampling of fine scale biological, chemical and physical structure in stratified waters. *Arch. Hydrobiol. Beih. Ergebn. Limnol.* 36:97-108.

46. Zhulin, I. B., Taylor, B. L. 1995. Chemotaxis in plant-associated bacteria: the search for the ecological niche, p. 451-459. In I. Fendrik (ed.), *Azospirillum VI and Related Microorganisms*. Springer-Verlag, Berlin.

47. Aravind, L., and C. P. Ponting. 1999. The cytoplasmic helical linker domain of receptor histidine kinase and methyl-accepting proteins is common to many prokaryotic signalling proteins. *FEMS Microbiol. Lett.* 176:111-6.

48. Ma, Q. 2001. HAMP domain and signaling mechanism of the Aer protein. Ph.D dissertation. Loma Linda University, Loma Linda, CA.

49. Maddock, J. R., and L. Shapiro. 1993. Polar location of the chemoreceptor complex in the *Escherichia coli* cell. *Science* 259:1717-23.

50. Lamanna, A. C., J. E. Gestwicki, L. E. Strong, S. L. Borchardt, R. M. Owen, and L. L. Kiessling. 2002. Conserved amplification of chemotactic responses through chemoreceptor interactions. *J. Bacteriol.* 184:4981-7.

51. Ames, P., C. A. Studdert, R. H. Reiser, and J. S. Parkinson. 2002. Collaborative signaling by mixed chemoreceptor teams in *Escherichia coli*. *Proc. Natl. Acad. Sci. USA* 99:7060-5.

52. Kim, K. K., H. Yokota, and S. H. Kim. 1999. Four-helical-bundle structure of the cytoplasmic domain of a serine chemotaxis receptor. *Nature* 400:787-92.

53. Bren, A., and M. Eisenbach. 2000. How signals are heard during bacterial chemotaxis: protein-protein interactions in sensory signal propagation. *J. Bacteriol.* 182:6865-73.

54. Bren, A., and M. Eisenbach. 2001. Changing the direction of flagellar rotation in bacteria by modulating the ratio between the rotational states of the switch protein FliM. *J. Mol. Biol.* 312:699-709.

55. Iuchi, S., and E. C. Lin. 1988. *arcA (dye)*, a global regulatory gene in *Escherichia coli* mediating repression of enzymes in aerobic pathways. *Proc. Natl. Acad. Sci. USA* 85:1888-92.

56. Iuchi, S., D. C. Cameron, and E. C. Lin. 1989. A second global regulator gene (arcB) mediating repression of enzymes in aerobic pathways of *Escherichia coli*. *J. Bacteriol.* 171:868-73.

57. Lynch, A. S., Lin, E. C. C. 1996. Responses to molecular oxygen, p. 1526-1538. In F. C. Neidhardt (ed.), *Escherichia coli* and *Salmonella*: Cellular and Molecular Biology, second ed, vol. 1. ASM Press, Washington, D.C.

58. Kwon, O., D. Georgellis, A. S. Lynch, D. Boyd, and E. C. Lin. 2000. The ArcB sensor kinase of *Escherichia coli:* genetic exploration of the transmembrane region. *J. Bacteriol.* 182:2960-6.

59. Georgellis, D., O. Kwon, and E. C. Lin. 2001. Quinones as the redox signal for the Arc two-component system of bacteria. *Science* 292:2314-6.

60. Iuchi, S., and E. C. Lin. 1992. Mutational analysis of signal transduction by ArcB, a membrane sensor protein responsible for anaerobic repression of operons involved in the central aerobic pathways in *Escherichia coli*. *J. Bacteriol.* 174:3972-80.

61. Iuchi, S. 1993. Phosphorylation/dephosphorylation of the receiver module at the conserved aspartate residue controls transphosphorylation activity of histidine kinase in sensor protein ArcB of *Escherichia coli*. *J. Biol. Chem.* 268:23972-80.

62. Pruss, B. M., J. W. Campbell, T. K. Van Dyk, C. Zhu, Y. Kogan, and P. Matsumura. 2003. FlhD/FlhC is a regulator of anaerobic respiration and the Entner-Doudoroff pathway through induction of the methyl-accepting chemotaxis protein Aer. *J. Bacteriol.* 185:534-543.

63. Gomelsky, M., I. M. Horne, H. J. Lee, J. M. Pemberton, A. G. McEwan, and S. Kaplan. 2000. Domain structure, oligomeric state, and mutational analysis of PpsR,

the *Rhodobacter sphaeroides* repressor of photosystem gene expression. *J. Bacteriol.* 182:2253-61.

64.   Penfold, R. J., and J. M. Pemberton. 1994. Sequencing, chromosomal inactivation, and functional expression in *Escherichia coli* of *ppsR*, a gene which represses carotenoid and bacteriochlorophyll synthesis in Rhodobacter sphaeroides. *J. Bacteriol.* 176:2869-76.

65.   Ponnampalam, S. N., J. J. Buggy, and C. E. Bauer. 1995. Characterization of an aerobic repressor that coordinately regulates bacteriochlorophyll, carotenoid, and light harvesting-II expression in *Rhodobacter capsulatus*. *J. Bacteriol.* 177:2990-7.

66.   Gomelsky, M., and S. Kaplan. 1997. Molecular genetic analysis suggesting interactions between AppA and PpsR in regulation of photosynthesis gene expression in *Rhodobacter sphaeroides* 2.4.1. *J. Bacteriol.* 179:128-34.

67.   Masuda, S., and C. Bauer. 2002. AppA Is a blue light photoreceptor that antirepresses photosynthesis gene expression in *Rhodobacter sphaeroides*. *Cell* 110:613.

68.   Oh, J. I., and S. Kaplan. 2000. Redox signaling: globalization of gene expression. *EMBO J.* 19:4237-47.

69.   Braatsch, S., M. Gomelsky, S. Kuphal, and G. Klug. 2002. A single flavoprotein, AppA, integrates both redox and light signals in *Rhodobacter sphaeroides*. *Mol. Microbiol.* 45:827-36.

70.   Delgado-Nixon, V. M., G. Gonzalez, and M. A. Gilles-Gonzalez. 2000. Dos, a heme-binding PAS protein from *Escherichia coli*, is a direct oxygen sensor. *Biochemistry* 39:2685-91.

71.   Gonzalez, G., E. M. Dioum, C. M. Bertolucci, T. Tomita, M. Ikeda-Saito, M. R. Cheesman, N. J. Watmough, and M. A. Gilles-Gonzalez. 2002. Nature of the displaceable heme-axial residue in the EcDos protein, a heme-based sensor from *Escherichia coli*. *Biochemistry* 41:8414-21.

72.   Quail, P. H. 1991. Phytochrome: a light-activated molecular switch that regulates plant gene expression. *Annu. Rev. Genet.* 25:389-409.

73.   Kehoe, D. M., and A. R. Grossman. 1996. Similarity of a chromatic adaptation sensor to phytochrome and ethylene receptors. *Science* 273:1409-12.

74.   Hughes, J., T. Lamparter, F. Mittmann, E. Hartmann, W. Gartner, A. Wilde, and T. Borner. 1997. A prokaryotic phytochrome. *Nature* 386:663.

75.   Yeh, K. C., S. H. Wu, J. T. Murphy, and J. C. Lagarias. 1997. A cyanobacterial phytochrome two-component light sensory system. *Science* 277:1505-8.

76.   Elich, T. D., and J. Chory. 1997. Phytochrome: if it looks and smells like a histidine kinase, is it a histidine kinase? *Cell* 91:713-6.

77.   Davis, S. J., A. V. Vener, and R. D. Vierstra. 1999. Bacteriophytochromes: phytochrome-like photoreceptors from nonphotosynthetic eubacteria. *Science* 286:2517-20.

78.   Bhoo, S. H., S. J. Davis, J. Walker, B. Karniol, and R. D. Vierstra. 2001. Bacteriophytochromes are photochromic histidine kinases using a biliverdin chromophore. *Nature* 414:776-9.

79.   Jiang, Z., L. R. Swem, B. G. Rushing, S. Devanathan, G. Tollin, and C. E. Bauer. 1999. Bacterial photoreceptor with similarity to photoactive yellow protein and plant phytochromes. *Science* 285:406-9.

80.   Giraud, E., J. Fardoux, N. Fourrier, L. Hannibal, B. Genty, P. Bouyer, B. Dreyfus, and A. Vermeglio. 2002. Bacteriophytochrome controls photosystem synthesis in anoxygenic bacteria. *Nature* 417:202-5.

81. Salomon, M., J. M. Christie, E. Knieb, U. Lempert, and W. R. Briggs. 2000. Photochemical and mutational analysis of the FMN-binding domains of the plant blue light receptor, phototropin. *Biochemistry* 39:9401-10.

82. Salomon, M., W. Eisenreich, H. Durr, E. Schleicher, E. Knieb, V. Massey, W. Rudiger, F. Muller, A. Bacher, and G. Richter. 2001. An optomechanical transducer in the blue light receptor phototropin from *Avena sativa. Proc. Natl. Acad. Sci. USA* 98:12357-61.

83. Losi, A., E. Polverini, B. Quest, and W. Gartner. 2002. First evidence for phototropin-related blue-light receptors in prokaryotes. *Biophys. J.* 82:2627-34.

84. Meyer, T. E. 1985. Isolation and characterization of soluble cytochromes, ferredoxins and other chromophoric proteins from the halophilic phototrophic bacterium *Ectothiorhodospira halophila. Biochim. Biophys. Acta.* 806:175-83.

85. Meyer, T. E., E. Yakali, M. A. Cusanovich, and G. Tollin. 1987. Properties of a water-soluble, yellow protein isolated from a halophilic phototrophic bacterium that has photochemical activity analogous to sensory rhodopsin. *Biochemistry* 26:418-23.

86. Baca, M., G. E. Borgstahl, M. Boissinot, P. M. Burke, D. R. Williams, K. A. Slater, and E. D. Getzoff. 1994. Complete chemical structure of photoactive yellow protein: novel thioester-linked 4-hydroxycinnamyl chromophore and photocycle chemistry. *Biochemistry* 33:14369-77.

87. Hoff, W. D., P. Dux, K. Hard, B. Devreese, I. M. Nugteren-Roodzant, W. Crielaard, R. Boelens, R. Kaptein, J. van Beeumen, and K. J. Hellingwerf. 1994. Thiol ester-linked p-coumaric acid as a new photoactive prosthetic group in a protein with rhodopsin-like photochemistry. *Biochemistry* 33:13959-62.

88. Kim, M., R. A. Mathies, W. D. Hoff, and K. J. Hellingwerf. 1995. Resonance Raman evidence that the thioester-linked 4-hydroxycinnamyl chromophore of photoactive yellow protein is deprotonated. *Biochemistry* 34:12669-72.

89. Groenhof, G., M. F. Lensink, H. J. Berendsen, and A. E. Mark. 2002. Signal transduction in the photoactive yellow protein. II. Proton transfer initiates conformational changes. *Proteins* 48:212-9.

90. Groenhof, G., M. F. Lensink, H. J. Berendsen, J. G. Snijders, and A. E. Mark. 2002. Signal transduction in the photoactive yellow protein. I. Photon absorption and the isomerization of the chromophore. *Proteins* 48:202-11.

91. Stragier, P., and R. Losick. 1996. Molecular genetics of sporulation in *Bacillus subtilis. Annu. Rev. Genet.* 30:297-41.

92. Fabret, C., V. A. Feher, and J. A. Hoch. 1999. Two-component signal transduction in *Bacillus subtilis*: how one organism sees its world. *J. Bacteriol.* 181:1975-83.

93. Perego, M., S. P. Cole, D. Burbulys, K. Trach, and J. A. Hoch. 1989. Characterization of the gene for a protein kinase which phosphorylates the sporulation-regulatory proteins SpoOA and SpoOF of *Bacillus subtilis. J. Bacteriol.* 171:6187-96.

94. Burbulys, D., K. A. Trach, and J. A. Hoch. 1991. Initiation of sporulation in *B. subtilis* is controlled by a multicomponent phosphorelay. *Cell* 64:545-52.

95. Perego, M., and J. A. Hoch. 1996. Protein aspartate phosphatases control the output of two-component signal transduction systems. *Trends Genet.* 12:97-101.

96. Wang, L., C. Fabret, K. Kanamaru, K. Stephenson, V. Dartois, M. Perego, and J. A. Hoch. 2001. Dissection of the functional and structural domains of phosphorelay histidine kinase A of *Bacillus subtilis. J. Bacteriol.* 183:2795-802.

97. LeDeaux, J. R., N. Yu, and A. D. Grossman. 1995. Different roles for KinA, KinB, and KinC in the initiation of sporulation in *Bacillus subtilis. J. Bacteriol.* 177:861-3.

98.   LeDeaux, J. R., and A. D. Grossman. 1995. Isolation and characterization of kinC, a gene that encodes a sensor kinase homologous to the sporulation sensor kinases KinA and KinB in *Bacillus subtilis*. *J. Bacteriol.* 177:166-75.

99.   Kobayashi, K., K. Shoji, T. Shimizu, K. Nakano, T. Sato, and Y. Kobayashi. 1995. Analysis of a suppressor mutation ssb (kinC) of surOB20 (spoOA) mutation in *Bacillus subtilis* reveals that kinC encodes a histidine protein kinase. *J. Bacteriol.* 177:176-82.

100. Fischer, H. M. 1994. Genetic regulation of nitrogen fixation in *rhizobia*. *Microbiol Rev* 58:352-86.

101. Gilles-Gonzalez, M. A., G. Gonzalez, M. F. Perutz, L. Kiger, M. C. Marden, and C. Poyart. 1994. Heme-based sensors, exemplified by the kinase FixL, are a new class of heme protein with distinctive ligand binding and autoxidation. *Biochemistry* 33:8067-73.

102. Perutz, M. F., M. Paoli, and A. M. Lesk. 1999. FixL, a haemoglobin that acts as an oxygen sensor: signalling mechanism and structural basis of its homology with PAS domains. *Chem. Biol.* 6:R291-7.

103. Dixon, R. 1998. The oxygen-responsive NIFL-NIFA complex: a novel two-component regulatory system controlling nitrogenase synthesis in gamma-proteobacteria. *Arch. Microbiol.* 169:371-80.

104. Rudnick, P., C. Kunz, M. K. Gunatilaka, E. R. Hines, and C. Kennedy. 2002. Role of GlnK in NifL-mediated regulation of NifA activity in *Azotobacter vinelandii*. *J. Bacteriol.* 184:812-20.

105. Blanco, G., M. Drummond, P. Woodley, and C. Kennedy. 1993. Sequence and molecular analysis of the *nifL* gene of *Azotobacter vinelandii*. *Mol. Microbiol.* 9:869-79.

106. Gropp, F., and M. C. Betlach. 1994. The *bat* gene of *Halobacterium halobium* encodes a trans-acting oxygen inducibility factor. *Proc. Natl. Acad. Sci. USA* 91:5475-9.

107. Woodley, P., and M. Drummond. 1994. Redundancy of the conserved His residue in *Azotobacter vinelandii* NifL, a histidine autokinase homologue which regulates transcription of nitrogen fixation genes. *Mol. Microbiol.* 13:619-26.

108. Narberhaus, F., H. S. Lee, R. A. Schmitz, L. He, and S. Kustu. 1995. The C-terminal domain of NifL is sufficient to inhibit NifA activity. *J. Bacteriol.* 177:5078-87.

109. Money, T., T. Jones, R. Dixon, and S. Austin. 1999. Isolation and properties of the complex between the enhancer binding protein NIFA and the sensor NIFL. *J. Bacteriol.* 181:4461-8.

110. Soderback, E., F. Reyes-Ramirez, T. Eydmann, S. Austin, S. Hill, and R. Dixon. 1998. The redox- and fixed nitrogen-responsive regulatory protein NIFL from *Azotobacter vinelandii* comprises discrete flavin and nucleotide-binding domains. *Mol. Microbiol.* 28:179-92.

111. Macheroux, P., S. Hill, S. Austin, T. Eydmann, T. Jones, S. O. Kim, R. Poole, and R. Dixon. 1998. Electron donation to the flavoprotein NifL, a redox-sensing transcriptional regulator. *Biochem. J.* 332 (Pt 2):413-9.

112. Klopprogge, K., R. Grabbe, M. Hoppert, and R. A. Schmitz. 2002. Membrane association of *Klebsiella pneumoniae* NifL is affected by molecular oxygen and combined nitrogen. *Arch. Microbiol.* 177:223-34.

113. Schmitz, R. A., K. Klopprogge, and R. Grabbe. 2002. Regulation of nitrogen fixation in *Klebsiella pneumoniae* and *Azotobacter vinelandii*: NifL, transducing two environmental signals to the nif transcriptional activator NifA. *J. Mol. Microbiol. Biotechnol.* 4:235-42.

114. Ninfa, A. J., Atkinson, M. R., Komberov, E. S., Feng, J., Ninfa, E. G. 1995. Control of nitrogen assimilation by the NRI-NrII two-component system of enteric bacteria, p. 67-88. In J. A. Hoch, Silhavy, T. J. (ed.), Two-Component Signal Transduction. American Society for Microbiology, Washington, D. C.

115. He, L., E. Soupene, and S. Kustu. 1997. NtrC is required for control of *Klebsiella pneumoniae* NifL activity. *J. Bacteriol.* 179:7446-55.

116. Zimmer, D. P., E. Soupene, H. L. Lee, V. F. Wendisch, A. B. Khodursky, B. J. Peter, R. A. Bender, and S. Kustu. 2000. Nitrogen regulatory protein C-controlled genes of *Escherichia coli:* scavenging as a defense against nitrogen limitation. *Proc. Natl. Acad. Sci. USA* 97:14674-9.

117. Jiang, P., M. R. Atkinson, C. Srisawat, Q. Sun, and A. J. Ninfa. 2000. Functional dissection of the dimerization and enzymatic activities of *Escherichia coli* nitrogen regulator II and their regulation by the PII protein. *Biochemistry* 39:13433-49.

118. Arcondeguy, T., R. Jack, and M. Merrick. 2001. P(II) signal transduction proteins, pivotal players in microbial nitrogen control. *Microbiol Mol Biol Rev* 65:80-105.

119. Jiang, P., and A. J. Ninfa. 1999. Regulation of autophosphorylation of *Escherichia coli* nitrogen regulator II by the PII signal transduction protein. *J. Bacteriol.* 181:1906-11.

120. Pioszak, A. A., P. Jiang, and A. J. Ninfa. 2000. The *Escherichia coli* PII signal transduction protein regulates the activities of the two-component system transmitter protein NRII by direct interaction with the kinase domain of the transmitter module. *Biochemistry* 39:13450-61.

121. Ninfa, E. G., M. R. Atkinson, E. S. Kamberov, and A. J. Ninfa. 1993. Mechanism of autophosphorylation of *Escherichia coli* nitrogen regulator II (NRII or NtrB): trans-phosphorylation between subunits. *J. Bacteriol.* 175:7024-32.

122. Martinez-Argudo, I., J. Martin-Nieto, P. Salinas, R. Maldonado, M. Drummond, and A. Contreras. 2001. Two-hybrid analysis of domain interactions involving NtrB and NtrC two-component regulators. *Mol. Microbiol.* 40:169-78.

123. Kramer, G., and V. Weiss. 1999. Functional dissection of the transmitter module of the histidine kinase NtrB in *Escherichia coli. Proc. Natl. Acad. Sci. USA* 96:604-9.

124. Ishida, M. L., M. C. Assumpcao, H. B. Machado, E. M. Benelli, E. M. Souza, and F. O. Pedrosa. 2002. Identification and characterization of the two-component NtrY/NtrX regulatory system in *Azospirillum brasilense. Braz. J. Med. Biol. Res.* 35:651-61.

125. Pawlowski, K., U. Klosse, and F. J. de Bruijn. 1991. Characterization of a novel *Azorhizobium caulinodans* ORS571 two-component regulatory system, NtrY/NtrX, involved in nitrogen fixation and metabolism. *Mol. Gen. Genet.* 231:124-38.

126. Arico, B., J. F. Miller, C. Roy, S. Stibitz, D. Monack, S. Falkow, R. Gross, and R. Rappuoli. 1989. Sequences required for expression of *Bordetella pertussis* virulence factors share homology with prokaryotic signal transduction proteins. *Proc. Natl. Acad. Sci. USA* 86:6671-5.

127. Stibitz, S., and M. S. Yang. 1991. Subcellular localization and immunological detection of proteins encoded by the vir locus of *Bordetella pertussis. J. Bacteriol.* 173:4288-96.

128. Karimova, G., J. Bellalou, and A. Ullmann. 1996. Phosphorylation-dependent binding of BvgA to the upstream region of the cyaA gene of *Bordetella pertussis. Mol. Microbiol.* 20:489-96.

129. Steffen, P., S. Goyard, and A. Ullmann. 1996. Phosphorylated BvgA is sufficient for transcriptional activation of virulence-regulated genes in *Bordetella pertussis*. *EMBO J.* 15:102-9.

130. Mattoo, S., A. K. Foreman-Wykert, P. A. Cotter, and J. F. Miller. 2001. Mechanisms of *Bordetella* pathogenesis. *Front. Biosci.* 6:E168-86.

131. Kinnear, S. M., R. R. Marques, and N. H. Carbonetti. 2001. Differential regulation of Bvg-activated virulence factors plays a role in *Bordetella pertussis* pathogenicity. *Infect. Immun.* 69:1983-93.

132. Uhl, M. A., and J. F. Miller. 1994. Autophosphorylation and phosphotransfer in the *Bordetella pertussis* BvgAS signal transduction cascade. *Proc. Natl. Acad. Sci. USA* 91:1163-7.

133. Beier, D., H. Deppisch, and R. Gross. 1996. Conserved sequence motifs in the unorthodox BvgS two-component sensor protein of *Bordetella pertussis*. *Mol. Gen. Genet.* 252:169-76.

134. Weiss, A. A., and S. Falkow. 1984. Genetic analysis of phase change in *Bordetella pertussis*. *Infect. Immun.* 43:263-9.

135. Miller, J. F., S. A. Johnson, W. J. Black, D. T. Beattie, J. J. Mekalanos, and S. Falkow. 1992. Constitutive sensory transduction mutations in the *Bordetella pertussis bvgS* gene. *J. Bacteriol.* 174:970-9.

136. Manetti, R., B. Arico, R. Rappuoli, and V. Scarlato. 1994. Mutations in the linker region of BvgS abolish response to environmental signals for the regulation of the virulence factors in *Bordetella pertussis*. *Gene* 150:123-7.

137. Bock, A., and R. Gross. 2002. The unorthodox histidine kinases BvgS and EvgS are responsive to the oxidation status of a quinone electron carrier. *Eur. J. Biochem.* 269:3479-84.

Chapter 3

bHLH-PAS PROTEINS IN *C. ELEGANS*

Jo Anne Powell-Coffman
Iowa State University, Ames, IA 50011

1. INTRODUCTION

During development and homeostasis, individual cells must divide, differentiate, migrate, adapt to the environment, or die at the appropriate times and places. A key to deciphering the molecular mechanisms by which cells make these decisions is to characterize the regulation and function of the proteins that regulate important changes in gene expression. The family of transcription factors that contain basic-helix-loop-helix and PAS motifs has been shown to control many critical developmental events and to mediate responses to certain environmental stimuli. For example, bHLH-PAS proteins play central roles in the development of specific neural tissues and vasculature, and they are core components of the molecular clock that govern circadian rhythms. bHLH-PAS proteins are also integral to the pathways that sense and respond to hypoxia (low oxygen) and certain xenobiotics (1). Phylogenetic analyses suggest that bHLH-PAS genes arose early in animal development, and in some cases, the functions of individual genes are largely conserved across phyla. This review describes the bHLH-PAS gene family in a genetic model organism, the nematode *Caenorhabditis elegans*.

1.1 Basic structure of bHLH-PAS proteins and complexes

bHLH-PAS proteins are characterized by common structural features. The basic domain is located at the N-terminus, and it is required for sequence-specific binding to DNA. The helix-loop-helix domain is juxtaposed to the basic domain, and it mediates dimerization with other bHLH-PAS proteins. Each bHLH-PAS protein includes two core PAS repeats, PAS-A and PAS-B, which are each around 100-120 amino acids in length (2). The PAS domains mediate binding to ligands, bHLH-PAS dimerization partners, chaperones, or transcriptional coregulatory proteins, depending on the particular protein. The aryl hydrocarbon receptor is the only bHLH-PAS protein that has been shown to bind a ligand. However, certain prokaryotic proteins that have a PAS domain (but not a bHLH motif) have been shown to bind low molecular weight ligands (2, 3). Generally, the transcriptional activation domain(s) are located at the C-termini of bHLH-PAS proteins.

bHLH-PAS transcriptional complexes are dimeric. The ARNT subfamily can dimerize with multiple bHLH-PAS partners, but the other bHLH-PAS proteins are each specialized for specific cellular or developmental functions. Each partner within the complex binds a specific DNA half-site.

2.   bHLH-PAS PROTEINS IN CAENORHABDITIS ELEGANS

2.1   *C. elegans* is a powerful genetic model system

The nematode *C. elegans* is a model system for molecular and genetic analyses of evolutionarily conserved cellular and developmental processes. *C. elegans* are transparent, and the adults are approximately one millimeter long. There are two sexes, self-fertile hermaphrodites and males, and a wild-type hermaphrodite consists of only 959 somatic cells. The wild-type embryonic and postembryonic lineages have been extensively documented and are essentially invariant, and this allows researchers to diagnose mutant phenotypes at the level of individual cells (4, 5). *C. elegans* have a short generation time (~3 days) and are easily cultivated. The sequence of the genome is known, and large-scale studies are underway to characterize the function of every gene by double-stranded RNA interference, gene knock-outs, full-genome microarrays, and semi-automated analyses of protein-protein interactions (6-11). These resources, coupled with directed genetic and molecular studies, have allowed researchers to identify and characterize complex regulatory networks (12).

2.2 Five bHLH-PAS proteins in *C. elegans*

Searches of the *C. elegans* genome have identified five gene products that contain both bHLH and PAS motifs (13-15) (Table 1). The five bHLH-PAS genes are: (i) *ahr-1*, the ortholog of the mammalian aryl hydrocarbon receptor; (ii) *aha-1*, the ortholog of mammalian ARNT; (iii) *hif-1*, the ortholog of the mammalian hypoxia inducible factor alpha subunits; (iv) *C15C8.2* a founding member of a novel class of bHLH-PAS genes; and (v) *T01D3.2*, a gene that has not been extensively characterized, but has an intriguing structure. Like its mammalian cognate, the AHA-1 protein is able to dimerize with multiple bHLH-PAS partners. (*C. elegans* genes are designated by lower case italics, while the names of proteins are in uppercase.).

3. AHR-1 AND AHA-1: THE *C. ELEGANS* ARYL HYDROCARBON RECEPTOR COMPLEX

3.1 The aryl hydrocarbon receptor

The mammalian aryl hydrocarbon receptor (AHR) (see chapter 8 by Bradfield) is a ligand-activated transcription factor. The presumptive endogenous ligands are not known, but a broad range of natural and synthetic planar, halogenated compounds have been shown to bind AHR (16). 2,3,7,8-tetrachlorodibenzo-*p*-dioxin (TCDD) is an unwanted byproduct that can form during the bleaching of paper pulp or the production of chlorinated herbicides, and it is the most potent known AHR ligand. Exposure to dioxin can cause birth defects, cancer, hepatotoxicity, immunological deficiencies, cognitive impairment, and death (17). A study of Vietnam War Veterans exposed to dioxin-contaminated herbicide suggests that the half-life of TCDD in human tissues is approximately 7 years (18). Other AHR ligands include polycyclic aromatic hydrocarbons such as benzo(a)pyrene, which is produced by cigarette smoking and other combustion processes. Unlike TCDD, benzo(a)pyrene is metabolized by genes that are upregulated by the AHR complex, and some of the resultant electrophilic intermediates directly mutate DNA (19, 20).

In the absence of ligand, AHR resides in the cytoplasm in a complex with 90 kDa heat shock proteins (HSP90) and additional chaperonins (21-25)Upon binding ligand, the receptor translocates to the nucleus, dissociates from HSP90, and binds a related transcription factor, the AHR nuclear translocator (ARNT) (26, 27). The AHR:ARNT complex binds to short

DNA sequences, termed xenobiotic response elements (XREs), in target genes. Biochemical studies indicate that the transcription factor complex is capable of recruiting multiple co-activators to regulate gene expression (30-34) AHR transcriptional targets include drug metabolizing enzymes and less well characterized genes that are thought to be responsible for the tumor-promoting effects of TCDD (for reviews, see (17, 35, 36).

Table 1. C. elegans bHLH-PAS proteins. The *C. elegans* genome encodes five bHLH-PAS genes. Co-immunoprecipitation assays and electrophoretic gel shift experiments using proteins translated in rabbit reticulocyte lysates have demonstrated that AHA-1 can bind to AHR-1, HIF-1, or CKY-1. AHA-1 can also bind DNA as a homodimer. AHA-1 binds the 5'GTG half site. The half sites shown for the other bHLH-PAS proteins are not necessarily the optimal binding sites. GFP reporter constructs have been used to investigate the expression pattern of each gene. The expression of AHA-1 has been confirmed with an AHA-1-specific antibody. The *ahr-1 (ia03), aha-1 (ia01)*, and *hif-1 (ia04)* deletion mutations are all predicted to be strong loss-of-function alleles. See text for further details.

bHLH-PAS complexes	Expression patterns	Functions & genetic evidence	References
AHR-1:AHA-1 5' TTGCGTG 3'	*ahr-1*:GFP is expression in a subset of neurons.	<u>Neural development</u> Mutants lacking *ahr-1* or *aha-1* exhibit specific defects in neuronal migration & differentiation.	(14) (Qin & Powell-Coffman, manuscript in prep.)
HIF-1:AHA-1 5' TACGTG 3'	*hif-1*:GFP is ubiquitously expressed. Proteasomal degradation of HIF-1 protein is inhibited by hypoxia.	<u>Adaptation to hypoxia</u> *hif-1* deficient mutants have decreased viability in hypoxic conditions (0.5% or 1% oxygen).	(15, 28, 29)
C15C8.2:AHA-1 5' TGCGTG 3'	*C15C8.2*:GFP is expressed in most non-neuronal cells in the pharynx.	<u>*aha-1* function is essential</u> Animals lacking *aha-1* function arrest during larval development. No *C15C8.2* mutants have been reported. C Current data support two non-exclusive models: i) the C15C8.2:AHA-1 complex has an essential developmental function; or ii) AHA-1 homodimers are required for developmental progression..	(15) (Jiang, Wu & Powell-Coffman, manuscript in prep.)
AHA-1:AHA-1 5' CACGTG 3'	AHA-1 protein is expressed in most, if not all, cells.		
T01D3.2	*T01D3.2*:GFP is expressed in two interneurons.	<u>Unknown</u> No mutations in this gene have been reported.	(Jiang, Wu & Powell-Coffman, manuscript in prep.)

In the absence of ligand, AHR resides in the cytoplasm in a complex with 90 kDa heat shock proteins (HSP90) and additional chaperonins (21-25)Upon binding ligand, the receptor translocates to the nucleus, dissociates from HSP90, and binds a related transcription factor, the AHR nuclear translocator (ARNT) (26, 27). The AHR:ARNT complex binds to short DNA sequences, termed xenobiotic response elements (XREs), in target genes. Biochemical studies indicate that the transcription factor complex is capable of recruiting multiple co-activators to regulate gene expression (30-34) AHR transcriptional targets include drug metabolizing enzymes and less well characterized genes that are thought to be responsible for the tumor-promoting effects of TCDD (for reviews, see (17, 35, 36).

AHR has important developmental functions. Mice that carry null mutations in the aryl hydrocarbon receptor (*Ah*) locus grow slowly and have decreased fertility (37-39). *Ah* null mice also exhibit liver defects. The hepatocyte cells in mice lacking AHR are small, and this appears to be due to a failure to remodel fetal vasculature (40). AHR is also expressed in other cell types, including the brain, but its function in most tissues is not known (41-43). As expected, AHR-deficient mice are resistant to many of the deleterious effects of AHR-activating pollutants (37-39, 44-46).

3.2 The *C. elegans* aryl hydrocarbon receptor complex

The *C. elegans* orthologs of AHR and ARNT were first identified by sequence similarity and are encoded by the genes *ahr-1* and *aha-1,* respectively (13, 14) The gene products, AHR-1 and AHA-1, share important biochemical properties with their mammalian cognates. Specifically AHR-1 forms a tight association with the 90 kDa heat shock protein (Hsp90), and AHR-1 and AHA-1 interact to bind DNA fragments containing the mammalian xenobiotic response element with sequence specificity. Biochemical analyses have also revealed differences between the *C. elegans* and human AHR proteins. AHR-1 does not bind to the mammalian AHR-associated chaperonin ARA9/AIP1/XAP2 (47), and the invertebrate AHR homologs that have been tested do not bind radiolabeled derivatives of TCDD or β-naphthoflavone (14, 48).

3.3 Regulation of *C. elegans* AHR-1

If *C. elegans* AHR-1 and other invertebrate AHR homologs aren't activated by TCDD, then how are their activities regulated? Two types of studies have provided some insight to this question. In the first study, the activities of *C. elegans* AHR-1 and human AHR were assayed in a yeast expression system. When the domains C-terminal to the PAS motifs were fused to LexA, both the human and the *C. elegans* chimeras were able to translocate to the nucleus and activate the transcription of a reporter gene containing LexA enhancers. This confirmed that *C. elegans* AHR-1 contains a transcriptional activation domain. When longer fusion proteins containing most of the PAS domains were assayed, neither the human nor the *C. elegans* chimera was able to activate transcription. This indicated that the PAS domains, which mediate binding to HSP90 and ligand in mammalian AHR, exert a repressive function that inhibited nuclear translocation or transcriptional activation. The addition of TCDD or β-naphthoflavone allowed activation of the human AHR fusion protein, but not the *C. elegans* AHR-1 fusion (14). These experiments suggest that *C. elegans* AHR-1 is not constitutively active and that it requires some form of post-translational activation.

Experiments in the fruit fly *Drosophila melanogaster* provided further data regarding the regulation of invertebrate AHR orthologs. The *Drosophila* orthologs of AHR and ARNT are the *spineless* and *tango* genes, respectively (49, 50). Ectopic expression of *spineless* is sufficient to cause nuclear localization of Tango and phenotypic abnormalities (50, 51). These data suggest that activation of Spineless is not restricted by a spatially restricted ligand. Other *Drosophila* and *C. elegans* bHLH-PAS proteins have been shown to localize to the nucleus more efficiently when co- expressed with a dimerization partner (15, 52). This supports a model in which co-localization of Spineless and Tango proteins may be sufficient for activation of the transcriptional complex (53).

Figure 1 illustrates 3 possible models for activation of *C. elegans* AHR-1. First, AHR-1 may be activated by a spectrum of ligands that does not include TCDD or β-naphthoflavone. Second, formation of a heterodimeric transcriptional complex may be a prerequisite for nuclear localization of AHR-1. Third, AHR-1 activity may be regulated by other signaling pathways, which could covalently modify AHR-1 by phosphorylation or by other means. The activity of mammalian AHR appears to be modulated by multiple mechanisms, including ligand binding, phosphorylation, and interaction with co-activators (1, 36). Further genetic and biochemical studies will eventually reveal the strategies utilized to regulate *C. elegans* AHR-1.

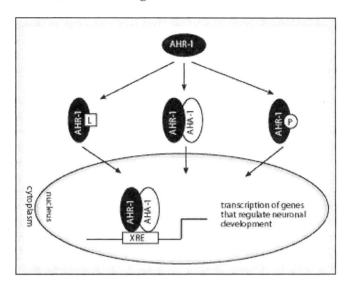

Figure 1. Models for regulation of AHR-1 activation. Like its mammalian cognate, C. elegans AHR-1 has been shown to bind HSP90, and it is likely to bind a chaperonin complex in the cytoplasm. Three possible mechanisms for AHR-1 activation and translocation to the nucleus are diagrammed here. (i) AHR-1 may be activated by a spectrum of ligands that does not include TCDD or β-naphthoflavone. (ii) Co-expression with its heterodimerization partner, AHA-1, may facilitate import of AHR-1 to the nucleus. (iii) AHR-1 may be activated by forms of post-translational modification other than ligand binding, such as phosphorylation.

3.4 *C. elegans* AHR-1 has a role in neuronal differentiation.

To investigate the function of *ahr-1*, we have used a combination of expression analyses and mutational studies. The *ahr-1 (ia03)* mutation is a ~1.5 kb deletion that removes part of the PAS domain and introduces an early translational stop. *C. elegans* homozygous for *ahr-1 (ia03)* are viable, and they exhibit subtle locomotive abnormalities. To understand the cellular basis for this phenotype, we constructed and assayed *ahr-1*:GFP reporter genes in transgenic animals. The *ahr-1*:GFP reporters are expressed in a subset of neurons. During larval development, *ahr-1*:GFP is expressed in specific touch receptor neurons and interneurons. These cells are not restricted to a single lineage, but they include the descendents of the Q neuroblasts. We are using cell-type-specific markers to examine the morphology, position, and differentiation of these neurons in animals lacking *ahr-1* function. We find that some cells, such as the SDQR interneuron, exhibit cell and axonal migration defects. Other neurons fail to express appropriate cell-type-specific markers. We conclude that *ahr-1* has a role in neuronal differentiation during normal development (Qin and Powell-Coffman, manuscript in preparation).

It will be important to identify the targets of the AHR-1 complex that direct neuronal development, and it will be interesting to explore the degree to which the neuronal functions of AHR-1 are conserved in other animals. A further challenge will be to identify regulators of AHR-1 activity, and this will require a more complete understanding of the *ahr-1*-defective phenotype.

4.   HIF-1 IS REQUIRED FOR ADAPTATION TO LOW ENVIRONMENTAL OXYGEN

4.1   Hypoxia Inducible Factors

Animals use both cellular and systemic strategies to adapt to changes in oxygen availability. During mammalian development and homeostasis, hypoxic tissues secrete growth factors to increase vascularization, and individual cells increase anaerobic metabolism in order to sustain basic cellular functions. Hypoxia-regulated genes also play central roles in tumor biology and in the recovery from strokes or cardiac failure (54). Many of the transcriptional responses to decreased oxygen (hypoxia) are mediated by the sequence-specific DNA binding hypoxia-inducible factors (HIF) (see chapter 8 by Semenza). HIF complexes consist of alpha and beta subunits, and both subunits are encoded by bHLH-PAS genes. The HIF-1β subunit is ARNT. The activity and the stability of HIF-1α is regulated by oxygen (for review see (55).

HIF transcription factors appear to have evolved in simple multicellular animals, and *C. elegans* has proven to be a valuable system for the study of hypoxia signaling and response. *C. elegans* have no apparent specialized respiratory structures, and they do not have a complex circulatory system. Any cell in the organism is only a few cell widths from the outer surface of the worm or the intestinal lumen (5). Thus, individual cells must sense oxygen availability and implement appropriate changes to survive.

4.2   C. elegans hif-1

The *C. elegans hif-1* gene is orthologous to the genes that encode HIF alpha subunits, and genetic and biochemical analyses have confirmed that the *C. elegans* gene product is similar to its mammalian cognates (15, 28).

hif-1 mRNA levels are not dramatically affected by environmental oxygen, but HIF-1 protein levels are induced by hypoxia and are rapidly decreased upon re-oxygenation (15, 28). The pathway that directs oxygen-dependent degradation of HIF-1α subunits is evolutionarily conserved. The proline in the conserved C-terminal motif LXXLAP in HIF-1α is hydroxylated in an oxygen-dependent manner. Once modified, HIF-1α has increased affinity for the von Hippel-Lindau tumor suppressor protein (VHL), and VHL targets HIF alpha subunits for polyubiquitination and proteasomal degradation (56-59). The *C. elegans* protein orthologous to VHL is encoded by the *vhl-1* gene, and *vhl-1* mutants fail to degrade HIF-1 protein in normoxic conditions (28)

Biochemical studies of mammalian HIF alpha subunits had described the characteristics of the enzyme that modified HIF-1α at the LXXLAP motif (60, 61), and Ratcliffe and colleagues used genetic strategies in *C. elegans* to identify the enzyme that modifies HIF-1 (28). They used a HIF-1-specific antibody to measure protein levels in an array of *C. elegans* mutants that were deficient in specific candidate proteins. They found that animals carrying loss-of-function mutations in the *egl-9* gene failed to downregulate HIF-1 expression in normoxic conditions. The *egl-9* gene was originally isolated as a gene required for normal egg-laying (62), and *egl-9* mutants also have decreased sensitivity to cyanide exposure (63). *egl-9* encodes a member of the 2-oxoglutarate-dependent oxygenase superfamily (64, 65). Ratcliffe and coworkers demonstrated that the EGL-9 protein acted directly on HIF–1 to hydroxylate the proline in the LXXLAP motif *in vitro*. Further, they identified mammalian homologs of EGL-9 (termed PHD 1, 2, & 3) and demonstrated that the mammalian PHDs hydroxylate human HIF-1α or HIF-2α *in vitro* (28). Thus, HIF-1, VHL-1, and EGL-9 appear to be part of an evolutionarily conserved regulatory network that senses hypoxia and implements appropriate transcriptional changes.

C. elegans HIF-1 binds AHA-1, the ortholog of ARNT/HIF-1β, to form a DNA binding complex that is similar to the HIF transcriptional complex found in mammals. In the intestine, which does not express any bHLH-PAS proteins other than *hif-1* and *aha-1*, nuclear localization of AHA-1 is dependent upon HIF-1 (15)(unpublished data). This confirms genetic interaction between the two proteins *in vivo*. A *hif-1*:GFP reporter gene, in which GFP expression is driven by predicted *hif-1* regulatory sequences, is expressed in every somatic cell. AHA-1 is also broadly expressed (15). This supports a model in which the HIF-1 complex regulates adaptation to oxygen availability at the level of individual cells.

Over 45 *hif-1* cDNAs have been completely or partially sequenced by the *C. elegans* genome project, and 4 alternatively spliced forms have been found (8, 66); data available at http://www.wormbase.org. Three *hif-1* mRNA species contain all of the previously defined functional domains, including the bHLH and PAS motifs, the oxygen-dependent degradation domain, and the putative transcriptional activation domain. Among these three forms of *hif-1*, the two most abundant mRNAs vary by only two codons, and a third form includes an additional ~200 codons 3' to the PAS domain. The fourth mRNA species is rare, and it appears to be transcribed from an alternative promoter downstream of the exons encoding the bHLH and PASA domains. The functional significance of the alternatively spliced forms is not known. The *hif-1 (ia04)* mutation (described below and in (15)) deletes three exons present in the three longer *hif-1* mRNAs and introduces premature stop codons.

4.3   *hif-1* is critical for survival in hypoxia, but not anoxia

C. *elegans hif-1* is essential for adaptation to hypoxia (0.5% or 1% oxygen). In the wild, C. *elegans* live in soil, where they encounter hypoxic microenvironments. When the environmental oxygen levels drop below 2%, C. *elegans* decrease their metabolic levels. They can continue to grow and reproduce in 1% oxygen (67), but adaptation to hypoxic conditions requires *hif-1* function. While ~97% of wild type animals can complete development and survive in 1% oxygen, the majority of *hif-1* mutants die during embryogenesis or larval development (15).

Surprisingly, *hif-1* mutants are not deficient in their ability to survive anoxia (0% oxygen). When C. *elegans* are completely deprived of oxygen, they enter a state of "suspended animation" that includes cessation of all movement and cell cycle arrest. Wild type animals of all stages can be incubated in anoxia for 24 hours at 20°C, and upon reoxygenation, greater than 85% survive (29, 67). Comparison of *hif-1*-deficient and wild-type embryos has demonstrated that *hif-1* function is not required for anoxia-induced suspended animation or for subsequent recovery upon re-oxygenation (29). Thus, when oxygen levels drop to 1% and C. *elegans* must lower their metabolic rates to survive, *hif-1* function is essential, but *hif-1* is not required for anoxia-induced arrest.

Crowder and colleagues have shown that adult C. *elegans* do not survive prolonged anoxia when incubated at 28°C, and they have termed this hypoxic death (68). When considered with the studies described above, this suggests that survival of anoxia is temperature-dependent. Certain loss-of-function mutations in the *daf-2* insulin-like receptor gene enable C. *elegans* to survive the hypoxic death regimen (68). These mutations also extend life span and cause constitutive dauer formation at 25°C. Dauers are a stress-resistant alternative third larval stage (69). It will be interesting to learn whether double mutants deficient in both *daf-2* and *hif-1* function are able to survive anoxia at 28°C, but it may prove challenging to uncouple the effects of oxygen deprivation and heat stress when interpreting the results.

4.4 *hif-1* mediates heat acclimation

Animals are better able to survive extreme heat if they are first exposed to intermediate temperatures at the upper end of their normal physiological range. This process of progressive adaptation to high temperatures is called heat acclimation. Horowitz, Treinin, and colleagues have developed a system for studying heat acclimation in *C. elegans*. They found that if wild type animals are shifted from 20°C to 35°C, then the entire population dies within 7 hours. However, if the animals are pre-incubated at 25°C for 18 hours, then most of the worms survive a 7 hour incubation at 35°C. *hif-1* mutants are apparently unable to undergo heat acclimation, as they do not benefit from pre-incubation at 25°C (Treinin, Jiang, Shleir, Powell-Coffman, and Horowitz, manuscript in preparation). Thus, HIF-1 has a critical role in adaptation to heat stresses. It will be interesting to learn whether HIF-α orthologs have an evolutionarily conserved role in heat acclimation.

5. ESSENTIAL DEVELOPMENTAL FUNCTION(S) OF BHLH-PAS PROTEINS IN *C. ELEGANS*.

5.1 *aha-1* function is essential

Although *ahr-1 (ia03); hif-1 (ia04)* double mutants are viable, animals that are homozygous for predicted null mutations in *aha-1* arrest during early larval development (Jiang, Wu, and Powell-Coffman, manuscript in preparation). Thus, AHA-1 has a function that is critical for viability or developmental progression that is independent of AHA-1 or HIF-1. Since bHLH-PAS proteins generally function as dimers, this suggests two non-exclusive models: (i) AHA-1 may form a dimer with one of the two remaining *C. elegans* bHLH-PAS proteins (T01D3.2 and C15C8.2/CKY-1); or (ii) AHA-1 may have essential functions as a homodimer.

5.2 *T01D3.2*

The *T01D3.2* gene has an intriguing structure and expression pattern, but it is unlikely to be essential. The predicted gene product has similarity to the subclass of bHLH-PAS proteins that include Single-minded and the hypoxia-inducible factor alpha subunits. Some features of the predicted protein are unusual. First, it is only 322 amino acids in length, and there is no apparent C-terminal transactivation domain. Second, the basic domain of T01D3.2 is unique. The N-terminal basic domains of bHLH-PAS genes mediate binding to DNA, and the protein sequence can be predictive of the DNA binding specificity. The T01D3.2 basic domain contains glutamine in a position that is usually occupied by basic residues (Table 2). It is not yet known whether the *T01D3.2* gene product binds DNA or whether it dimerizes with AHA-1. Some mammalian bHLH-PAS proteins function as negative regulators of other bHLH-PAS complexes (70, 71), and it is possible that T01D3.2 may inhibit the function of the HIF-1:AHA-1 or AHR-1:AHA-1 complexes. A reporter gene containing predicted *T01D3.2* regulatory sequences and N-terminal coding sequence fused to GFP is expressed in the AVH interneurons (unpublished data). No essential function has been described for these neurons.

Table 2. Basic domains of the C. elegans bHLH-PAS proteins and homologous proteins. The basic domains of the C. elegans bHLH-PAS proteins and representative mammalian orthologs are aligned. The predicted T01D3.2 protein is similar to SIM and HIF proteins, but the basic domain includes a glutamine in a position that is characteristically occupied by a basic residue, and it is labeled with a star. Abbreviations: *Caenorhabditis elegans* (C.e.); *Drosophila melanogaster* (D.m.); *Homo sapiens* (H.s.); *Mus musculus* (M.m.).

		★
C.e.	T01D3.2	----------METNLSEEKQKP**SKSQA**QQ**RR**
D.m.	Sim	-----------------MKEK**SKNAARTRR**
H.s.	SIM1	MKEK**SKNAARTRR**
C.e.	HIF-1	-------MEDNRKRNME**RRR**RET**S**RH**AARDRR**
H.s.	HIF-1α	-MEGAGGANDKKKISSE**RRKEKSRDAARSRR**
C.e.	AHA-1	GKY**AR**MEDEMGE-NKERF**ARENHSEIERRRR**
H.s.	ARNT	ERF**AR**SDDEQSSAD**KER**L**ARENHSEIERRRR**
C.e.	AHR-1	**YA**SKRRQRNFKRVRDP-PKQLTNT**NPSKRHR**
H.s.	AHR	**YA**SRKRRKPVQKTVKPIPAEGIKS**NPSKRHR**
C.e.	C15C8.2	TMGMSSAGSSNGSNLVNGQQ**RSTRGASK**QRR
H.s.	NXF	MY**RSTKGASKARR**

5.3   Evidence for a C15C8.2:AHA-1 complex in the pharynx

The fifth bHLH-PAS gene, *C15C8.2*, is closely related to the predicted products of mammalian Nerve X Factor, and *Drosophila* Dysfusion (Dys) (see Chapter 5 by Crews). It will be interesting to decipher the cellular or developmental functions of this subfamily of bHLH-PAS genes and to determine whether they have evolutionarily conserved roles.

C15C8.2 likely has a function in the pharynx, a neuromuscular feeding organ at the anterior end of the worm. A *C15C8.2*:GFP reporter construct is expressed in most non-neuronal pharyngeal cells. Interestingly, AHA-1 is localized to the nuclei of these cells (Jiang, Wu, and Powell-Coffman, manuscript in preparation). Studies with *hif-1* in other tissues have shown that co-expression of a heterodimerization partner correlates with nuclear enrichment of AHA-1 (15). Thus, AHA-1 and the C15C8.2 gene product may interact in the pharynx. In support of this, when AHA-1 and C15C8.2 proteins are expressed in rabbit reticulocyte lysates, they can interact to form a DNA binding complex. To test the hypothesis that *aha-1* mutants arrest because they require C15C8.2 and AHA-1 function in the pharynx, we constructed a chimeric gene in which *aha-1* expression is directed by *C15C8.2* upstream regulatory sequences. The *C15C8.2:aha-1* chimera rescues the *aha-1 (ia01)* larval lethal phenotype (Jiang, Wu, and Powell-Coffman, manuscript in preparation). The final test of this model will be phenotypic analysis of animals lacking *C15C8.2* function. Unfortunately, the pharynx is unusually resistant to RNAi, and no mutations in *C15C8.2* have been reported. Thus, the existing data supports a model in which C15C8.2 and AHA-1 have an essential function in the pharynx, but it does not preclude an important role for AHA-1 homodimers in the pharynx, or in other tissues.

ACKNOWLEDGEMENTS

The Powell-Coffman Lab is supported by grants from the National Science Foundation (#9874456) and from the American Heart Association (Established Investigator Award to J.A.P.-C.).

REFERENCES

1. Gu, Y. Z., J. B. Hogenesch, and C. A. Bradfield. 2000. The PAS superfamily: sensors of environmental and developmental signals. *Annu. Rev. Pharmacol. Toxicol.* 40:519-61.

2. Taylor, B. L., and I. B. Zhulin. 1999. PAS domains: internal sensors of oxygen, redox potential, and light. *Microbiol. Mol. Biol. Rev.* 63:479-506.

3. Pellequer, J.-L., R. Brudler, and E. D. Getzoff. 1999. Biological sensors: more than one way to sense oxygen. *Curr. Biol.* 9:R416-418.

4. Sulston, J. E., and H. R. Horvitz. 1977. Post-embryonic cell lineages of the nematode, Caenorhabditis elegans. *Dev. Biol.* 56:110-156.

5. Sulston, J. E., E. Schierenberg, J. G. White, and J. N. Thomson. 1983. The embryonic cell lineage of the nematode Caenorhabditis elegans. *Dev. Biol.* 100:64-119.

6.   Consortium, T. C. e. S. 1998. Genome sequence of the nematode C. elegans: a platform for investigating biology. *Science* 282:2012-8.

7.   Hill, A. A., C. P. Hunter, B. T. Tsung, G. Tucker-Kellogg, and E. L. Brown. 2000. Genomic analysis of gene expression in C. elegans. *Science* 290:809-812.

8.   Reboul, J., P. Vaglio, N. Tzellas, N. Thierry-Mieg, T. Moore, C. Jackson, T. Shin-i, Y. Kohara, D. Thierry-Mieg, J. Thierry-Mieg, et al. 2001. Open-reading-frame sequence tags (OSTs) support the existence of at least 17,300 genes in C. elegans. *Nat. Genet.* 27:332-6.

9.   Walhout, A. J., S. J. Boulton, and M. Vidal. 2000. Yeast two-hybrid systems and protein interaction mapping projects for yeast and worm. *Yeast* 17:88-94.

10.   Jiang, M., J. Ryu, M. Kiraly, K. Duke, V. Reinke, and S. K. Kim. 2001. Genome-wide analysis of developmental and sex-regulated gene expression profiles in Caenorhabditis elegans. *Proc. Nat. Acad. Sci. USA* 98:218-223.

11.   Kim, S. K., J. Lund, M. Kiraly, K. Duke, M. Jiang, J. M. Stuart, A. Eizinger, B. N. Wylie, and G. S. Davidson. 2001. A gene expression map for Caenorhabditis elegans. *Science* 293:2087-2092.

12.   Boulton, S. J., A. Gartner, J. Reboul, P. Vaglio, N. Dyson, D. E. Hill, and M. Vidal. 2002. Combined functional genomic maps of the C. elegans DNA damage response. *Science* 295:127-131.

13.   Hahn, M. E., S. I. Karchner, M. A. Shapiro, and S. A. Perera. 1997. Molecular evolution of two vertebrate aryl hydrocarbon (dioxin) receptors (AHR1 and AHR2) and the PAS family. *Proc. Natl. Acad. Sci. USA* 94:13743-13748.

14.   Powell-Coffman, J. A., C. A. Bradfield, and W. B. Wood. 1998. *Caenorhabditis elegans* orthologs of the aryl hydrocarbon receptor and its heterodimerization partner the aryl hydrocarbon receptor nuclear translocator. *Proc. Natl. Acad. Sci. USA* 95:2844-2849.

15.   Jiang, H., R. Guo, and J. A. Powell-Coffman. 2001. The Caenorhabditis elegans hif-1 gene encodes a bHLH-PAS protein that is required for adaptation to hypoxia. *Proc. Natl. Acad. Sci. USA* 98:7916-7921.

16.   Denison, M. S., S. D. Seidel, W. J. Rogers, M. Ziccardi, G. M. Winter, and S. Health-Pagliuso (ed.). 1998. Natural and synthetic ligands for the Ah receptor. Taylor & Francis, Philadelphia, PA.

17.   Schmidt, J. V., and C. A. Bradfield. 1996. Ah receptor signaling pathways. *Annu. Rev. Cell Dev. Biol.* 12:55-89.

18.   Pirkle, J. L., W. H. Wolfe, D. G. Patterson, L. L. Needham, J. E. Michalek, J. C. Miner, M. R. Peterson, and D. L. Phillips. 1989. Estimates of the half-life of 2,3,7,8-tetrachlorodibenzo-p-dioxin in Vietnam Veterans of Operation Ranch Hand. *J. Toxicol. Environ. Health* 27:165-171.

19.   Thorgeirsson, S. S., and D. W. Nebert. 1977. The Ah locus and the metabolism of chemical carcinogens and other foreign compounds. *Adv. Cancer Res.* 25:149-193.

20.   Denissenko, M. F., A. Pao, M. Tang, and G. P. Pfeifer. 1996. Preferential formation of benzo[a]pyrene adducts at lung cancer mutational hotspots in P53. *Science* 274:430-432.

21.   Denis, M., S. Cuthill, A. C. Wikstrom, L. Poellinger, and J.-A. Gustafsson. 1988. Association of the dioxin receptor with the Mr 90,000 heat shock protein: a structural kinship with the glucocorticoid receptor. *Biochem. Biophys. Res. Commun.* 155:801-807.

22.   Perdew, G. H. 1988. Association of the Ah receptor with the 90-kDa heat shock protein. *J. Biol. Chem.* 263:13802-13805.

23. Carver, L. A., and C. A. Bradfield. 1997. Ligand-dependent interaction of the aryl hydrocarbon receptor with a novel immunophilin homolog *in vivo. J. Biol. Chem.* 272:11452-11456.

24. Ma, Q., and J. J. P. Whitlock. 1997. A Novel Cytoplasmic Protein that Interacts with the Ah Receptor, contains Tetratricopeptide Repeat Motifs, and augments the transcriptional Response to 2, 3, 7, 8-Tetrachlorodibenzo-p-dioxin. *J. Biol. Chem.* 272:8878-8884.

25. Meyer, B. K., J. R. Petrulis, and G. H. Perdew. 2000. Aryl hydrocarbon (Ah) receptor levels are selectively modulated by hsp90-associated immunophilin homolog XAP2. *Cell Stress Chaperones* 5:243-254.

26. Probst, M. R., S. Reisz-Porszasz, R. V. Agbunag, M. S. Ong, and O. Hankinson. 1993. Role of the aryl hydrocarbon receptor nuclear translocator protein in aryl hydrocarbon (dioxin) receptor action. *Mol. Pharmacol.* 44:511-518.

27. Reyes, H., S. Reisz-Porszasz, and O. Hankinson. 1992. Identification of the Ah receptor nuclear translocator protein (Arnt) as a component of the DNA binding form of the Ah receptor. *Science* 256:1193-1195.

28. Epstein, A. C., J. M. Gleadle, L. A. McNeill, K. S. Hewitson, J. O'Rourke, D. R. Mole, M. Mukherji, E. Metzen, M. I. Wilson, A. Dhanda, et al. 2001. C. elegans EGL-9 and mammalian homologs define a family of dioxygenases that regulate HIF by prolyl hydroxylation. *Cell* 107:43-54.

29. Padilla, P. A., T. G. Nystul, R. A. Zager, A. C. Johnson, and M. B. Roth. 2002. Dephosphorylation of cell cycle-regulated proteins correlates with anoxia-induced suspended animation in Caenorhabditis elegans. *Mol. Biol. Cell* 13:1473-1483.

30. Kobayashi, A., K. Numayama-Tsuruta, K. Sogawa, and Y. Fujii-Kuriyama. 1997. CBP/p300 functions as a possible transcriptional coactivator of Ah receptor nuclear translocator (Arnt). *J. Biochem. (Tokyo)* 122:703-710.

31. Kumar, M. B., and G. H. Perdew. 1999. Nuclear receptor coactivator SRC-1 interacts with the Q-rich subdomain of the AhR and modulates its transactivation potential. *Gene Expr.* 8:273-286.

32. Nguyen, T. A., D. Hoivik, J. E. Lee, and S. Safe. 1999. Interactions of nuclear receptor coactivator/corepressor proteins with the aryl hydrocarbon receptor complex. *Arch. Biochem. Biophys.* 367:250-257.

33. Beischlag, T. V., S. Wang, D. W. Rose, J. Torchia, S. Reisz-Porszasz, K. Muhammad, W. E. Nelson, M. R. Probst, M. G. Rosenfeld, and O. Hankinson. 2002. Recruitment of the NCoA/SRC-1/p160 family of transcriptional coactivators by the aryl hydrocarbon receptor/aryl hydrocarbon receptor nuclear translocator complex. *Mol. Cell. Biol.* 22:4319-4333.

34. Tohkin, M., M. Fukuhara, G. Elizondo, S. Tomita, and F. J. Gonzalez. 2000. Aryl hydrocarbon receptor is required for p300-mediated induction of DNA synthesis by adenovirus E1A. *Mol. Pharmacol.* 58:845-851.

35. Hankinson, O. 1995. The Aryl hydrocarbon receptor complex. *Ann. Rev. Biochem.* 35:307-40.

36. Whitlock, J. P., Jr. 1999. Induction of cytochrome P4501A1. *Annu. Rev. Pharmacol. Toxicol.* 39:103-125.

37. Fernandez-Salguero, P., T. Pineau, D. M. Hilbert, T. McPhail, S. S. Lee, S. Kimura, D. W. Nebert, S. Rudikoff, J. M. Ward, and F. J. Gonzalez. 1995. Immune system impairment and hepatic fibrosis in mice lacking the dioxin-binding Ah receptor. *Science* 268:722-726.

38.   Schmidt, J. V., G. H.-T. Su, J. K. Reddy, M. C. Simon, and C. A. Bradfield. 1996. Characterization of a murine Ahr null allele: involvement of the Ah receptor in hepatic growth and development. *Proc. Natl. Acad. Sci. USA* 93:6731-6736.

39.   Mimura, J., K. Yamashita, K. Nakamura, M. Morita, T. N. Takagi, K. Nakao, M. Ema, K. Sogawa, M. Yasuda, M. Katsuki, et al. 1997. Loss of teratogenic response to 2,3,7,8-tetrachlorodibenzo-p-dioxin (TCDD) in mice lacking the Ah (dioxin) receptor. *Genes Cells* 2.

40.   Lahvis, G. P., S. L. Lindell, R. S. Thomas, R. S. McCuskey, C. Murphy, E. Glover, M. Bentz, J. Southard, and C. A. Bradfield. 2000. Portosystemic shunting and persistent fetal vascular structures in aryl hydrocarbon receptor-deficient mice. *Proc. Natl. Acad. Sci. USA* 97:10442-10447.

41.   Abbott, B. D., L. S. Birnbaum, and G. H. Perdew. 1995. Developmental expression of two members of a new class of transcription factors: I. Expression of aryl hydrocarbon receptor in the C57BL/6N mouse embryo. *Dev. Dyn.* 204:133-43.

42.   Jain, S., E. Maltepe, M. M. Lu, C. Simon, and C. A. Bradfield. 1998. Expression of ARNT, ARNT2, HIF1 alpha, HIF2 alpha and Ah receptor mRNAs in the developing mouse. *Mech. Dev.* 73:117-123.

43.   Petersen, S. L., M. A. Curran, S. A. Marconi, C. D. Carpenter, L. S. Lubbers, and M. D. McAbee. 2000. Distribution of mRNAs encoding the arylhydrocarbon receptor, arylhydrocarbon receptor nuclear translocator, and arylhydrocarbon receptor nuclear translocator-2 in the rat brain and brainstem. *J. Comp. Neurol.* 427:428-439.

44.   Shimizu, Y., Y. Nakatsuru, M. Ichinose, Y. Takahashi, H. Kume, J. Mimura, Y. Fujii-Kuriyama, and T. Ishikawa. 2000. Benzo[a]pyrene carcinogenicity is lost in mice lacking the aryl hydrocarbon receptor. *Proc. Natl. Acad. Sci. USA* 97:779-782.

45.   Matikainen, T., G. I. Perez, A. Jurisicova, J. K. Pru, J. J. Schlezinger, H. Y. Ryu, J. Laine, T. Sakai, S. J. Korsmeyer, R. F. Casper, et al. 2001. Aromatic hydrocarbon receptor-driven Bax gene expression is required for premature ovarian failure caused by biohazardous environmental chemicals. *Nat. Genet.* 28:355-360.

46.   Vorderstrasse, B. A., L. B. Steppan, A. E. Silverstone, and N. I. Kerkvliet. 2001. Aryl hydrocarbon receptor-deficient mice generate normal immune responses to model antigens and are resistant to TCDD-induced immune suppression. *Toxicol. Appl. Pharmacol.* 171:157-164.

47.   Bell, D. R., and A. Poland. 2000. Binding of aryl hydrocarbon receptor (AhR) to AhR-interacting protein. The role of hsp90. *J. Biol. Chem.* 275:36407-14.

48.   Butler, R. A., M. L. Kelley, W. H. Powell, M. E. Hahn, and R. J. Van Beneden. 2001. An aryl hydrocarbon receptor (AHR) homologue from the soft-shell clam, Mya arenaria: evidence that invertebrate AHR homologues lack 2,3,7,8-tetrachlorodibenzo-p-dioxin and beta-naphthoflavone binding. *Gene* 278:223-234.

49.   Sonnenfeld, M., M. Ward, G. Nystrom, J. Mosher, S. Stahl, and S. Crews. 1997. The *Drosophila tango* gene encodes a bHLH-PAS protein that is orthologous to mammalian Arnt and controls CNS midline and tracheal development. *Development* 124:4583-4594.

50.   Duncan, D. M., E. A. Burgess, and I. Duncan. 1998. Control of distal antennal identity and tarsal development in *Drosophila* by *spineless-aristapedia*, a homolog of the mammalian dioxin receptor. *Genes Dev.* 12:1290-1303.

51.   Emmons, R. B., D. Duncan, P. A. Estes, P. Kiefel, J. T. Mosher, M. Sonnenfeld, M. P. Ward, I. Duncan, and S. T. Crews. 1999. The Spineless-Aristapedia and Tango bHLH-PAS proteins interact to control antennal and tarsal development in *Drosophila*. *Development* 126:3937-3945.

52. Ward, M. P., J. T. Mosher, and S. T. Crews. 1998. Regulation of *Drosophila* bHLH-PAS protein cellular localization during embryogenesis. *Development* 125:1599-1608.

53. Crews, S. T., and C.-M. Fan. 1999. Remembrance of things PAS: regulation of development by bHLH-PAS proteins. *Curr. Opin. Genet. Dev.* 9:580-587.

54. Semenza, G. L. 2000. HIF-1 and human disease: one highly involved factor. *Genes Dev.* 14:1983-1991.

55. Wenger, R. H. 2002. Cellular adaptation to hypoxia: O2-sensing protein hydroxylases, hypoxia-inducible transcription factors, and O2-regulated gene expression. *FASEB J.* 16:1151-1162.

56. Maxwell, P. H., M. S. Wiesener, G. W. Chang, S. C. Clifford, E. C. Vaux, M. E. Cockman, C. C. Wykoff, C. W. Pugh, E. R. Maher, and P. J. Ratcliffe. 1999. The tumour suppressor protein VHL targets hypoxia-inducible factors for oxygen-dependent proteolysis. *Nature* 399:271-275.

57. Kamura, T., S. Sato, K. Iwai, M. Czyzyk-Krzeska, R. C. Conaway, and J. W. Conaway. 2000. Activation of HIF1alpha ubiquitination by a reconstituted von Hippel-Lindau (VHL) tumor suppressor complex. *Proc. Natl. Acad. Sci. USA* 97.

58. Ohh, M., C. W. Park, M. Ivan, M. A. Hoffman, T. Y. Kim, L. E. Huang, N. Pavletich, V. Chau, and W. G. Kaelin. 2000. Ubiquitination of hypoxia-inducible factor requires direct binding to the beta-domain of the von Hippel-Lindau protein. *Nat. Cell Biol.* 2:423-427.

59. Tanimoto, K., Y. Makino, T. Pereira, and L. Poellinger. 2000. Mechanism of regulation of the hypoxia-inducible factor-1 alpha by the von Hippel-Lindau tumor suppressor protein. *EMBO J.* 19:4298-4309.

60. Ivan, M., K. Kondo, H. Yang, W. Kim, J. Valiando, M. Ohh, A. Salic, J. M. Asara, W. S. Lane, and W. G. Kaelin, Jr. 2001. HIFalpha targeted for VHL-mediated destruction by proline hydroxylation: implications for O2 sensing. *Science* 292:464-468.

61. Jaakkola, P., D. R. Mole, Y. M. Tian, M. I. Wilson, J. Gielbert, S. J. Gaskell, A. Kriegsheim, H. F. Hebestreit, M. Mukherji, C. J. Schofield, et al. 2001. Targeting of HIF-alpha to the von Hippel-Lindau ubiquitylation complex by O2-regulated prolyl hydroxylation. *Science* 292:468-72.

62. Trent, C., N. Tsung, and H. R. Horvitz. 1983. Egg-laying defective mutants of the nematode Caenorhabditis elegans. *Genetics* 104:619-647.

63. Gallagher, L. A., and C. Manoil. 2001. Pseudomonas aeruginosa PAO1 Kills Caenorhabditis elegans by Cyanide Poisoning. *J. Bacteriol.* 183:6207-6214.

64. Aravind, L., and E. V. Koonin. 2001. The DNA-repair protein AlkB, EGL-9, and leprecan define new families of 2-oxoglutarate- and iron-dependent dioxygenases. *Genome Biol.* 2:RESEARCH0007.

65. Darby, C., C. L. Cosma, J. H. Thomas, and C. Manoil. 1999. Lethal paralysis of Caenorhabditis elegans by Pseudomonas aeruginosa. *Proc. Natl. Acad. Sci. USA* 96:15202-15207.

66. Stein, L., P. Sternberg, R. Durbin, J. Thierry-Mieg, and J. Spieth. 2001. WormBase: network access to the genome and biology of Caenorhabditis elegans. *Nucleic Acids Res.* 29:82-86.

67. Van Voorhies, W. A., and S. Ward. 2000. Broad oxygen tolerance in the nematode Caenorhabditis elegans. *J. Exp. Biol.* 203 Pt 16:2467-2478.

68. Scott, B. A., M. S. Avidan, and C. M. Crowder. 2002. Regulation of hypoxic death in C. elegans by the insulin/IGF receptor homolog DAF-2. *Science* 296:2388-2391.

69. Riddle, D. L., and P. S. Albert (ed.). 1997. Genetic and environmental regulation of dauer larva development. Cold Spring Harbor Laboratory Press.

70.   Mimura, J., M. Ema, K. Sogawa, and Y. Fujii-Kuriyama. 1999. Identification of a novel mechanism of regulation of Ah (dioxin) receptor function. *Genes Dev.* 13:20-25.

71.   Makino, Y., R. Cao, K. Svensson, G. Bertilsson, M. Asman, H. Tanaka, Y. Cao, A. Berkenstam, and L. Poellinger. 2001. Inhibitory PAS domain protein is a negative regulator of hypoxia-inducible gene expression. *Nature* 414:550-554.

Chapter 4

DROSOPHILA bHLH-PAS DEVELOPMENTAL REGULATORY PROTEINS

Stephen T. Crews
The University of North Carolina at Chapel Hill, Chapel Hill, NC 27599

1. INTRODUCTION

Drosophila bHLH-PAS proteins play important roles in development and physiology. They can be divided into three groups: (i) Tgo and its bHLH-PAS dimerization partners, (ii) circadian rhythms, and (iii) hormone function. The Tgo dimerization group carries-out many developmental roles, and includes: Dysfusion (Dys), Similar (Sima), Single-minded (Sim), Spineless (Ss), and Trachealess (Trh). They are the subject of this chapter (the other *Drosophila* PAS genes are reviewed in the chapters by Hogenesch and Kay, Montell, and Wilson). Members of the Tgo dimerization group share a number of common features. Foremost, they form DNA-binding heterodimers with Tgo. They are well-conserved between nematodes, insects, and mammals. Each carries-out multiple developmental roles, and some have roles as master regulators of tissue formation. The tissues, cell types, and biological processes whose development and function are influenced by bHLH-PAS proteins are diverse. There is little redundancy or overlap of function, although, there is one example of a bHLH-PAS protein regulating levels of another.

2.   TANGO: THE DIMERIZATION PARTNER

2.1   Structure and biochemistry

Tgo (1, 2) is the *Drosophila* ortholog of the Aryl hydrocarbon nuclear receptor (Arnt), the first vertebrate bHLH-PAS protein identified (3), and its close relative, Arnt2. The evolving realization that *Drosophila* had orthologs of mammalian Ahr and HIF-1α, which both use Arnt as a dimerization partner; the ability of *Drosophila* Sim to dimerize with Human Arnt; and the identification of a CNS midline enhancer element, whose sequence was consistent with binding to a Sim:Arnt heterodimer (4), strongly suggested that a fly *Arnt* existed. Use of human *Arnt* probes to screen *Drosophila* clone libraries (1, 2) were employed to identify *tgo*. The sequence structure of Tgo resembles Arnt, and both can dimerize with either insect or mammalian bHLH-PAS proteins (1), but there are differences. Vertebrate Arnt has a functional N-terminal nuclear localization sequence absent in Tgo (5), and Tgo has a Paired repeat near its C-terminus (1). The Paired repeat consists of alternating His-Pro residues and is found on a number of interesting transcription factors, including Bicoid and Paired (6). While its function in Tgo is unknown, interestingly, three *tgo* mutants have stop codons just before the Paired repeat.

Tgo forms DNA-binding heterodimers with its partners, and there is no evidence that it can function as a homodimer. However, if it does function as a homodimer or monomer, then it must be in an unconventional manner unrelated to DNA binding, since Tgo resides in the cytoplasm in the absence of a bHLH-PAS dimerization partner, and in the nucleus in the presence of a partner (7). Both cell culture transfection and in vivo experiments have indicated that Sim:Tgo, Trh:Tgo, and Sima:Tgo preferentially bind an ACGTG core sequence that is referred to as a CNS midline element (CME) (1, 2, 4, 8), and Ss:Tgo binds GCGTG (9). Tgo binds the GTG half-site and the partner binds the other AC or GC half-site. The recognition sequence of Dys:Tgo is unknown. The transient transfection results also indicate that each heterodimer (Dys:Tgo has not been tested) functions as a transcriptional activator.

2.2   Genetics

Given the striking phenotypes of *sim*, *trh*, and *ss* mutants, it was surprising that mutants in *tgo* had not been identified before it was cloned. Nevertheless, two approaches yielded *tgo* mutants. One utilized reverse genetics (1). In this scheme, a P-element transposon was mobilized into the *tgo* gene creating a lethal mutation. Four EMS *tgo* mutants were then isolated that failed to complement the P-element mutant strain. Another approach screened for dominant enhancers of a weak *ss* mutant phenotype (9). In this latter method, it was anticipated that mutations in genes that function in the same developmental pathways as *ss* would enhance the weak *ss* phenotype. Three alleles of *tgo* were identified in this manner. These results also provided genetic evidence that Ss interacts with Tgo in vivo.

Mutations of *tgo* are embryonic lethal, and phenotypic analysis of mutant embryos revealed CNS midline and tracheal defects (1), consistent with a role as a partner of Sim in controlling CNS midline cell development and Trh in controlling tracheal development. However, all *tgo* mutants analyzed showed weaker embryonic defects than observed in *sim* and *trh* null mutants. Since the evidence is strong that Tgo is a required partner for Sim and Trh, and the *Drosophila* genome contains no other *tgo*-like genes, the weak phenotypes are likely due to the presence of maternally-contributed *tgo* RNA and the possibility that none of the *tgo* mutants are null.

Mutant strains of *ss* produce viable adult flies and show defects in adult bristle, antennal, and leg morphology (10). These tissues are all derived from imaginal discs that develop during larval growth and metamorphosis. Since *tgo* mutants are embryonic lethal, testing *tgo* mutants for *ss* defects required generating mosaic *tgo* flies (9). In this manner, embryos heterozygous for *tgo* are allowed to develop, the homozygous *tgo* mutant cells are induced during postembryonic development, and flies assayed for adult morphological defects. The three *tgo* mutants analyzed showed *ss*-like phenotypes, and animals with *tgo*[5] mutant tissue showed defects nearly as severe as *ss* null mutants. In summary, the genetic analysis of *tgo* is consistent with it being a dimerization partner for Sim, Ss, and Trh.

2.3 bHLH-PAS protein interactions and Tgo subcellular localization

Both *tgo* RNA and protein are found in all embryonic cells (1, 2, 7). In most cells, Tgo protein is present in the cytoplasm and excluded from the nucleus (7). However, in a number of cells Tgo protein is localized to nuclei. These cells include the CNS midline, trachea, salivary duct, sensory cells, and larval antennal primordia – all sites of Sim, Ss and Trh protein localization. This led to the idea that Tgo is cytoplasmic if no bHLH-PAS partner protein is present, but in the presence of a partner protein the two dimerize and translocate into the nucleus (Figure 1). This was confirmed by ectopically expressing *sim*, *ss*, or *trh* and showing that both the partner protein and Tgo colocalized to nuclei in those cells (7, 9). Since nuclear localization was observed in all cell types and developmental times investigated, this indicated that dimerization and nuclear localization were not under developmental regulation. However, this does not imply that other factors are unimportant in dimerization and localization.

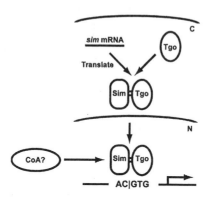

Figure 1. Model for Tgo interactions with partner bHLH-PAS proteins. Tgo is localized to the cytoplasm (C) in the absence of a partner bHLH-PAS protein. When a bHLH-PAS gene, such as *sim*, is expressed in the cell, the mRNA is translated into protein, dimerizes with Tgo, and the complex translocates into the nucleus (N), where it binds DNA and activates transcription. Sim:Tgo binds an ACGTG-containing binding site. Target gene expression likely also requires interactions between Sim:Tgo and transcriptional coactivators (CoA), although regulation by corepressors remains a possibility. This model holds for all partners of Tgo, although nuclear translocation of Trh:Tgo requires phosphorylation, and the appearance of Sima protein is due to hypoxia-induced inhibition of protein degradation rather than transcriptional control.

3.   SINGLE-MINDED – MASTER REGULATOR OF CNS MIDLINE CELL DEVELOPMENT

The *sim* gene functions as a master regulator of CNS midline cell development. The lethal gene designated as *l(3)S8* was first reported in 1964 (11, 12). Later it was shown that *l(3)S8* mutations had a severely disorganized embryonic CNS (13). The gene was renamed "*single-minded*" because the two longitudinal axonal connectives that run along the length of the wild-type *Drosophila* CNS were now fused into a single connective in the mutant (Figure 2A, B). Further analysis indicated that *sim* mutant embryos were missing the cells that lie along the midline of the CNS (13). *sim* mutants were also isolated based on an absence of the cuticular ventral midline denticles (14), a defect later shown to be due to an absence of a midline-to-epidermis signaling pathway (15, 16). The region around *sim* was genetically and molecularly well-characterized. This facilitated the cloning of *sim*, and its identification was based on its prominent expression in the CNS midline cells (13) (Figure 2C). The *sim* gene is also expressed in a number of cells besides the midline cells.

Figure 2. sim phenotypes and expression. Anterior is to the left in all panels. (A) Wild-type *Drosophila* embryo stained with an antibody that reacts with all nerve cells and axons. The axon scaffold consists of two longitudinal connectives running along the A/P axis and two commissures/ganglion that cross the midline. (B) *sim* mutant embryo showing a collapsed axon phenotype with the appearance of a single axon bundle running along the A/P axis, instead of the two characteristic longitudinal connectives. (C) Wild type embryo stained with anti-Sim showing the appearance of Sim in CNS midline cell nuclei. Magnification is higher in (C) than in (A, B).

3.1   *Drosophila* CNS midline cells

3.1.1   Development of the CNS midline cells

The *Drosophila* embryonic CNS consists of a brain and ventral nerve cord. The nerve cord is comprised of segmentally-repeated ganglia. Each ganglion has ~1000 neurons and glia. The ganglion is bilaterally symmetrical, and has a distinct set of cells at the midline. The mature embryonic *Drosophila* midline cells consist of ~15 neurons and 2-3 glia (13, 17, 18) (Figure 3). The midline neurons are: (i) 2 midline precursor 1 (MP1) interneurons, (ii) 2 MP3 interneurons, and (iii) ~10 progeny of the median neuroblast that include interneurons and neurosecretory motoneurons. The midline glia enwrap the two commissural axon bundles that cross the midline, and also participate in signaling pathways that control a variety of developmental processes.

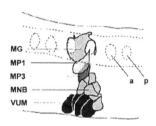

Figure 3. Mature *Drosophila* CNS midline cells. The different CNS midline cells of a mature embryo are labeled to the left (see text). The anterior commissure (a) and posterior commissures (p) that cross the midline are shown enwrapped by the midline glia (MG). The midline neurons lie below the MG. Adapted from Bossing and Technau (17).

The mature CNS midline cells are derived from a small set of midline precursor cells, which are often referred to as the mesectoderm (reviewed in (19, 20). The mesectoderm consists of two single cell-wide strips of cells that run along the anterior/posterior axis of the blastoderm embryo adjacent to the mesoderm. As gastrulation takes place, the two mesectodermal stripes join together at the ventral midline. Initially, there are ~8 midline precursor cells/segment. These cells undergo a synchronous cell division to give rise to 16 precursors/segment. This is followed by changes in cellular morphology in which the midline nuclei migrate inward and the cell maintains a cytoplasmic extension to the embryo surface. The midline precursors go on to divide and differentiate into mature midline neurons and glia, and the specific fates of individual midline cells are determined, at least in part, by the functions of the segment polarity genes, including *engrailed*, *hedgehog*, *patched*, and *wingless* (21). Within the midline glial lineage, apoptosis acts to establish the final number of mature glia in each segment (22). Given their simplicity and extensive study, the *Drosophila* CNS midline cells have the potential to be one of the premier neurogenomic systems for studying how neural precursor cells develop into a diverse set of motoneurons, interneurons and glial cell types. Over 200 genes have been identified that are expressed or function in the CNS midline cells.

3.1.2   Control of development and differentiation by CNS midline cell signaling

The midline cells constitute an important signaling center during *Drosophila* embryogenesis that influences the differentiation and migration of a number of neighboring cell types. CNS midline-dependent cellular processes include the: (i) formation and patterning of the underlying ventral epidermis and salivary tissue, (ii) proliferation of brain neuroblasts and formation of axonal connections between brain and ventral nerve cord, (iii) control of midline axon crossing and organization, (iv) muscle cell migration, and (v) formation of nearby mesodermal and CNS cells. These processes are mediated in most cases by morphogens secreted by the midline cells.

Spitz signaling control of cell fate and proliferation. One such morphogen is Spitz, a TGF-α-like protein that acts through the *Drosophila* Epidermal growth factor receptor (Egfr). Spitz is secreted by the midline cells and patterns the underlying ventral epidermis in a graded manner (15, 16). The Spitz signaling pathway is also involved in directing the ventral-most salivary primordia to become duct cells (see 4.6.1.2 below) (23), and for the formation of the mesodermally-derived dorsal median cells, an unusual set of cells that lie above the CNS (24, 25). Recent work has shown that Sim and Spitz are required for neuroblast proliferation and subsequent formation of the midbrain (26). What is particularly noteworthy about this observation is that this signal emanates from Sim+ cells in the foregut, not the CNS. This is anatomically consistent, since the Sim+ foregut cells constitute a "midline" within the midbrain primordia (i.e. the foregut passes through the embryonic brain), and *sim* blastoderm expression in the ectodermal foregut cells is an anterior extension of *sim* expression in the mesectoderm. Nevertheless, it is striking that this function is carried-out in two distinct tissue types. This foregut-derived signaling pathway also likely mediates the formation of axonal connections between the embryonic brain and nerve cord (27). It has been hypothesized that the ancestral brain of arthropods was located dorsal to the foregut and connected to the ventral nerve cord by axonal connectives without an intervening midbrain. Page (26) has proposed that an evolutionary extension of Sim/Spitz signaling from the CNS midline cells into the foregut cells promoted the formation of the now existent midbrain.

Midline control of axonogenesis. The CNS midline cells also play an important role in formation of the commissural axon bundles that connect the lateral halves of the CNS. Most nerve cells send their axons across the midline via commissural bundles where they join distinct longitudinal pathways that run along the longitudinal axis of the CNS (28). The Netrins are secreted by the midline and attract axons to the midline. Conversely, the Slit protein is also secreted by the midline and acts to both repel axons that do not cross and to inhibit recrossing of commissural axons. In addition, Slit acts as a morphogen to direct axons into distinct longitudinal bundles.

Slit and FGF control of cell migration and fate. There is a subpopulation of somatic muscle cells that originate above the CNS. Midline-derived Slit acts as a repellent to guide these muscle cell precursors outwards from the interior of the embryo to the body wall (29, 30). Both Slit and Fibroblast Growth Factor (FGF) pathways are involved in directing the fates and differentiation of some non-midline nervous system cells. *slit* is required for the formation of specific neurons from the ganglion mother cell (GMC) precursors (31). Normally, GMC-1 of the RP2/sib lineage divides asymmetrically to generate two distinct neurons: RP2 and RP2sib. In *sim* or *slit* mutants, GMC-1 fails to divide asymmetrically bout instead generates two identical RP2 neurons. The insect midline also has an FGF-like activity that influences the differentiation of serotonergic neurons upon crossing the midline (32). Finally, analysis of *sim* mutants indicated that 15% of the lateral CNS neurons were absent (33), and it was proposed that this effect is due to the influence of multiple midline-derived signaling pathways, consistent with the other reports described in this section. These results all point to the midline as an important developmental signaling center. Interestingly, this function is a role shared with the vertebrate floor plate, a group of specialized neuroepithelial cells, which resides at the ventral midline of the developing spinal cord.

3.2   Sim midline genetics

Examination of *sim* mutant embryos revealed that the collapsed CNS phenotype is due to a severe disruption in CNS midline cell development (13). Gastrulation is normal and the mesectodermal cells come together at the ventral midline. However, the midline precursor cells never properly form, and all subsequent developmental events fail to take place, including cell division, cell shape changes, and differentiation into midline neurons and glia (34). Ultimately, the cells die (22). Most, if not all, genes that are expressed in the CNS midline precursor cells require *sim* to initiate or maintain their expression (35). Conversely, genes that are expressed in the adjacent, lateral CNS cells, but not the midline cells, are expressed in the midline in *sim* mutants (36-38). Misexpression of *sim* throughout the neuroectoderm is able to convert the entire CNS to midline cells (34). This indicates that *sim* acts as a genetic switch to activate midline transcription and to repress midline expression of genes normally expressed in the adjacent lateral CNS. Presumably, the combination of these two functions directs a neuroectodermal cell to become CNS midline and not lateral CNS. The data are consistent with the idea that the *sim* gene acts as a master regulator of CNS midline cell development.

3.3   *sim* midline expression

The *sim* gene is prominently expressed in the CNS midline cells throughout development (13, 39). Initial expression is in midline precursors (mesectoderm) at the cellular blastoderm stage just before gastrulation. It remains on in all midline precursors until they begin to differentiate into neurons and glia. At that time, *sim* is expressed at high levels in glia, but low levels in neurons. The biphasic expression of *sim*, initially in midline precursors and later in midline glia, is reflected in the organization of the gene. The *sim* gene consists of 8 exons spanning over 20.5 kb (40), and contains two promoters, an early promoter (P_E) and a late promoter (P_L) (34). P_E governs initial expression of *sim* in the mesectoderm and subsequent midline precursors. P_L directs *sim* transcription later in midline precursors and then in the midline glia. *sim* remains on in midline cells of the larval ventral nerve cord (41). Functionally, there are three major modes of *sim* expression: (i) initial activation in the mesectoderm, (ii) maintenance of expression in the midline precursors, and (iii) strong expression in midline glia and weak expression in midline neurons.

3.4   Flipping the switch: activating *sim* expression in the midline cells

The mesectoderm lies between the mesoderm and neuroectoderm along the blastoderm embryonic dorsal/ventral (D/V) axis. The key event in dictating whether a cell will become mesectoderm is activation of *sim* gene expression. Thus, specification of the CNS midline cells is essentially an issue of how D/V patterning genes activate *sim* expression in the two single-cell wide mesectodermal stripes. Since this D/V patterning process requires the specification of only a single cell diameter, it is not surprising that a complex set of developmental processes are required including: morphogenetic gradients, transcriptional activation and repression, combinatorial actions of transcription factors, and cell signaling (Figure 4A). The results of genetics and biochemical experiments indicate that the Dorsal and Twist transcription factors act on the *sim* promoter in conjunction with a bHLH (Daughterless:Scute) heterodimer and activate *sim* transcription broadly in the ventral region of the blastoderm (19, 40). The ventral boundary of *sim* expression is established by the Snail zinc finger protein, which is expressed at high levels in the mesoderm but is off in the mesectoderm. Sna represses *sim* transcription, thereby establishing a sharp ventral *sim* expression boundary between the mesoderm and mesectoderm (19, 42). The dorsal boundary of *sim* expression is established via the

Suppressor of Hairless [Su(H)] transcription factor. Su(H) represses *sim* transcription in the neuroectoderm forming the dorsal *sim* expression on-off boundary (43). In addition, Notch signaling, positioned by Sna promotion of Notch-Delta endocytosis (44, 45), converts Su(H) from a repressor in the mesectoderm to a direct activator of *sim*. The result is that *sim* is expressed at high levels in the mesectoderm, but not at all in adjacent mesodermal or neuroectodermal cells. All of these transcription factors act directly on the *sim* gene, and *sim* P_E has a dense and complex arrangement of transcription factor binding sites (Figure 4B).

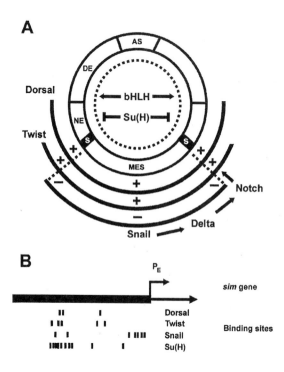

Figure 4. Activation of *sim* expression along the D/V axis of the blastoderm embryo. See text for details. (A) Schematic of cross-section of a blastoderm embryo showing the distribution of transcription factors that regulate initial sim expression. Ventral is at bottom. Filled box with white "s" represents mesectoderm – the site of initial *sim* expression. AS – amnioserosa; DE – dorsal ectoderm; NE – neuroectoderm; MES – mesoderm; "+" and arrows indicate positive regulation, and (-) and blocked lines indicate negative regulation of the *sim* gene. Dotted lines indicate that bHLH and Su(H) are expressed throughout the blastoderm. (B) The arrangement of transcription factor binding sites in the *sim* early regulatory region is shown below a schematic of the *sim* gene. Shown is 3.7 kb of DNA that flanks the *sim* early promoter (P_E).

3.5 Structure and biochemistry

Drosophila Sim belongs to a subfamily of bHLH-PAS proteins highly conserved between insects and vertebrates. Mammals have two *Sim* genes, *Sim1* and *Sim2* (see chapter by Fan). *Drosophila* Sim has four major regions, which, from N-terminus to C-terminus, are: (i) bHLH DNA domain, (ii) PAS-1 and PAS-2 domains, (iii) Ala-Ala-Gln repeats, and (iv) homopolymeric stretches (Figure 5). The basic region mediates DNA binding in combination with the Tgo basic region. Biochemically, Sim heterodimerizes with Tgo (1), and this interaction is dependent on HLH and PAS-1 domains (Nystrom and Crews, unpubl.). Sim does not homodimerize (1). Together, Sim:Tgo binds to CMEs, which contain a core ACGTG recognition sequence (1, 4). The PAS domain-containing region binds to the chaperone Hsp90 (46), and also interact with unidentified cofactors required for transcriptional specificity (8, 47). Closely following the PAS domains is a stretch of 10 Ala-Ala-Gln repeats. Despite its striking sequence structure, deletion of the Ala-Ala-Gln region does not affect the ability of Sim to function in transcriptional activation in vivo (48). The C-terminal region has at least three discrete transcriptional activation regions consisting of stretches rich in asparagine, glutamine, glycine, histidine, and serine (49).

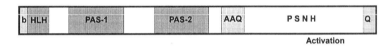

Figure 5. Sequence structure of the Sim protein. Shown are the bHLH, PAS-1, and PAS-2 domains, along with regions rich in various amino acids. These include a stretch of Ala-Ala-Gln (AAQ), Pro, Ser, His, Asn (PSNH), and Gln (Q). There are multiple transcriptional activation sequences in the C-terminal section of the protein.

3.6 Midline precursor gene expression – Sim activation, autoregulation, and repression

Activation. The critical role of *sim* in CNS midline cell development is indicated by the finding that genes that are normally expressed in the CNS midline cells require *sim* function for either the initiation or maintenance of their midline expression. The cis-regulatory regions of three genes, *Toll*, *rhomboid*, and *breathless*, have been studied in vivo using germline transformation and in vitro mutagenesis techniques. Analysis of the *Toll* gene identified 4 putative Sim:Tgo binding sites (CMEs) in an 0.9 kb fragment that drove high levels of midline precursor transcription (4). The

CMEs were within a 662 bp region. Mutation of all 4 CMEs abolished midline transcription. The role of the CME was further demonstrated by multimerizing a 20 bp fragment containing *Toll* CME-4, and showing it could drive strong midline precursor transcription in vivo. Similar analysis of *rhomboid* revealed two CMEs within 60 bp that are required for midline precursor transcription (50), and study of *breathless* revealed three essential CMEs within 144 bp (2). It is likely that every gene expressed in the midline precursor cells is directly regulated by Sim:Tgo heterodimers, and their regulatory regions contain multiple, clustered Sim:Tgo binding sites.

However, binding of Sim:Tgo is not sufficient for activation of midline precursor transcription. Ectopic expression of *sim* throughout the ectoderm reveals that activation of midline precursor-expressed genes can occur only in the ventral ectoderm, but not dorsal ectoderm (8, 34, 48). This occurs despite the observation that Sim:Tgo can enter nuclei and bind DNA in the dorsal ectoderm (7). Additional analysis of the *rhomboid* gene indicated that additional cis-regulatory sequences were required in addition to the CMEs for midline transcription (8). Most likely, midline precursor gene expression normally involves Sim:Tgo as well as either a ventral coactivator or a dorsal corepressor. Transgenic swap experiments have revealed that the unidentified factor(s) functions through the Sim PAS domains (8).

Autoregulation. Genetic and molecular studies demonstrated that *sim* is required for its own continued expression in midline precursors. Expression of a transgene containing the *sim* P_E regulatory region driving *lacZ* expression showed a more rapid reduction in levels in a *sim* mutant compared to wild-type (34). In addition, levels of *sim* RNA declined more rapidly in *sim* mutant embryos (34). Analysis of the *sim* P_E regulatory region revealed 4 Sim:Tgo binding sites within 560 bp, and mutation of these sites abolished *sim*-dependent midline precursor expression (4). This indicates that autoregulation is due to direct action of Sim:Tgo on the *sim* regulatory region. Midline precursor expression is also driven from the *sim* P_L (34). In *sim* mutant embryos, P_L-mediated expression is absent. Thus, DV patterning proteins initially activate *sim* in the mesectodermal cells, and then Sim maintains its own expression via a positive feedback loop acting on both promoters. Additional factors must be responsible for P_L transcription in midline glia and the absence of midline glial P_E expression.

Repression. Mesectodermal cells are initially fated to become *ventral nerve cord defective* (*vnd*)-positive neuroectodermal cells, but the expression of *sim* switches them to a midline fate [*vnd* is a key regulator of ventral neuroectodermal neural specification in much the way *sim* controls midline development (51, 52)]. *sim* represses midline expression of genes, including *tartan*, *wingless*, and *vnd*, that are expressed in the ventral neuroectoderm (36-38). The mechanism of *sim* repression was revealed in studies on the *vnd* gene (48). Three lines of evidence demonstrated that Sim is a pure transcriptional activator that represses indirectly. (i) Deletion of Sim:Tgo binding sites from the *vnd* regulatory region does not affect midline repression when tested in vivo. (ii) Analysis of mutant forms of Sim, tested using an in vivo misexpression-repression assay, indicated that mutations that abolished the ability of Sim to activate transcription (e.g. removal of the DNA binding or activation domains) also abolished repression. (iii) Substitution of the Sim activation domain with the VP16 activation domain restored the ability of Sim to both activate and repress transcription in vivo. These results indicate that Sim represses midline transcription indirectly by activating transcription of distinct repressive factors. Complete understanding of the role of transcriptional repression in CNS midline cell development will require identification of these repression factors.

3.7   Midline glial regulation

The Sim protein is prominently expressed in midline glia, both in the embryo and larva. Since distinct, but overlapping, sets of genes are expressed in midline precursors and midline glia, does Sim:Tgo directly control midline glial expression, and do the two modes of regulation require the use of different regulatory cofactors? Midline glial expression is derived from the *sim* late promoter, P_L (34). The *slit* gene, which plays important roles in midline-directed axon guidance, is expressed in the midline glia, and was studied as a representative midline glial-expressed gene. Analysis of fragments of *slit* genomic DNA by germline transformation discovered a 380 bp fragment of *slit* from intron 1 that drove *lacZ* in the midline glia (53). This fragment has a single CME, and mutation of the CME results in a loss of midline glial expression (4). This indicates that Sim:Tgo directly regulates *slit* expression.

Two genes, *fish-hook* (*fish*; also called *Dichaete*) and *drifter* (*dfr*), both encode transcription factors expressed in midline glia. *fish* encodes a Sox HMG domain transcription factor and *dfr* encodes a POU-homeobox transcription factor. Mutations in either gene result in weak phenotypes in which expression of midline glial markers is relatively normal (47, 54). Although midline glia are present, they fail to migrate properly. However, when *fish dfr* double mutants were analyzed, midline glia fail to form and gene expression, including *slit-lacZ*, is greatly reduced. Biochemical experiments revealed that Fish binds Sim via the Sim PAS domain and Fish Sox domain. Dfr binds Fish via the Dfr POU domain (47). These results indicate that midline glial expression is dependent on a transcription factor complex that includes Tgo:Sim:Fish:Dfr. Neither Fish nor Dfr have been implicated in controlling midline precursor cell transcription. Thus, the ability of Sim:Tgo to regulate gene expression in both CNS midline precursor cells and midline glia is dependent, at least in part, on interactions of Sim:Tgo with different coregulatory proteins and distinct cis-regulatory sequences.

Sim is able to control the transcription of different gene sets in midline cells and other diverse cell types (also see below). Current results suggest that transcriptional specificity arises from the interaction of Sim:Tgo with different coregulators. The identification of these coregulatory proteins and biochemical study of how they interact with Sim (particularly the role of the PAS domain) are important areas of future research. In addition, it is of great interest to determine how Sim, which dictates general CNS midline identity, and the segment polarity proteins, which promote different midline neural and glial cell fates, interact at the molecular level to promote specific transcription patterns in individual midline cells.

3.8   Non-midline functions of Sim

3.8.1   Postembryonic brain function – control of locomotion

There are two sites of *sim* expression in the larval brain: (i) the lamina and medulla of the optic lobes, and (ii) clusters of neurons in the central complex (41). The optic lobes mediate processing of visual information that is received from the retina. In the 3rd instar larva, *sim* is expressed in most or all of the neurons of the lamina and many medullary neurons. *sim* expression is not in the optic lobe proliferative zones, but precedes neural differentiation. Analysis of flies heterozygous for a *sim* temperature sensitive mutant allele (sim^{J1-47}) and a *sim* null mutant (sim^{H9}) reared at the permissive temperature (17°C) revealed defects in axonal organization (41). At the level of the inner optic chiasm in which the medullary axons connect with the lobula and lobular plate (additional sites of higher order visual processing), there were axonal fibers entering the lobula from the medulla via abnormal paths. These results suggest that *sim* may be controlling aspects of medullary neuron axon guidance.

The larval brain central complex expression of *sim* is in three paired clusters of neurons that lie on either side of the midline. One function of the central complex is the coordination of movement (55). Analysis of sim^{J1-47}/sim^{H9} flies showed adult behavioral and morphological brain defects consistent with the central complex expression (41). When wild-type flies were tested in a behavioral paradigm consisting of a circular stage and two opposing visual cues, they walked in straight lines back and forth between the cues. However, sim^{J1-47}/sim^{H9} flies did not walk in straight lines, but walked in circles. An individual fly could turn left or turn right, but not both, nor walk straight. Although basic locomotion appeared normal, its coordination was defective. Male courtship behavior was also affected. Examination of the adult brain revealed that interhemispheric axonal connections were defective; the neuropil was thinner in *sim* mutants than in wild-type and there was disorganization at sites of axonal crossing. Thus, mutant defects in the Sim-positive cells may affect interhemispheric communication, leading to a split-brain fly that cannot properly coordinate its movement. Future questions concern the development and function of the Sim-positive central complex cells, and the developmental roles of *sim* in the central complex and optic lobes. Does *sim* control neurogenesis in the central brain as it does in the CNS midline cells, or does it influence other neurodevelopmental processes, such as axon guidance?

3.8.2   Genital structures and sterility

During embryonic development, the midline expression of *sim* extends past the presumptive CNS into segments A9-10, and terminates at the proctodeal (anal) opening (35, 41). The staining in segments A9 and A10 overlaps with the sites of the genital disc primordia. These are the presumptive imaginal structures that will give rise to the adult genitalia. Analysis of *sim* mutant embryos indicates that the genital discs form, but are misplaced (41). The misplacement could be due to either a defective genital disc or a consequence of improper condensation of the CNS. The proctodeum is also abnormal. Analysis of the cuticle indicates that the anal slit, which constitutes the proctodeal opening, is absent and the adjacent anal pads have fused. Thus, *sim* contributes to the formation of the genital disc and midline structures associated with the proctodeum.

sim^{J1-47} mutants kept at the permissive temperature or *sim*^{H9}/*Df(3R)ry*^{75} mutants survive into adulthood, but exhibit male and female sterility (41). A small percentage of *sim*$^{J1-47}$ mutant adults lack the genitalia and anus. Males lack the clasper and penis, and females lack the vulva. Both sexes lack an anus and the anal plates, and the flies are closed at the posterior end. Male and female gonads were only rudiments, and unattached to the gut via the internal genital structures. Thus, defects in the *sim* mutant genital discs result in severe defects in the genital structures and sterility. While gut structure appeared normal in newly emerged adults, since the hindgut was not connected to an anal opening, it became swollen. This resulted in the premature death of the flies.

3.8.3   Muscle precursors

There is a small cluster of 3-5 *sim*-expressing cells/hemisegment that arise just above the CNS (30). These cells migrate laterally to the body wall, where they differentiate into ventral oblique somatic muscles. The proper migration of these mesodermal cells requires Slit repulsive signals derived from the CNS midline cells (29, 30). The expression of *sim* in these cells is transient, occurring initially as the cells first appear as pre-migratory muscle precursors before migration, and largely disappearing before they differentiate as muscles. Despite the expression of *sim* at a critical time in the development of these cells, genetic analysis of *sim* mutants specifically lacking expression in the muscle precursor cells did not reveal any obvious abnormalities in somatic musculature (30). This suggests that, unlike its major role in midline neurogenesis, *sim* does not play a major role in myogenesis.

4.   DROSOPHILA TRACHEALESS – REGULATOR OF TRACHEAL AND SALIVARY DUCT DEVELOPMENT

The *Drosophila trh* gene was first discovered in the Nobel Prize-winning genetic screen of cuticle phenotypes by Nusslein-Volhard and Wieschaus as a mutant embryo devoid of trachea and a defective with a defective filzkörper (56). Two groups, one interested in tracheal development and the other in salivary gland development, identified the *trh* gene by virtue of P-element enhancer trap insertions that showed expression in both cell types (57, 58). Subsequent cloning and expression analysis of the *trh* gene revealed that it is expressed in the embryonic trachea, salivary duct, and subset of CNS cells. *trh* plays important developmental roles in the trachea and salivary duct. The CNS function is unknown, and it is expressed after their development into mature cell types (57) (Ward and Crews, unpubl.), suggesting a function in axonogenesis, synaptic connectivity, or neural function. Identification of *trh* in other insects and arthropods has revealed other additional sites of expression and potential functions, including silk gland development in the silk moth, and osmoregulation in brine shrimp.

4.1   Tracheal development and *trh* genetics

Since insects do not have an oxygen-carrying circulatory system, they depend on a diffuse, multi-branched trachea to deliver oxygen. Tracheal development (59) begins with the formation of segmentally repeated placodes. These cells invaginate and branch. Tracheal branches from different segments fuse to form the tubular, air-filled trachea that is closed except at the anterior and posterior spiracular openings. The posterior spiracle is connected to the trachea by the filzkörper, an elaborate structure which acts as a tracheal air filter.

The tracheal placode fails to invaginate in *trh* mutants, and the cells do not differentiate and form trachea (57, 58, 60). This is accompanied by an absence of expression (57, 58) of a number of tracheal-expressed genes, including *trh*, which undergoes positive autoregulation (58). One important tracheal-expressed gene whose expression is dependent on *trh* is *breathless* (*btl*), which is required for tracheal migration (2). Ectopic expression of *trh* resulted in additional tracheal placodes forming at two distinct sites in the dorsal ectoderm (58). These results suggested that *trh* is a master regulator of tracheal development. However, the situation is more complex. While *trh* clearly plays an important role in tracheal development, there are tracheal-expressed genes that are not dependent on *trh* function (61). In addition, while the filzkörper fail to elongate in *trh* mutants, they are able to secrete cuticle, suggesting that the cells have the correct identity, but cannot form tubes. Since *trh* is required for formation of trachea and salivary duct, both tubular cell types, it has been suggested that *trh* comprises a regulatory cassette that functions in tubule formation (57). Since known tracheal target genes of *trh* include *btl* and *rho*, which both participate in tracheal migration and invagination, the role of *trh* may be more in morphogenesis than tracheal precursor cell fate.

4.2 Trh biochemistry and tracheal expression

The sequence of Trh is similar to the other bHLH-PAS partners of Tgo: a bHLH domain near the N-terminus followed by PAS-1 and PAS-2, and ending with a large unconserved C-terminal region that functions as a transcriptional activation domain (57, 58). There are two splice variants that affect the sequence of PAS-1 and the spacer between PAS-1 and PAS-2 (58) - the significance of these variant proteins is unknown. Trh:Tgo binds to CMEs with ACGTG core sequences, and activates transcription (1, 2, 8). It is not surprising that Trh:Tgo recognizes the same binding site as Sim:Tgo and Sima:Tgo, since the Trh basic region has 9/13 aa identity with Sim and 11/13 aa identify with Sima. In vivo analysis has been carried-out on two Trh:Tgo target genes, *btl* and *rho* (2, 50). Both have multiple CMEs in close proximity that are required for tracheal expression. Both genes are expressed in the CNS midline cells, and the same CMEs are required for both tracheal expression by Trh and midline expression by Sim.

There is prominent *trh* expression in all embryonic tracheal cells and the posterior spiracle (57, 58). *trh* expression appears just as the tracheal primordia form, and remains on throughout embryogenesis and larval development in most tracheal cells. One exception is tracheal fusion cells, in which Trh levels decline due to negative regulation by Dys (62). It will be important to determine what function *trh* plays during late embryonic and postembryonic development, and whether Sima (see below) may control Trh levels and function under hypoxic conditions, akin to Dys regulation of Trh.

4.3   Trh nuclear localization and phosphorylation

Examination of Trh subcellular localization shows it to be an exclusively nuclear protein (7, 58). Misexpression of *trh* results in ectopic Trh nuclear localization regardless of the cell type it is expressed (7). While this suggests that Trh dimerization and nuclear localization are unregulated, additional factors are required. Nuclear localization of Trh is dependent on phosphorylation by the *Dakt1* protein kinase B (PKB) (63). Trh is phosphorylated by Dakt1 at S665, which lies in the C-terminal region thought to be involved in transcriptional activation and nuclear localization (and also outside of the bHLH-and PAS domains). Mutants of *Dakt1* result in reduced levels of *trh* and *btl*, consistent with a role in regulating Trh function. How Dakt1 influences Trh nuclear localization is unknown, but could be due to effects on: (i) nuclear import by allowing access to the Trh nuclear localization sequence, (ii) export and removal of an inhibitor to nuclear export, or (iii) ability to dimerize with Tgo. The zygotic expression of *Dakt1* includes tracheal cells and sites in which ectopic trachea are formed when *trh* is expressed throughout the dorsal ectoderm. *Dakt1* corresponds to the vertebrate *akt* oncogene, which is regulated by phosphatidylinositol signaling. This suggests that Trh nuclear localization and function is dependent on a signaling pathway, although its biological roles remain unknown. Dakt1 may also be a component of the accessory factors proposed to regulate Trh function post-transcriptionally in the dorsal and dorsolateral ectoderm (8).

4.4   Transcriptional specificity: Drifter-Trh interactions

Ectopic expression of *trh* throughout the ectoderm results in ectopic tracheal pits and gene expression (58). However, the additional tracheal tissue forms at only two sites within the dorsal ectoderm. This suggests that additional factors are required for *trh* function and at least one of them is spatially restricted to the sites of ectopic *trh*-induced tracheal cells. Another aspect of transcriptional specificity concerns Sim and Trh. Even though

Sim:Tgo binds the same DNA sequences as Trh:Tgo it is unable to induce tracheal gene expression when ectopically expressed (8). Transgenic domain swap experiments revealed that a Trh protein with Sim PAS domains behaved like Sim and unlike Trh, indicating that transcriptional specificity resided within the PAS domains (8). Two factors, Dakt1 and Drifter (Dfr) have emerged as cofactors for Trh function. One component of Trh tissue specificity is phosphorylation by Dakt1, since it is required for the formation of ectopic trachea when *trh* is misexpressed. The zygotic transcription of *Dakt1* is spatially restricted to tracheal regions. However, the *Dakt1* maternal component is broadly expressed early in embryogenesis (63), and. the S665 PKB phosphorylation site also lies well outside of the PAS domains, indicating that other spatially-restricted factors are required for Trh function.

The Drifter (Dfr) POU-homeobox gene influences tracheal development and is expressed early in tracheal formation, similar to *trh* expression (61, 64). Dfr is also expressed at the sites in the dorsal ectoderm where ectopic trachea form when *trh* is misexpressed (50). Misexpression of both *trh* and *dfr* revealed the presence of ectopic trachea at sites in the ectoderm and head, beyond those observed for ectopic expression of only *trh*. These results demonstrate that Dfr is a coactivator of Trh in vivo. Biochemical experiments revealed that the Dfr POU domain directly binds to the Trh PAS domain (50) (Figure 6), consistent with the requirement of the Trh PAS domain for transcriptional specificity. Thus, Trh function requires the presence of both the Dfr coactivator and phosphorylation for proper function, and the coactivator clearly restricts Trh target gene activation. It is interesting that Dfr functions as a coactivator with both Trh and Sim, but biochemically behaves differently: Dfr binds directly to Trh via its PAS domain, but indirectly with Sim, which requires an additional cofactor, Fish, that binds to both Sim and Dfr (47) (Figure 6).

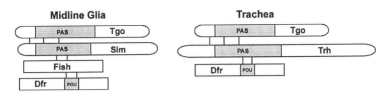

Figure 6. Dfr employs multiple modes of biochemical interactions with bHLH-PAS proteins. In the midline glia, Dfr interacts with Sim indirectly by binding to Fish via the Dfr POU domain, which binds to Sim via the Sim PAS domain. In the trachea, the Dfr POU domain binds directly to Trh via the Trh PAS domain.

4.5   Initiation of *trh* transcription

The expression of *trh* and corresponding formation of tracheal placodes occurs at precise positions in the dorsal ectoderm. It is convenient to think of the tracheal placodes as coordinates along the D/V and A/P axes, specified by axis patterning genes. Along the D/V axis, the dorsal extent of *trh* expression is established by repression by the Decapentaplegic (Dpp) TGF-β signaling pathway (57, 58), and the ventral extent is set by repression by the epidermal growth factor (EGF) pathway (16). The cues governing A/P positioning are not well known although Wingless has been implicated (58, 65). The tracheal placodes are relegated to 10 thoracic and abdominal segments and are not found in the terminal segments. This is due to terminal region repression by the *spalt* gene (66). Initial expression of *dfr* is independent of *trh* and likely governed directly by the same genes that regulate *trh* expression. This is consistent with their co-equal roles in tracheal development. After *trh* and *dfr* are activated in tracheal placode cells, they combine to autoregulate their own expression, since the initial patterning cues fade out. Trh and Dfr also combine to activate transcription of some tracheal target genes, and they may individually combine with other proteins to regulate expression of other target genes (61). Further insight into the respective developmental roles of *trh* and *dfr* will emerge as the identities of additional target genes for each transcription factor are discovered.

4.6   Non-tracheal roles of *trh*

4.6.1   Drosophila salivary duct development

4.6.1.1   Expression and genetics

The *Drosophila trh* gene plays a prominent role in controlling embryonic salivary duct development. The formation of the salivary tissue has emerged as an excellent system for studying tissue formation (67, 68). The two salivary glands are connected to the pharynx by ducts. Each gland is joined to its own duct, and these two ducts merge at their anterior ends to form a common duct. The salivary primordia consist of precursors to both duct and gland cells. *trh* is initially expressed in the entire primordia, but is later restricted to the salivary duct cells (57, 58). Despite the expression of *trh* in the primordia, only the duct cells are affected in *trh* mutant embryos (57, 58, 69). The *trh* mutant salivary glands appear relatively normal, but are closed at the site where they would normally join with the ducts. In contrast, the salivary duct cells fail to invaginate, and remain on the surface of the embryo. Trh regulates expression of several genes in the duct cells, including *eyegone*, which is required for formation of the individual ducts (70). However, *trh* may be carrying-out only a subset of duct cell developmental functions, such as morphogenesis, since some duct-expressed genes are not dependent on *trh* (67). Trh and Tgo colocalize to nuclei in both salivary primordia and duct. However, the Trh coactivator, Dfr, is absent in the salivary duct, so that Trh:Tgo presumably interacts with other coregulatory proteins to control duct cell transcription.

4.6.1.2   Control of *trh* salivary duct expression

Since *trh* plays an important role in salivary duct development, understanding the factors that control its expression provide insight into how regulatory proteins dictate cell fates. There are also interesting evolutionary implications since similar factors may control the formation of tissues in other species (see below). Initial expression of *trh* is in the ventral ectoderm of parasegment 2 (corresponds to the posterior maxillary segment plus anterior labial segment) (23, 57), which is the domain for the salivary primordia. *trh* expression is then restricted to the duct primordia, which lies ventral to the gland cells. Thus, regulation of *trh* involves three issues: (i) how D/V patterning restricts expression to the ventral ectoderm, (ii) how homeotic genes, which control parasegmental identity, restrict expression to parasegment 2, and (iii) how *trh* expression is restricted to the duct cells, and is absent from the gland cells.

The *Sex combs reduced* (*Scr*) homeotic gene controls parasegment 2 identity, and *Scr* is required, along with the *extradenticle* and *homothorax* transcription factor genes, to restrict *trh* expression to only parasegment 2 (23, 57). Thus, mutants of *Scr* result in a loss of *trh* expression and ubiquitous expression of *Scr* results in additional *trh* expression in more anterior segments (57). The *Dpp* gene functions along with *brinker* in the dorsal ectoderm to repress *trh* expression, thus restricting *trh* expression to the salivary primordia within the ventral ectoderm (23, 57). Within the salivary primordia, the ventral-most cells become duct cells, and the more dorsal cells form the glands. *trh* expression is present in the duct cells and is absent from the gland cells. The transcriptional repressor, Fork head (Fkh) functions to repress *trh* in the gland cells (69). *fkh* expression is complementary to *trh*: it is off in the duct cells, but present in the gland cells. Thus, a key issue is how is *fkh* expression restricted to the gland cells, but not duct cells. This is due to midline-directed Sim/Spitz signaling.

The ventral-most midline cells in parasegment 2 express *sim*, which likely controls *spitz/EGFR* signaling (23). Spitz acts as a morphogen emanating from the midline that is high in concentration ventrally and low dorsally. Mutants in *sim* or other *spitz* class genes result in a loss of duct cell fate and a corresponding expansion of gland cell fate (23, 69). One of the consequences of Spitz signaling is repression of *fkh* in the duct cell primordia. Since Fkh represses *trh* expression, this results in restriction of *trh* to the duct cells.

4.6.2   *Bombyx trh* and silk gland development

The silkworm, *Bombyx mori*, is well known for its ability to produce silk. It has been proposed that the silk-producing gland is a modified salivary gland, since it is derived from the labial segment, as is the salivary gland. The silk gland consists of three distinct regions: the anterior silk gland (ASG), middle silk gland (MSG), and posterior silk gland (PSG). The ASG is a duct for the silk proteins secreted by the MSG and PSG. This functional specialization of the silk gland resembles that of the salivary gland in which an anterior salivary duct connects the two salivary glands to the foregut.

Bombyx possesses a *trh* gene (*Bm-trh*) (71) that is highly related to *Drosophila trh*. Embryonic expression of *Bm-trh* was observed in the silk gland, trachea, and supracolonic trachea (homolog of the *Drosophila* posterior spiracle). Tracheal and supracolonic tracheal expression begins in the placodes and continues throughout development. This is analogous to *trh* expression in *Drosophila* and suggests a similar function as a regulator of tracheal development. Initial *Bm-trh* expression is present in the primordia of the entire silk gland, but then becomes restricted to the ASG, the silk gland duct. This also resembles *trh* expression in the *Drosophila* salivary primordia, in which *trh* is initially expressed throughout the primordia, and is then restricted to the salivary duct.

The *Bombyx silk gland factor-3* (*SGF-3*) gene (72) encodes a POU-homeobox protein highly related to *Drosophila* Dfr. *SGF-3* is expressed in the developing trachea and silk gland. Initially, *SGF-3* is expressed throughout the entire silk gland primordia, but is later restricted to the ASG and part of the MSG. This expression pattern overlaps with *Bm-trh*, which is also expressed in the trachea, silk gland primordia, and then ASG. As in *Drosophila* and *Artemia* (see below), Bm-Trh and SGF-3 may interact as a regulatory cassette to control transcription and development. The one phylogenetic difference is that *Drosophila* Trh and Dfr do not interact to control salivary duct development.

Further molecular analysis of silk gland gene expression, revealed additional similarities between silk gland and salivary gland development. *Drosophila trh* is initially expressed in cells that will give rise to both salivary duct and glands, but is later expressed in only the duct cells. The *Drosophila fkh* gene is expressed in the gland cells and required for their development (73). *fkh* represses *trh* in the gland cells, thus restricting *trh* expression to the duct cells. The *Bombyx fkh* gene (*Salivary Gland Factor-*

1; *SGF-1*) is expressed only in the MSG and PSG, but not ASG (74). *Bm-trh* initially is expressed throughout the silk gland primordia, and is later restricted to the ASG (71). Thus, *Bombyx* Fkh may be acting to repress *Bm-trh* and *SGF-3* in the MSG and PSG, similar to its role in the *Drosophila* salivary primordia.

4.6.3   *Artemia trh* and osmoregulation

The branchiopod crustacean brine shrimp, *Artemia franciscana*, resides in salt pools as cysts. Upon hydration, the cysts hatch and larvae develop and live in the hyperosmotic salt ponds. Nauplius larvae, the first emerging larval forms, possess a specialized organ, the salt gland, which regulates osmolarity. Later in development, the salt gland is resorbed, and osmoregulation is carried-out by the thoracic epipods that reside on the appendages (75). Given its lifestyle, osmoregulation is a critical element of brine shrimp physiology.

Artemia possess a *trh* gene (*Af-trh*) highly related to insect *trh* (76). *In situ* hybridization of *Artemia* larvae with an *Af-trh* probe revealed expression in the salt gland of the nauplius and the thoracic epipods of older larvae – both sites of osmoregulation (Figure 7). Thus, another function of *trh* may be the development and function of crustacean osmoregulatory organs. An *Artemia drifter* gene, *APH-1*, was identified and its expression pattern analyzed (77). Interestingly, it is also expressed in the salt gland (epipod expression was not reported). This suggests that Af-Trh and APH-1 interact to control transcription in *Artemia* similar to their role in insect tracheal development, and represent an evolutionarily-conserved regulatory protein cassette.

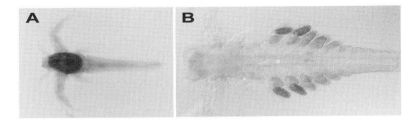

Figure7. Artemia trh is expressed in sites of osmoregulation. (A) Expression of *Af-trh* in the naupliar salt gland. (B) Later expression of *Af-trh* in the larval epipods. Adpated from (76).

Drosophila osmoregulation occurs in a subset of cells in the hindgut (78), a site in which *trh* expression has not been reported. Thus, this aspect of *trh* function may not be conserved between crustaceans and insects. *Artemia* do not possess identifiable respiratory organs, and respiration is likely due to diffusion over the entire body surface, including the epipods. It remains an interesting, but open question, whether the role of *trh* in respiratory system development is conserved in the crustacean subphylum.

5.   DYSFUSION: REGULATOR OF TRACHEAL FUSION

5.1   Dys structure and expression

The Dys bHLH-PAS protein was identified using a bioinformatic screen of the *Drosophila* genome (62). It is a member of well-conserved subgroup of bHLH-PAS proteins conserved in *C. elegans* (C15C8.2) (see chapter by Powell-Coffman) and mammals (NXF). Dys is a nuclear protein and overlaps with nuclear Tgo, strongly suggesting that Dys and Tgo heterodimerize. Expression of *dys* is found in a variety of cell types including: (i) tracheal fusion cells, (ii) epidermal leading edge, (iii) foregut atrium, (iv) nervous system, (v) hindgut, and (vi) anal pad. Work on the *dys* gene has focused on its role in tracheal fusion cells.

The insect trachea are derived from a series of segmentally-reiterated placodes. Signals from adjacent cells induce the tracheal cells to migrate in multiple directions (59). In most cases, these branches meet and fuse. There are four branch types that fuse, including the: (i) dorsal branch, (ii) dorsal trunk, (iii) lateral trunk, and (iv) ganglionic branch. Each migrating branch has a tip cell that guides branching and then mediates fusion (79). In this manner, the mature, diffuse tracheal system is created. The fusion process is complex, involving recognition, adhesion, cytoskeletal rearrangement, formation of adherens junctions, and formation of a continuous tubule. The fusion cell has a distinct tracheal cell identity indicated by a unique pattern of gene expression and cellular function.

The *dys* gene is expressed in all tracheal fusion cells, but no other tracheal cells. The *escargot* (*esg*) gene encodes a zinc finger transcription factor that is also expressed in tracheal fusion cells. *esg* expression occurs prior to fusion, and plays an important role in fusion cell fate and function (80, 81). *dys* is expressed in fusion cells after the appearance of *esg*, but also before fusion occurs (62). Analysis of *dys* expression in *esg* mutant embryos showed that *esg* regulates *dys* expression in all branches, except the dorsal trunk. The expression of *dys* in fusion cells suggested a role in the fusion cell process.

5.2   Dys function and Dys-Trh interactions

The function of *dys* was examined using RNAi (62). Embryos injected with *dys* RNAi had a high percentage of tracheal fusion defects. The dorsal branch, lateral trunk, and ganglionic branch showed an absence of fusion. The dorsal trunk was unaffected. These results were similar to those for *esg*, which showed defects in all branches except the dorsal trunk.

Initially, the Trh protein exists at uniformly high levels in all tracheal cells (7). As Dys protein levels rise in fusion cells during wild-type tracheal development, the levels of Trh decline in fusion cells (62). The decline of Trh is dependent on *dys*, since Trh levels remain high when *dys* expression is absent due to injection of *dys* RNAi. The mechanism in which Dys downregulates Trh is unknown. One model is that Dys outcompetes Trh for their common dimerization partner, Tgo, and the reduction of Trh:Tgo heterodimers causes a corresponding reduction in *trh* transcription, since *trh* is autoregulatory.

The developmental significance of the reduction in Trh by Dys is unknown, but may facilitate the transition of the tip cell from a migratory role to a fusion role. Since there are a number of examples in vertebrates and invertebrates in which bHLH-PAS genes overlap in expression (82, 83), the study of Dys-Trh interactions may prove to be an important model. There are a number of important issues to be addressed. What are the identities of Dys target genes and what are their roles in tracheal fusion and related processes. Does *dys* carry-out a subset of *esg* functions? Does Dys activate or repress transcription? What is the mechanism by which Dys reduces Trh levels, and what is the biological significance of this reduction? Does Dys regulate levels of other bHLH-PAS proteins?

6.   SIMILAR: HYPOXIA REGULATION

The *similar* (*sima*) gene was identified by low stringency hybridization with a *sim* probe (84). When the human *hypoxia inducible factor-1α* (*HIF-1α*) (see chapter by Semenza) gene was cloned and sequenced (85), it became apparent that Sima was most related to HIF-1α, and represented a potential candidate for an insect HIF. Subsequent work (86, 87) has confirmed that Sima is a HIF-1α ortholog, and functions in regulating the response to hypoxia, much the way HIF-1α regulates the mammalian response to hypoxia.

6.1   *Drosophila* hypoxia responsiveness

Insight into the *Drosophila* physiological, developmental, and behavioral responses to low oxygen levels have appeared in several studies. Use of a transgenic fly strain containing a hypoxia-sensitive reporter indicated that the transcriptional response to hypoxia occurs throughout development (86). Responsiveness increases during embryogenesis with a peak in late-stage embryos, and decreases afterwards. However, it remains relatively high during larval development. This is particularly relevant since larvae crawl through rotting food where oxygen may be limiting. Consistent with this, hypoxia alters the feeding behavior of larvae through a nitric oxide/cyclic GMP pathway, such that oxygen-deprived larvae stop feeding and begin moving (presumably with the desire to find an environment with higher oxygen levels) (88). Induction of the reporter peaked at oxygen concentrations around 3% (normoxia is 20%). Induction was absent at 0% oxygen (anoxia), consistent with work showing that anoxia results in cell cycle and metabolic arrest.

During hypoxic conditions, the trachea is the main cell type showing high hypoxia-sensitive reporter induction (86). The significance of this is unclear. One proposal is that the trachea may act as a sensor for oxygen and convey that information to the nervous system, which influences larval behavior. However, the main effects of hypoxia have been observed in non-tracheal cells, as they seek a source of oxygen (59). Hypoxic larval cells secrete Branchless (Bnl; a fly fibroblast growth factor-like protein) and induce new tracheal branches to migrate towards the Bnl-producing cells (89). Another mechanism involves the extension of cytoplasmic projections from hypoxic non-tracheal cells that attach to tracheoles and pull them towards the oxygen-starved cells (59, 90). Consistent with the ability of non-tracheal cells to respond to hypoxia, induction of the hypoxia-dependent reporter can be seen outside of the trachea when conditions of hypoxia are relatively severe (86). It will be important to analyze *sima* mutants to understand its role in hypoxia regulation and additional, unforeseen, biological processes. Does *sima* regulate Bnl-mediated tracheal branching and the ability of non-tracheal cells to extend cytoplasmic processes and grab existing tracheoles? What is the relevance of the strong induction of hypoxia-dependent gene regulation in the trachea? Does *sima* mediate long-term changes in morphology, physiology, and behavior?

6.2   Structure and expression

The Sima protein is one of the longest *Drosophila* bHLH-PAS proteins (1505 aa). Sima heterodimerizes with Tgo (1) and is able to bind and activate transcription from an ACGTG CME-like DNA sequence (Estes and Crews, unpubl.) (8, 86). As with other bHLH-PAS proteins, the bHLH domain is near the N-terminus followed by PAS-1 and PAS-2 domains. The C-terminal half of Sima has a large number of homopolymeric stretches, including 13 stretches of poly[glutamine]. These homopolymeric stretches likely contribute to the transcriptional activation function that resides in the C-terminal region. Within the C-terminal region is also an evolutionarily-conserved oxygen-dependent degradation domain (ODDD), which regulates the levels of Sima protein with respect to oxygen concentration (86). In addition, the ODDD contains sequences required for cytoplasmic localization under normoxia (see below).

RNA levels of *sima* appear uniform in most, if not all, cells of the embryo under normoxia (84), and the levels change little under hypoxia (86). Sima protein levels are low at normal oxygen levels, and the protein is localized to the cytoplasm (86). Protein levels dramatically increase under low oxygen tension, and the protein is localized to nuclei (86). As with mammalian HIF-1α protein, this suggests that Sima protein levels and function are predominantly regulated at the level of protein degradation.

6.3 Biochemistry of hypoxia regulation

The regulation of HIF-1α stability under varying oxygen conditions has been well-studied in vertebrates (see chapter by Semenza), and the same basic model applies to Sima. Under normoxia, HIF-1α is hydroxylated on two prolines in the ODDD by HIF prolyl hydroxylases. These hydroxylated compounds bind von Hippel Lindau (VHL) factor, which mediates degradation by the ubiquitin/proteasome. Under hypoxia, the hydroxylase is inhibited, VHL fails to bind, and the protein is no longer degraded. While the biochemical details have not been described, *Drosophila* proteins related to those involved in hypoxia control of HIF-1α have been identified. These include *Drosophila* von Hippel-Lindau (D-VHL) factor (91) and HIF prolyl hydroxylase (Hph; CG1114) (86).

D-VHL RNAi experiments resulted in tracheal defects, such as excessive branching, looping, and breakage (91), its role in hypoxia and mediating Sima degradation has not been studied. Its expression is predominantly tracheal. The *Drosophila Hph* gene is ubiquitously expressed in the embryo. *Hph* RNAi and genetic experiments carried-out under normoxia resulted in upregulation of Sima protein and its localization to the nucleus (86), accompanied by activation of the hypoxia-sensitive reporter. Expression of *Hph* was upregulated under hypoxia, similar to affects seen with mammalian prolyl hydroxylases. These results indicate that prolyl 4-hydroxylase functions in hypoxia regulation. Mutational analysis of Sima is consistent with Hph acting directly on Sima. The ODDD contains two prolines within sequences conserved between Sima and HIF-1α. Deletion of the ODDD resulted in nuclear accumulation of Sima under normoxia and activation of the hypoxia-sensitive reporter (86). This confirms that the ODDD controls Sima degradation and cytoplasmic retention under normoxia, and that it is regulated by prolyl 4-hydroxylase. While, the role of D-VHL and proteins of the ubiquitin degradation pathway have not been analyzed for their roles in hypoxia regulation, the evidence is strong that the regulatory pathway is highly related to that utilized in mammals and *C. elegans*. Nevertheless, it will be interesting to see whether there exist features of hypoxia regulation idiosyncratic to *Drosophila*.

6.4   Transcriptional specificity

Sima:Tgo is able to bind and activate transcription from a multimerized ACGTG (CME) core sequence, just as Sim:Tgo and Trh:Tgo. Since each protein complex regulates different developmental and physiological processes, this raises the issue of transcriptional specificity and how the different complexes control distinct gene batteries. Insight into this issue has emerged from analysis of two related, but distinct, in vivo gene reporters that were tested for their response to hypoxia (86). When a pentamer of the murine *erythropoietin* hypoxia response element (HRE), which contains HIF-1α:Arnt (and presumably Sima:Tgo) CME binding sites, was tested in vivo, it failed to be activated under hypoxia. When a fragment of the mammalian *lactate dehydrogenase A* (*LDH-A*) gene was tested, it was activated under hypoxia. The *LDH-A* fragment has two HREs and a cyclic AMP responsive element (CRE). These results suggest that a CRE-binding protein acts as a coactivator with Sima:Tgo for hypoxia induction. Since this reporter is not expressed at significant levels in CNS midline cells or trachea, this regulatory region is specific for Sima:Tgo activation , but not for Sim:Tgo or Trh:Tgo. Further progress will emerge when target genes of Sima:Tgo are identified and analyzed.

7.   *SPINELESS*: APPENDAGE IDENTITY

Mutations in the *ss* gene were first reported in 1923 by Bridges and Morgan (92). Null mutants of *ss* are viable, and possess a number of interesting phenotypes that include: (i) reduction in bristle size, (ii) deletion of much of the tarsal segment of the leg, and (iii) transformation of the distal antenna to distal leg (Figure 8). The latter two phenotypes have established *ss* as a key gene in controlling appendage identity. The *ss* gene was cloned based on the identification of a P-element transposon insertion that had a weak *ss* phenotype (10). Molecular analysis revealed that *Drosophila ss* gene is orthologous to the mammalian *aryl hydrocarbon receptor* (*ahr*; dioxin receptor) and *C. elegans ahr-1* (93), indicating that these genes encode a highly-conserved subfamily of bHLH-PAS proteins. Unlike *ahr*, which is best known for its role in toxin metabolism (see chapter by Yao et al.), the known roles of *ss* are developmental. Direct tests have shown that *Drosophila* Ss does not bind dioxin (94).

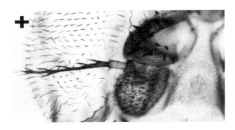 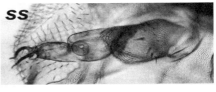

Figure 8. ss mutant shows a transformation of antenna to leg. Wild-type adult (+) showing antenna including the distal arista attached to segment A3. In *ss* null mutant, the arista is transformed to distal leg. Adapted from (9).

7.1   Antennal and leg anatomy and development

The adult antenna and leg are derived from imaginal discs that are specified during embryogenesis and develop during the larval and pupal stages. The mature antenna is an olfactory organ that consists of 6 segments (proximal to distal): A1-5, and the arista (A6). Olfactory neurons reside on A3. Below the antenna are the maxillary palps, which are also olfactory organs. The adult leg is derived from the leg imaginal disc and consists of the coxa, trochanter, femur, tibia, and 5 tarsal segments (proximal to distal). The last tarsal segment terminates in a claw.

7.2   Biochemistry

Like Sim and Trh, Ss heterodimerizes with Tgo, likely migrates to the nucleus, binds DNA and activates transcription (9). Ss:Tgo prefers GCGTG binding sites, as do Ahr:Arnt heterodimers. This differs from the ACGTG preference of Sim:Tgo, Trh:Tgo, and Sima:Tgo, and provides a partial explanation for how Ss:Tgo regulates different genes and biological processes distinct from the other bHLH-PAS proteins. When *ss* is ectopically expressed, Tgo localizes in the nucleus at the ectopic sites, suggesting that a Ss:Tgo nuclear complex is formed at those sites (9). Since no differences have been observed with respect to cell type, this suggests that nuclear entry is not dependent on localized or tissue-specific ligand binding, as is required for vertebrate Ahr. However, this observation does not rule out that some Ss function may still require ligand binding or modification.

7.3   Expression

ss is generally expressed at sites of *ss* genetic function (10). Embryonic sites include: (i) the larval antennal sense organ primordia, (ii) the gnathal segments (maxillary, mandibular, and labial), (iii) in patches within the thoracic segments that correspond to the presumptive legs, and (iv) sensory cells. All of these sites colocalized with nuclear Tgo, except the leg primordia. Corresponding to the sites of embryonic expression are two larval *ss* phenotypes: a deformed antennal sensory organ and mislocalized dorsomedial papilla of the maxillary sense organ.

Postembryonically, *ss* is expressed in the leg, antennal, wing, genital, and eye discs and all bristle precursor cells (10). *ss* mutants delete the distal region of leg segment 1 and segments 2-4, resulting in a fusion of 1 and 5. Leg disc expression of *ss* likely corresponds to those tarsal cells deleted in *ss* mutants. Expression in the antennal disc is predominantly in antennal segments A2, A3 and the aristae, consistent with the *ss* mutant defects in A2, A3 and the aristae, and the absence of *ss* defects in antennal segments A1 (10, 95). During pupariation, *ss* is expressed in bristle sensory organ precursors in most discs, which also correlates with the *ss* bristle defects. Later it can be seen that *ss* expression is restricted to the bristle cells, but not to the socket cells. The postembryonic expression of *ss* correlates nicely with the known *ss* defects, although expression in the genital, labial, and eye discs does not correlate with any known *ss* phenotype. The *ss* gene is dynamically expressed in a variety of cell types; critical to its function is understanding its regulation and the nature of the target genes it regulates.

7.4   *ss* genetics and regulation

7.4.1   Antennal development

ss null mutants result in a transformation of the antennal arista into distal leg (see (10). The antennal third segment is converted to smooth cuticle of unspecified identity. In addition misexpression of *ss* in imaginal discs is able to transform maxillary palps, rostral membrane (which lies between the antenna and maxillary palp), and distal leg to antennal-like structures (10).

These results indicate that *ss* plays an important role in specifying the identity of the distal antenna. Since ectopic expression of *ss* only converts certain cell types to antennal fates, Ss presumably functions with cell-type specific coregulators to carry-out its functions.

The *Distalless* (*Dll*) and *homothorax* (*hth*) homeobox genes regulate both appendage identity and proximodistal positional information (95). Mutants in both genes result in a transformation of antenna to leg. Mutations in *Dll* also result in distal deletions of both leg and antenna, and *hth* mutants result in proximal deletions of both appendages. Genetic studies indicate that both *Dll* and *hth* activate *ss* expression in the antenna and leg (10, 95). Both *Dll* and *hth* are spatially restricted in the developing appendages, and their expression and activities are likely required for the spatial distribution of *ss* in the antennal and leg discs. While *ss* plays an important role in antennal identify, it is not a master regulator of antennal gene expression, since numerous genes involved in antennal differentiation and function are not regulated by *ss (95)*. Two roles for *ss* are repression of tarsal and tarsal claw in the antenna, and the formation of olfactory sense organs in A3. Thus, *ss* is a downstream target of Dll and Hth and carries-out a subset of their functions.

7.4.2   Leg development

ss null mutants result in a loss of the distal part of tarsal segment 1 (T1) and T2-4. When *ss* is misexpressed in distal and proximal leg segments, it results in a deletion of the medial femur and tibia segments and a transformation of the claw into arista (10). Regulation of *ss* in the leg resembles antennal *ss* regulation, since *Dll* controls ss expression in both appendages (10). One function of *ss* is to regulate the expression of *bric-a-brac* (*bab*). *bab* is required for the formation of several tarsal joints (96). Ss positively regulates *bab* in the leg (10), and *dachshund* is involved in repressing *bab* (97), resulting in stripes of *bab* expression at the tarsal joints.

7.5   Evolutionary considerations

Despite seemingly different biological roles for *ss* and *Ahr*, it has been proposed that an ancestral *ss* gene was involved in chemosensory function (9). This is consistent with the expression of *C. elegans Ahr-1* in sensory neurons (see chapter by Powell-Coffman), *Drosophila ss* in formation of sensory organs, and mammalian Ahr in binding (sensing) toxic molecules. Future work may reveal additional similarities. It cannot be excluded that the nematode and insect proteins also function as ligand-binding receptors and may have physiological roles. Similarly, further work on *Ahr* may reveal conserved developmental functions.

8.   SUMMARY

The *Drosophila* bHLH-PAS proteins that partner with Tgo share a number of properties. It is convenient to group Dys, Sim, Ss, and Trh together, since their regulation and mode of action differs from Sima. Each member of this "DSST" subgroup is specifically expressed in a number of cell types during embryonic and postembryonic development. There is little redundancy or overlap among these proteins, and this lack of redundancy likely contributes to the numerous mutant phenotypes associated with these genes. Several DSST genes play fundamental roles in tissue-specific development. The *sim* gene can reasonably be called a master regulator of CNS midline cell development, since it is required for all midline cell developmental events (directly or indirectly), and the initiation or maintenance of expression of every midline-expressed gene assayed. The *ss* and *trh* genes also play important roles in the formation of tissue types, since like *sim*, their ectopic expression results in formation of their corresponding cell types (antenna, trachea, and CNS midline cells) at ectopic sites. However, even in the antenna and trachea, ss and trh, respectively, do not control all aspects of development and transcription. For all three genes, the misexpression experiments that result in additional sites of tissue generation are not ubiquitous but found only at specific locations. This reinforces that idea that these proteins require additional factors for function. These may be coregulators that directly interact with or modify the bHLH-PAS proteins or additional factors that act independently and carry-out functions complementary to those of the bHLH-PAS proteins.

The ability of each bHLH-PAS gene to regulate different target genes in different cell types, as well as the fact that multiple bHLH-PAS proteins dimerize with Tgo and bind the same DNA sequences raises issues of transcriptional specificity. While other mechanisms may exist, current evidence suggests that specificity arises from interactions between bHLH-PAS proteins and different coregulatory proteins (Trh-Dfr, Sim-Fish-Dfr). The PAS domain mediates these interactions, but more biochemical work needs to be carried-out to better define the PAS domain interactions, as well as understand the entire range of PAS domain functions. Why are there two PAS domains/protein, and why is the sequence so evolutionarily diverse? Even though regulatory proteins, such as Sim and Trh, regulate different cell types (CNS midline and trachea) there exist many genes that are expressed in both cell types and utilize the same binding sites for both modes of expression. Is there an evolutionary significance to this, such that an ancestral organism possessed a cell type or tissue that is the predecessor to both existing *Drosophila* cell types? Does this extend to other cell types regulated by bHLH-PAS proteins?

The *Drosophila* bHLH-PAS partners of Tgo are expressed in a diverse array of cell types, although there are similarities. Three of the five (Dys, Sima, and Trh) play roles in tracheal development. Is this fortuitous or purposeful? Evidence for the latter derives from the observation that Dys downregulates Trh in tracheal fusion cells. Since many prokaryotic and eukaryotic PAS proteins function in sensory roles and the PAS domain often mediates ligand binding, it is tempting to think that the *Drosophila* developmental bHLH-PAS proteins may also respond to developmental signals. Evidence to data has not turned up much evidence for this, although Trh function is dependent on phosphorylation. Instead the spatial and temporal control over function derives primarily from the transcriptional regulation of these genes. The *sim* gene is a large gene with multiple promoters and a sophisticated array of cis-control elements that direct its highly specific expression. Although only *sim* has been studied in detail, the *dys*, *ss*, and *trh* genes are also large and likely to contain a large number of cis-control elements. Nevertheless, the developmental, physiological, and environmental regulation of function via cofactor binding (protein or small molecule), or modification via the PAS domain or other protein region, should be considered. As an example, phosphorylation of Trh function by PKB was unexpected. The PAS domain proteins remain central to understanding the regulation of developmental and physiological processes in *Drosophila*.

ACKNOWLEDGEMENTS

I would like to thank Debbie Andrew, Ian Duncan, Dan Lau, and John Nambu for thoughtful comments on the manuscript.

REFERENCES

1. Sonnenfeld, M., M. Ward, G. Nystrom, J. Mosher, S. Stahl, and S. Crews. 1997. The *Drosophila tango* gene encodes a bHLH-PAS protein that is orthologous to mammalian Arnt and controls CNS midline and tracheal development. *Development* 124:4583-4594.

2.   Ohshiro, T., and K. Saigo. 1997. Transcriptional regulation of *breathless* FGF receptor gene by binding of TRACHEALESS/dARNT heterodimers to three central midline elements in *Drosophila* developing trachea. *Development* 124:3975-3986.

3.   Hoffman, E. C., H. Reyes, F. Chu, F. Sander, L. H. Conley, B. A. Brooks, and O. Hankinson. 1991. Cloning of a subunit of the DNA-binding form of the Ah (dioxin) receptor. *Science* 252:954-958.

4.   Wharton, J., K. A., R. G. Franks, Y. Kasai, and S. T. Crews. 1994. Control of CNS midline transcription by asymmetric E-box elements: similarity to xenobiotic responsive regulation. *Development* 120:3563-3569.

5.   Eguchi, H., T. Ikuta, T. Tachibana, Y. Yoneda, and K. Kawajiri. 1997. A nuclear localization signal of human aryl hydrocarbon receptor nuclear translocator/hypoxia-inducible factor 1β is a novel bipartite type recognized by the two components of nuclear pore-targeting complex. *J. Biol. Chem.* 272:17640-17647.

6.   Frigerio, G., M. Burri, D. Bopp, S. Baumgartner, and M. Noll. 1986. Structure of the segmentation gene *paired* and the Drosophila PRD gene set as part of a gene network. *Cell* 47:735-746.

7.   Ward, M. P., J. T. Mosher, and S. T. Crews. 1998. Regulation of *Drosophila* bHLH-PAS protein cellular localization during embryogenesis. *Development* 125:1599-1608.

8.   Zelzer, E., P. Wappner, and B.-Z. Shilo. 1997. The PAS domain confers target gene specificity of *Drosophila* bHLH/PAS proteins. *Genes Dev.* 11:2079-2089.

9.   Emmons, R. B., D. Duncan, P. A. Estes, P. Kiefel, J. T. Mosher, M. Sonnenfeld, M. P. Ward, I. Duncan, and S. T. Crews. 1999. The Spineless-Aristapedia and Tango bHLH-PAS proteins interact to control antennal and tarsal development in *Drosophila*. *Development* 126:3937-3945.

10.   Duncan, D. M., E. A. Burgess, and I. Duncan. 1998. Control of distal antennal identity and tarsal development in *Drosophila* by *spineless-aristapedia*, a homolog of the mammalian dioxin receptor. *Genes Dev.* 12:1290-1303.

11.   Hilliker, A. J., S. H. Clark, W. M. Gelbart, and A. Chovnick. 1981. Cytogenetic analysis of the rosy micro-region, polytene chromosome interval 87D2-4; 87E12-F1, of *Drosophila melanogaster. Drosophila Inform. Serv.* 56:65-72.

12.   Schalet, A., R. P. Kernaghan, and A. Chovnick. 1964. Structural and phenotypic definition of the rosy cistron in Drosophila melanogaster. *Genetics* 50:1261-1268.

13.   Thomas, J. B., S. T. Crews, and C. S. Goodman. 1988. Molecular genetics of the *single-minded* locus: a gene involved in the development of the Drosophila nervous system. *Cell* 52:133-141.

14.   Mayer, U., and C. Nüsslein-Volhard. 1988. A Group of genes required for pattern formation in the ventral ectoderm of the Drosophila embryo. *Genes Dev.* 2:1496-1511.

15.   Kim, S. H., and S. T. Crews. 1993. Influence of Drosophila ventral epidermal development by the CNS midline cells and spitz class genes. *Development* 118:893-901.

16.   Raz, E., and B. Z. Shilo. 1993. Establishment of ventral cell fates in the Drosophila embryonic ectoderm requires DER, the EGF receptor homolog. *Genes Dev.* 10:1937-1948.

17. Bossing, T., and G. M. Technau. 1994. The fate of the CNS midline progenitors in *Drosophila* as revealed by a new method for single cell labelling. *Development* 120:1895-1906.

18. Schmid, A., A. Chiba, and C. Q. Doe. 1999. Clonal analysis of Drosophila embryonic neuroblasts: neural cell types, axon projections and muscle targets. *Development* 126:4653-4689.

19. Crews, S. T. 1998. Control of cell lineage-specific development and transcription by bHLH-PAS proteins. *Genes Dev.* 12:607-620.

20. Jacobs, J. R. 2000. The midline glia of Drosophila: a molecular genetic model for the developmental functions of glia. *Prog. Neurobiol.* 62:475-508.

21. Hummel, T., K. Schimmelpfeng, and C. Klambt. 1999. Commissure formation in the embryonic CNS of *Drosophiila* I. Identification of the required gene functions. *Dev. Biol.* 209:381-398.

22. Sonnenfeld, M. J., and J. R. Jacobs. 1994. Mesectodermal cell fate analysis in *Drosophila* midline mutants. *Mech. Dev.* 46:3-13.

23. Panzer, S., D. Weigel, and S. K. Beckendorf. 1992. Organogenesis in Drosophila melanogaster: embryonic salivary gland determination is controlled by homeotic and dorsoventral patterning genes. *Development* 114:49-57.

24. Luer, K., J. Urban, C. Klambt, and G. M. Technau. 1997. Induction of identified mesodermal cells by CNS midline progenitors in *Drosophila. Development* 124:2681-2690.

25. Zhou, L., H. Xiao, and J. R. Nambu. 1997. CNS midline to mesoderm signaling in Drosophila. *Mech. Dev.* 67:59-68.

26. Page, D. T. 2003. A function for EGF receptor signaling in expanding the developing brain in Drosophila. *Curr. Biol.* 13:474-82.

27. Therianos, S., S. Leuzinger, F. Hirth, C. S. Goodman, and H. Reichert. 1995. Embryonic development of the Drosophila brain: formation of commissural and descending pathways. *Development* 121:3849-60.

28. Dickson, B. J. 2002. Molecular mechanisms of axon guidance. *Science* 298:1959-64.

29. Kidd, T., C. Russell, C. S. Goodman, and G. Tear. 1998. Dosage sensitive and complementary functions of roundabout and commissureless control axon crossing of the CNS midline. *Neuron* 20:25-33.

30. Lewis, J. O., and S. T. Crews. 1994. Genetic analysis of the Drosophila single-minded gene reveals a CNS influence on muscle patterning. *Mech. Dev.* 48:81-91.

31. Mehta, B., and K. M. Bhat. 2001. Slit signaling promotes the terminal asymmetric division of neural precursor cells in the Drosophila CNS. *Development* 128:3161-8.

32. Condron, B. G. 1999. Serotonergic neurons transiently require a midline-derived FGF signal. *Neuron* 24:531-40.

33. Menne, T. V., K. Luer, G. M. Technau, and C. Klambt. 1997. CNS midline cells in *Drosophila* induce the differentiation of lateral neural cells. *Development* 124:4949-4958.

34. Nambu, J. R., J. L. Lewis, K. A. Wharton, and S. T. Crews. 1991. The Drosophila *single-minded* gene encodes a helix-loop-helix protein which acts as a master regulator of CNS midline development. *Cell* 67:1157-1167.

35.   Nambu, J. R., R. G. Franks, S. Hu, and S. T. Crews. 1990. The *single-minded* gene of Drosophila is required for the expression of genes important for the development of CNS midline cells. *Cell* 63:63-75.

36.   Xiao, H., L. A. Hrdlicka, and J. R. Nambu. 1996. Alternate function of the *single-minded* and *rhomboid* genes in development of the *Drosophila* ventral neuroectoderm. *Mech. Dev.* 58:65-74.

37.   Chang, Z., B. D. Price, S. Bockheim, M. J. Boedigheimer, R. Smith, and A. Laughon. 1993. Molecular and genetic characterization of the *Drosophila tartan* gene. *Dev. Biol.* 160:315-332.

38.   Mellerick, D. M., and M. Nirenberg. 1995. Dorsal-ventral patterning genes restrict NK-2 homeobox gene expression to the ventral half of the central nervous system of Drosophila embryos. *Dev. Biol.* 171:306-316.

39.   Crews, S. T., J. B. Thomas, and C. S. Goodman. 1988. The Drosophila *single-minded* gene encodes a nuclear protein with sequence similarity to the *per* gene product. *Cell* 52:143-151.

40.   Kasai, Y., S. Stahl, and S. Crews. 1998. Specification of the *Drosophila* CNS midline cell lineage: direct control of *single-minded* transcription by dorsal/ventral patterning genes. *Gene Expression* 7:171-189.

41.   Pielage, J., G. Steffes, D. C. Lau, B. A. Parente, S. T. Crews, R. Strauss, and C. Klambt. 2002. Novel behavioral and developmental defects associated with Drosophila single-minded. *Dev. Biol.* 249:283-99.

42.   Kasai, Y., J. R. Nambu, P. M. Lieberman, and S. T. Crews. 1992. Dorsal-ventral patterning in Drosophila: DNA binding of snail protein to the single-minded gene. *Proc. Natl. Acad. Sci. USA* 89:3414-3418.

43.   Morel, V., and F. Schweisguth. 2000. Repression by suppressor of hairless and activation by Notch are required to define a single row of single-minded expressing cells in the Drosophila embryo. *Genes Dev.* 14:377-88.

44.   Cowden, J., and M. Levine. 2002. The Snail repressor positions Notch signaling in the Drosophila embryo. *Development* 129:1785-93.

45.   Morel, V., R. Le Borgne, and F. Schweisguth. 2003. Snail is required for Delta endocytosis and Notch-dependent activation of single-minded expression. *Dev. Genes Evol.* 213:65-72.

46.   McGuire, J., P. Coumailleau, M. L. Whitelaw, J. A. Gustafsson, and L. Poellinger. 1995. The basic helix-loop-helix/PAS factor Sim is associated with hsp90. Implications for regulation by interaction with partner factors. *J. Biol. Chem.* 270:31353-7.

47.   Ma, Y., K. Certel, Y. Gao, E. Niemitz, J. Mosher, A. Mukherjee, M. Mutsuddi, N. Huseinovic, S. T. Crews, W. A. Johnson, et al. 2000. Functional interactions between Drosophila bHLH/PAS, Sox, and POU transcription factors regulate CNS midline expression of the slit gene. *J. Neurosci.* 20:4596-605.

48.   Estes, P., J. Mosher, and S. T. Crews. 2001. *Drosophila single-minded* represses gene transcription by activating the expression of repressive factors. *Dev. Biol.* 232:157-175.

49. Franks, R. G., and S. T. Crews. 1994. Transcriptional activation domains of the Single-minded bHLH protein are required for CNS midline cell development. *Mech. Dev.* 45:269-277.

50. Zelzer, E., and B. Shilo. 2000. Interaction between the bHLH-PAS protein Trachealess and the POU-domain protein Drifter, specifies tracheal cell fates. *Mech. Dev.* 19:163-173.

51. McDonald, J. A., S. Holbrook, T. Isshiki, J. Weiss, C. Q. Doe, and D. M. Mellerick. 1998. Dorsoventral patterning in the Drosophila central nervous system: the vnd homeobox gene specifies ventral column identity. *Genes Dev.* 12:3603-3612.

52. Chu, H., C. Parras, K. White, and F. Jimenez. 1998. Formation and specification of ventral neuroblasts is controlled by vnd in Drosophila neurogenesis. *Genes Dev.* 12:3613-3624.

53. Wharton, J., K. A., and S. T. Crews. 1993. CNS midline enhancers of the *Drosophila slit* and *Toll* genes. *Mech. Dev.* 40:141-154.

54. Sanchez-Soriano, N., and S. Russell. 1998. The Drosophila SOX-domain protein Dichaete is required for the development of the central nervous system midline. *Development* 125:3989-3996.

55. Strauss, R. 2002. The central complex and the genetic dissection of locomotor behaviour. *Curr. Opin. Neurobiol.* 12:633-8.

56. Jurgens, G., E. Wieschaus, C. Nusslein-Volhard, and H. Kluding. 1984. Mutations affecting the pattern of the larval cuticle in Drosophila melanogaster. *Roux Arch. Dev. Biol.* 193:283-295.

57. Isaac, D. D., and D. J. Andrew. 1996. Tubulogenesis in *Drosophila*: a requirement for the *trachealess* gene product. *Genes Dev.* 10:103-117.

58. Wilk, R., I. Weizman, L. Glazer, and B.-Z. Shilo. 1996. *trachealess* encodes a bHLH-PAS protein and is a master regulator gene in the *Drosophila* tracheal system. *Genes Dev.* 10:93-102.

59. Manning, G., and M. A. Krasnow. 1993. Development of the *Drosophila* tracheal system, p. 609-685. In M. Bate and A. Martinez Arias (ed.), The Development of *Drosophila melanogaster*. Cold Spring Harbor Laboratory Press, Cold Spring Harbor, N. Y.

60. Younossi-Hartenstein, A., and V. Hartenstein. 1993. The role of the tracheae and musculature during pathfinding of Drosophila embryonic sensory axons. *Dev. Biol.* 158:430-47.

61. Boube, M., M. Llimargas, and J. Casanova. 2000. Cross-regulatory interactions among tracheal genes support a co-operative model for the induction of tracheal fates in the Drosophila embryo. *Mech. Dev.* 91:271-8.

62. Jiang, L., and S. T. Crews. 2003. The *Drosophila dysfusion* bHLH-PAS gene controls tracheal fusion and levels of the Trachealess bHLH-PAS protein. *Mol. Cell. Biol.* In press.

63. Jin, J., N. Anthopoulos, B. Wetsch, R. C. Binari, D. D. Isaac, D. J. Andrew, J. R. Woodgett, and A. S. Manoukian. 2001. Regulation of Drosophila tracheal system development by protein kinase B. *Dev Cell* 1:817-27.

64. Anderson, M. G., G. L. Perkins, P. Chittick, R. J. Shrigley, and W. A. Johnson. 1995. *drifter*, a *Drosophila* POU-domain transcription factor, is required for correct differentiation and migration of tracheal cells and midline glia. *Genes Dev.* 9:123-137.

65. de Celis, J. F., M. Llimargas, and J. Casanova. 1995. Ventral veinless, the gene encoding the Cf1a transcription factor, links positional information and cell differentiation during embryonic and imaginal development in Drosophila melanogaster. *Development* 121:3405-16.

66. Kuhnlein, R. P., and R. Schuh. 1996. Dual function of the region-specific homeotic gene spalt during Drosophila tracheal system development. *Development* 122:2215-23.

67. Andrew, D. J., K. D. Henderson, and P. Seshaiah. 2000. Salivary gland development in Drosophila melanogaster. *Mech. Dev.* 92:5-17.

68. Bradley, P. L., A. S. Haberman, and D. J. Andrew. 2001. Organ formation in Drosophila: specification and morphogenesis of the salivary gland. *Bioessays* 23:901-11.

69. Kuo, Y. M., N. Jones, B. Zhou, S. Panzer, V. Larson, and S. K. Beckendorf. 1996. Salivary duct determination in *Drosophila*: roles of the EGF receptor signaling pathway and the transcription factors fork head and trachealess. *Development* 122:1909-17.

70. Jones, N. A., Y. M. Kuo, Y. H. Sun, and S. K. Beckendorf. 1998. The Drosophila Pax gene eye gone is required for embryonic salivary duct development. *Development* 125:4163-74.

71. Matsunami, K., H. Kokubo, K. Ohno, P. Xu, K. Ueno, and Y. Suzuki. 1999. Embryonic silk gland development in Bombyx: molecular cloning and expression of the Bombyx trachealess gene. *Dev. Genes Evol.* 209:507-14.

72. Matsunami, K., H. Kokubo, K. Ohno, and Y. Suzuki. 1998. Expression pattern analysis of SGF-3/POU-M1 in relation to sericin-1 gene expression in the silk gland. *Dev. Growth Differ.* 40:591-7.

73. Weigel, D., H. Bellen, G. Jurgens, and H. Jackle. 1989. Primordium-specific requirement of the homeotic gene *fork head* in the developing gut of the *Drosophila* embryo. *Wilhelm Roux's Arch. Dev. Biol.* 198:201-201.

74. Kokubo, H., S. Takiya, V. Mach, and Y. Suzuki. 1996. Spatial and temporal expression pattern of *Bombyx fork head*/SGF-1 gene in embryogenesis. *Dev. Genes Evol.* 206:80-85.

75. Martin, J. W. 1992. Branchiopoda, Microscopic Anatomy of the Invertebrates, Volume 9: Crustacea. Wiley-Liss Inc.

76. Mitchell, B., and S. T. Crews. 2002. Expression of the *Artemia trachealess* gene in the salt gland and epipod. *Evol. Dev.* 4:1-10.

77. Chavez, M., C. Landry, S. Loret, M. Muller, J. Figueroa, B. Peers, F. Rentier-Delrue, G. G. Rousseau, M. Krauskopf, and J. A. Martial. 1999. APH-1, a POU homeobox gene expressed in the salt gland of the crustacean Artemia franciscana. *Mech. Dev.* 87:207-12.

78. Lengyel, J. A., and D. D. Iwaki. 2002. It takes guts: the Drosophila hindgut as a model system for organogenesis. *Dev. Biol.* 243:1-19.

79. Samakovlis, C., N. Hacohen, G. Manning, D. Sutherland, K. Guillemin, and M. A. Krasnow. 1996. Development of the *Drosophila* tracheal system occurs by a series of morphologically distinct but genetically coupled branching events. *Development* 122:1395-1407.

80. Samakovlis, C., G. Manning, P. Steneberg, N. Hacohen, R. Cantera, and M. A. Krasnow. 1996. Genetic control of epithelial tube fusion during *Drosophila* tracheal development. *Development* 122:3531-6.

81. Tanaka-Matakatsu, M., T. Uemura, H. Oda, M. Takeichi, and S. Hayashi. 1996. Cadherin-mediated cell adhesion and cell motility in *Drosophila* trachea regulated by the transcription factor Escargot. *Development* 122:3697-705.

82. Gradin, K., J. McGuire, R. H. Wenger, I. Kvietikova, M. L. Whitelaw, R. Toftgard, L. Tora, M. Gassmann, and L. Poellinger. 1996. Functional interference between hypoxia and Dioxin signal transduction pathways: competition for recruitment of the Arnt transcription factor. *Mol. Cell. Biol.* 16:5221-5231.

83. Woods, S. L., and M. L. Whitelaw. 2002. Differential activities of murine single minded 1 (SIM1) and SIM2 on a hypoxic response element. Cross-talk between basic helix-loop-helix/per-Arnt-Sim homology transcription factors. *J. Biol. Chem.* 277:10236-43.

84. Nambu, P. A., and J. R. Nambu. 1996. The Drosophila fish-hook gene encodes a HMG domain protein essential for segmentation and CNS development. *Development* 122:3467-75.

85. Wang, G. L., B.-H. Jiang, E. Rue, and G. L. Semenza. 1995. Hypoxia-inducible factor 1 is a basic helix-loop-helix-PAS heterodimer regulated by cellular O2 tension. *Proc. Natl. Acad. Sci. USA* 92:5510-5514.

86. Lavista-Llanos, S., L. Centanin, M. Irisarri, D. M. Russo, J. M. Gleadle, S. N. Bocca, M. Muzzopappa, P. J. Ratcliffe, and P. Wappner. 2002. Control of the hypoxic response in Drosophila melanogaster by the basic helix-loop-helix PAS protein similar. *Mol. Cell. Biol.* 22:6842-53.

87. Bacon, N. C. M., P. Wappner, J. F. O'Rourke, S. M. Bartlett, B. Shilo, C. W. Pugh, and P. J. Ratcliffe. 1998. Regulation of the *Drosophila* bHLH-PAS protein Sima by hypoxia: functional evidence for homology with mammalian HIF-1α. *Bioch. Biophys. Res. Comm.* 249:811-816.

88. Wingrove, J. A., and P. H. O'Farrell. 1999. Nitric oxide contributes to behavioral, cellular, and developmental responses to low oxygen in Drosophila. *Cell* 98:105-14.

89. Jarecki, J., E. Johnson, and M. A. Krasnow. 1999. Oxygen regulation of airway branching in Drosophila is mediated by branchless FGF. *Cell* 99:211-20.

90. Wigglesworth, V. B. 1977. Structural changes in the epidermal cells of Rhodnius during tracheole capture. *J. Cell Sci.* 26:161-74.

91. Adryan, B., H. J. Decker, T. S. Papas, and T. Hsu. 2000. Tracheal development and the von Hippel-Lindau tumor suppressor homolog in Drosophila. *Oncogene* 19:2803-11.

92. Bridges, C., and T. H. Morgan. 1923. The third-chromosome group of mutant characters of Drosophila melanogaster. *Carnegie Inst. Wash. Publ.*:1-251.

93. Powell-Coffman, J. A., C. A. Bradfield, and W. B. Wood. 1998. *Caenorhabditis elegans* orthologs of the aryl hydrocarbon receptor and its heterodimerization partner

the aryl hydrocarbon receptor nuclear translocator. *Proc. Nat. Acad. Sci. USA* 95:2844-2849.

94. Butler, R. A., M. L. Kelley, W. H. Powell, M. E. Hahn, and R. J. Van Beneden. 2001. An aryl hydrocarbon receptor (AHR) homologue from the soft-shell clam, Mya arenaria: evidence that invertebrate AHR homologues lack 2,3,7,8-tetrachlorodibenzo-p-dioxin and beta-naphthoflavone binding. *Gene* 278:223-34.

95. Dong, P. D., J. S. Dicks, and G. Panganiban. 2002. Distal-less and homothorax regulate multiple targets to pattern the Drosophila antenna. *Development* 129:1967-74.

96. Godt, D., J. L. Couderc, S. E. Cramton, and F. A. Laski. 1993. Pattern formation in the limbs of Drosophila: bric a brac is expressed in both a gradient and a wave-like pattern and is required for specification and proper segmentation of the tarsus. *Development* 119:799-812.

97. Chu, J., P. D. Dong, and G. Panganiban. 2002. Limb type-specific regulation of bric a brac contributes to morphological diversity. *Development* 129:695-704.

Chapter 5

METHOPRENE-TOLERANT, A PAS GENE CRITICAL FOR JUVENILE HORMONE SIGNALING IN DROSOPHILA MELANOGASTER

Thomas G. Wilson
The Ohio State University, Columbus, OH 43210

1. INTRODUCTION

Since their discovery, bHLH-PAS genes have been found to be widespread among organisms, ranging from prokaryotes to humans (1, 2). Proteins encoded by genes in the PAS family function to sense environmental and developmental signals that reach an organism or a tissue within an organism, and they initiate a response that usually involves the regulation of target gene(s). As discussed in these chapters, the environmental signals range from chemical (planar aromatic hydrocarbons; oxygen tension) to physical (light). The developmental signals are less well-understood, but PAS mutants have shown the importance of PAS genes during development (3-10). The availability of the genomic sequence of humans, *Drosophila melanogaster*, and *Caenorhabditis elegans* within the past several years has allowed the constellation of PAS genes to be identified in these organisms, and it seems unlikely that new PAS genes having functions that are radically different from those of the present family members will be uncovered. So, our task at present is to elucidate the functions and molecular mechanisms of these genes.

In this chapter I describe the *Methoprene-tolerant* (*Met*) PAS gene from *D. melanogaster* and related genes from dipteran insects. *Met* shares functional characteristics with other PAS genes in that it senses not only a developmental signal, specifically the insect juvenile hormone (JH), but also a class of environmental chemicals, juvenile hormone analog (JHA) insecticides (11). The insect then responds to these signals with actions that affect development and reproduction. This response seems to involve gene regulation, although not to the extent seen for some other PAS genes. *Met* may be specific to invertebrates, perhaps limited to Insecta, since a vertebrate gene with strong homology to *Met-l* has not been discovered in GenBank searches. This situation is unlike, for example, the aryl hydrocarbon nuclear translocator (ARNT) and hypoxia-inducible factor (HIF) vertebrate genes whose invertebrate homologs *tango* (12) and *similar* (13) are evident. Since JH is restricted to Insecta (14, 15), *Met* may be primarily involved in the endocrinology of JH. I begin with a brief description of insect endocrinology, then follow with an account of the discovery of *Met* and its characteristics.

2.   ENDOCRINE CONTROL OF INSECT DEVELOPMENT

Insect development and reproduction are under the primary control of two hormones, the steroid 20-OH ecdysone (20-HE) and the sesquiterpenoid juvenile hormone (JH) (Figure 1). We know much about 20-HE; it is responsible for molting, metamorphosis, and oogenesis, and appears to act in a typical steroid hormone mechanism of action regulating gene expression (16, 17). In contrast, JH is poorly understood. This hormone acts to control the character of the molt (18). In this role a high JH titer acts in concert with 20-HE to promote a larval-larval molt, maintaining the "status quo" of larval development. Prior to metamorphosis, the JH titer drops, and 20-HE secretion triggers a larval-pupal molt and initiates the sweeping developmental and physiological changes that occur during metamorphosis. JH reappears during adult life to control vitellogenic oocyte development, and this hormone is also involved in a poorly understood role in male accessory gland protein synthesis (15).

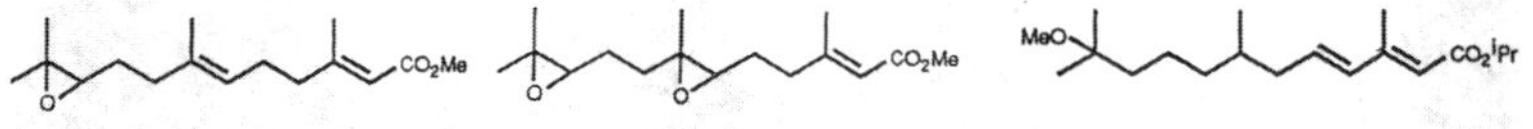

JHIII   JH bisepoxide   Methoprene

Figure 1. Chemical structures for the two JHs found in D. melanogaster and for the JHA methoprene

The 20-HE receptor is comprised of two subunits, products of the *Ecdysone-receptor (EcR)* and the *ultraspiracle (usp)* genes (16, 19, 20). *EcR* is a member of the steroid hormone receptor superfamily (16) and *usp* shares homology with the retinoic acid receptor-X (RXR) component of vertebrate nuclear receptors (19, 21). During metamorphosis a large number of genes involved in the metamorphic change are activated (22). In contrast, the JH receptor has not been described for any insect. This deficiency has been a major reason for our poor understanding of the mechanism of action of JH. Although it seems clear that transcriptional regulation of certain genes is under JH control, the number of genes is small (relative to 20-HE), and the mechanism of transcriptional regulation is unclear.

Several agrochemical companies have synthesized JH chemical analogs (JHA) for pest insect control, and the best known of these is methoprene (Figure 1). JHAs act as JH agonists and have proven useful in insect endocrinology research in addition to control of certain insect pests (23). Insects are masters at evolving resistance to a variety of insecticides, and their response to JHAs has been no exception (24, 25). Work in my lab has focused on understanding the genetics of resistance of *D. melanogaster* to methoprene. Results from this work have led to the identification of a gene, termed *Methoprene-tolerant (Met)*, which, when mutated, is responsible for resistance. Subsequent isolation and sequencing of *Met* showed it to belong to the bHLH-PAS family of transcriptional regulators. Our previous biochemical work indicated that MET is involved in the action of JH, perhaps as a component of the receptor, and the discovery that *Met* is a member of the PAS gene family has reinforced this conclusion. Subsequent study of the phenotypic characteristics of *Met* mutants and biochemical properties of MET should continue to offer insight not only into the biology of this gene and how it fits into the PAS family, but also into the mechanism of action of JH.

3.   DISCOVERY OF *MET* MUTANTS

The structures of the three major JH molecules found in insects were worked out by 1970, and most insects, including *D. melanogaster*, secrete only JH III (26). *D. melanogaster* glands in culture also produce a unique JH bisepoxide (27), but this hormone cannot be detected *in vivo*, has less biological activity than JH III, and its role in this insect is unclear.

Methoprene   (isopropyl-(2E,4E)-11-methoxy-3,7,11-trimethyl-2,4-dodecadienonate), has good insecticidal activity against certain insects, particularly dipteran insects (23). One early prediction stated that insects would have difficulty evolving resistance to an analog of their own hormone

(28), but resistance in a beetle and fly species was soon demonstrated (24). I became intrigued with this resistance, and I believe that a better understanding of it can provide information not only about the genetics of resistance evolution but also about the mechanism of JH signaling in insects. I turned to *D. melanogaster* to identify the genetic basis of methoprene resistance.

We mutagenized susceptible *D. melanogaster*, selected for resistant lines (29), and in the initial screens recovered two that showed high (50-100-fold) resistance. Genetic mapping and complementation tests quickly showed that both mutants mapped to the X-chromosome and were allelic. We termed the resistance gene Methoprene-tolerant (*Met*) and its wild-type allele *Met$^+$*. We have characterized *Met* as follows (11): (i) *Met* results in 50-100-fold resistance to both the toxic and morphogenetic effects of methoprene, (ii) *Met* is expressed as a semidominant mutation; resistance is present in heterozygotes, although not as strongly as in homozygotes. (iii) *Met* acts in a cell-autonomous manner in genetic mosaics, ruling out a circulating factor such as either a degradative enzyme or carrier protein as the resistance mechanism, (iv) *Met* results in resistance to the natural hormone, JH III, to the homolog JH bisepoxide (unpublished), and to each of two additional JH analogs that have been tested (30). However, *Met* flies are not resistant to other classes of insecticides. This is an important distinction: it shows that *Met* is not a general insecticide resistance gene but is specific for JH and JH analog insecticides.

4.   BIOCHEMISTRY OF *MET* RESISTANCE

Elucidating the biochemistry of the resistance of *Met* mutants became essential for an understanding of this gene. Possible insecticide resistance mechanisms that have been demonstrated for other insects include enhanced secretion or metabolism, tissue sequestration, reduced cuticular penetration of JH, and target-site insensitivity (31-33). The first three of these mechanisms were ruled out by direct experimentation (34). The last mechanism, termed target-site insensitivity, involves a protein that is the molecular "target" of the insecticide and physically binds it. Susceptible insects bind the insecticide with high affinity and are readily killed; while resistant ones have an altered target protein (due to a mutation in the encoding gene) that binds the insecticide more poorly, resulting in resistance. This mechanism of resistance is widespread and has been documented for a variety of insecticides (35). We examined this mechanism by measuring JH binding to broken cell preparations of a JH target tissue, larval fat body. Preparations from *Met* flies showed a 10-fold lower binding affinity (poorer binding) for JH III than did similar extracts from Met^+ flies (34). We have found consistent binding results with three *Met* alleles and with other *Met* and Met^+ stock constructions (34) and are very confident of these results. This binding protein has characteristics expected of a JH receptor (36), but direct evidence is lacking. We believe that Met^+ encodes either a component of the receptor or a protein that must be intimately and stoichiometrically involved in JH reception.

A similar toxin-resistance mechanism in vertebrates has been described, and work elucidating this resistance has been important in the identification of the *Ah* (*aryl hydrocarbon responsiveness*) locus (see Ch. 7 by Gao et al.). At least one strain of mice, DBA/2J, shows resistance to the toxic compound dioxin, and toxicity was found to genetically segregate with the *Ah* locus (37). When binding of radiolabeled dioxin was measured with liver cytosolic preparations from these mice, binding affinity was found to be poorer relative to preparations from a susceptible strain (37). The authors interpreted the resistance as due to altered toxin binding to AHR in this strain.

Since the other mechanism of resistance, increased metabolism of toxins (31, 33), is also widespread, we had to ensure that *Met* was not acting to enhance the degradation of methoprene/JH. Increased metabolism usually confers resistance to a variety of insecticides (32, 38, 39). We unambiguously ruled out this possibility with the following experiments/observations: (i) a direct demonstration of altered JH III binding in *Met* extracts (34), (ii) a failure to detect increased JH III metabolism in *Met* flies (34, 40), (iii) resistance of *Met* mutants restricted to JH/JH analog insecticides (11; unpublished), and (iv) GC-MS measurements of JH III titers in *Met* flies showed essentially wild-type levels (Rembold and Wilson, unpublished); one would expect the endogenous titer to be low if JH III metabolism was enhanced. Therefore, I remain confident in our assessment of a target-site insensitivity mechanism of resistance conferred by *Met* mutants.

5.   MOLECULAR CLONING AND SEQUENCE OF *MET*

We cloned Met^+ using the technology of *P*-element transposon-tagging (41). We recovered two *P*-element alleles, Met^{K17} and Met^{43} from a mutant screen and used them to establish a physical map (Figure 1) of the genomic region (42). Bi-directional DNA sequencing from the points of each of the two *P*-element insertion sites revealed these sites to be located 273 bp (Met^{43}) and 424 bp (Met^{K17}) 5'-upstream of an open reading frame (ORF).

Germ-line transformation (43) of *Met* flies with either (i) a genomic fragment (fragment St-H, Figure 2) that included this ORF or (ii) a 3.3 kb *Met⁺* cDNA (Figure 2) showed unambiguously that this ORF encodes a functional Met^+ gene (44).

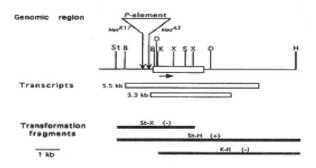

Figure 2. Met+gene region. Genomic organization of an 8 kb region including the Met+ gene ORF (boxed). P-element insertional sites in the MetA3 and MetK17 alleles are shown at the arrows. The solid arrow represents the direction of transcription of Met+. The locations of the transcripts as deduced from cDNA sequencing (3.3-kb transcript) and RT-PCR analysis (5.5-kb transcript) are noted below the map. The genomic transformation fragments are indicated below the transcription units; those that did not rescue the resistance phenotype are noted (-) and one (St-H) producing methoprene susceptibility in transformant flies is noted (+). The fragments are designated by the restriction enzyme sites. D=Hind III; S=Sal I; K=Kpn I; St=Stu I; B=Bam HI; X=Xho I; H=Hpa I. Taken from (45).

Northern analysis showed that the Met^+ gene is expressed as two transcripts of sizes 5.5 and 3.3 kb (Figure 2); both transcripts include the entire ORF but different lengths of untranslated sequences. The larger transcript is expressed throughout development, including the methoprene-sensitive period at the onset of metamorphosis, while the smaller transcript is expressed only in adult ovaries and (presumably contributed as a maternal mRNA) in the first half of embryonic development. Since a role for JH during embryonic development is unknown, the function of MET produced from this smaller transcript is likewise unknown.

Sequence analysis readily demonstrated that MET has considerable sequence similarity to the bHLH-PAS family of transcription factors (44). PAS domains have been found to function in protein-protein interaction (46) and target gene specificity (47). Additionally, the ligand-binding region of AHR includes a PAS region (48). PAS proteins usually function as heterodimers either with other PAS proteins or with selected nuclear receptors, acting as co-activators/co-repressors in the latter function (49).

We isolated and sequenced the homologous gene from *D. virilis* (Kearing and Wilson, unpublished), a Drosophilid that is thought to have separated from *D. melanogaster* 50-60 million years ago (50). Due to this relatively long evolutionary time period, *D. virilis* is useful for gene sequence comparison to *D. melanogaster*, aiding in the identification of the conserved portions. Before isolating *Met⁺* from *D. virilis*, it was first necessary to ensure that this species would react to methoprene treatment in a similar manner as does *D. melanogaster*, because insects from many other orders do not react to methoprene treatment. We incorporated methoprene in varying doses in the food in a manner analogous to our standard treatment of *D. melanogaster* (11, 29) and introduced wild-type *D. virilis* larvae. Examination of the pupae and emerging adults demonstrated that methoprene is toxic to *D. virilis* in a dose-dependent manner and additionally results in morphogenetic defects similar to those seen with *D. melanogaster* (unpublished). Thus, as expected, methoprene affects this species in a similar manner as *D. melanogaster*, implicates *Met⁺* in the mechanism of action, and validates comparison of the *Met⁺* genes from both these two species.

Met⁺ is similar between *D. melanogaster* (78 kD) and *D. virilis* (Figure 3). The arrangement of the bHLH and PAS domains is similar. There is a single intron at precisely the same location in the PAS-B region of the gene as in *D. melanogaster*, although its size is different (69 bp in *D. melanogaster*; 1044 bp in *D. virilis*). Sequence homology is very high (>90%) in the bHLH and two PAS domains (Fig. 3) but lower (overall 71% identity) in other regions of the gene. There is one LXXLL motif at the beginning of the PAS-A domain that is conserved between the two species.

BLAST searches reveal similarity with another *D. melanogaster* gene that we term *Met-l* and with a mosquito gene, both to be described. These genes are 68-86% identical in the bHLH and PAS domains to *Met⁺* (Figure 3), and have smaller islands of conserved sequence elsewhere in the gene. BLAST searches with either bHLH or PAS domains from *Met⁺* reveals other PAS genes, both vertebrate and invertebrate, that have lower similarity (about 35% identical) in the conserved domains. The vertebrate PAS gene with perhaps the highest similarity to these domain sequences is ARNT, having about 35% identity in the conserved domains. So, unlike the *D. melanogaster* genes *similar, tango,* and *spineless,* whose vertebrate homologs were evident from their amino acid sequence comparisons (12, 51, 52), no vertebrate PAS family member is sufficiently similar to suggest homology.

PAS Proteins

bHLH:

1. REARNRAEKNRRDKLNGSIQELSTMVPHVAIEPRRVDKTAVLRFAAHALR
   REARNRAEKNRRDKLNGSIQELS MVPHVA PRRVDKTAVLRF AH LR
2. REARNRAEKNRRDKLNGSIQELSAMVPHVAESPRRVDKTAVLRFSAHGLR
   REARN AEK RRDKLN SIQ L MVPH AES RR DKTAVLRF HGLR
3. REARNLAEKQRRDKLNASIQILATMVPHAAESSRRLDKTAVLRFATHGLR
   REARNLAEKQRRDKLNASIQELATMVPHAAES RRLDKTAVLRFATHGLR
4. REARNLAEKQRRDKLNASIQELATMVPHAAESTRRLDKTAVLRFATHGLR

PAS-A:

1. LMDMLDSFFLTLTCHGHILLISASIEQHLGHCQSDLYGQSIMQITHPEDQNMLKQQL
   L ML F LT TC G I L S S EQ LGHCQ DLYGQ   THP D LKQQL
2. LFRMLNGFLLTVTCRGQIVLVSPSVEQFLGHCQTDLYGQNLFNLTHPDDHALLKQQL
   L L   LT TC GQIVLVS SVEQ LGHCQ DLYGQNL THPDD LL QQL
3. LMQLLDCCFLTLTCSGQIVLVSTSVEQLLGHCQSDLYGQNLLQITHPDDQDLLRQQL
   LMQLLD FLTLTC GQIVLVS SVEQLLGHCQSDLYGQNLLQITHPDDQDLLRQQL   .
4. LMQLLDSFFLTLTCNGQIVLVSGSVEQLLGHCQSDLYGQNLLQITHPDDQDLLRQQL

PAS-B:

1. EYKTRHLIDGRIIDCDQRIGIVAGYMTDEVRNLSPFTFMHNDD
   EYKTRHLIDGRI CDQRI IVAGY TDEV LSPFTFMH DD
2. EYKTRHLIDGRIVQCDQRISIVAGYLTDEVSGLSPFTFMHRDD
   EY TRHLIDG I CDQRI VAGY DEV LSPF FMH DD
3. EYHTRHLIDGSIIDCDQRIGLVAGYMKDEVRNLSPFCFMHLDD
   EY TRHLIDGSIIDCDQRIG VAGYMKDEVRNLSPF FMHLDD
4. EYRTRHLIDGSIIDCDQRIGIVAGYMKDEVRNLSPFSFMHLDD

Figure 3. Comparison of MET and MET-L amino acid sequences for the bHLH, PAS-A, and PAS-B domains. 1. MET-L from *D. melanogaster*; 2. ORF from *A. gambiae*; 3. MET from *D. melanogaster*; 4. MET from *D. virilis*. Identical amino acids are noted between sequences.

6. *MET²⁷*, A NULL ALLELE

What is the phenotype of flies that lack a functional *Met⁺*? Since the precise function of MET is unknown, identification of a *Met* null allele by screening for a predicted phenotype (e.g., larval lethality) in a mutant screen is difficult if the phenotype is uncertain. Therefore, alleles recovered in our standard mutant screens for methoprene resistance (29) were examined for any that showed the most apparent indication of a null allele: lack of Met^+ transcript. We have recovered about a dozen alleles of *Met* using ethylmethane sulfonate, X-rays, and *P*-transposable genetic elements as mutagenic agents (11, 53). When examined by Northern analysis, one allele, Met^{27}, totally lacked either of the two Met^+ transcripts (3). RT-PCR of mRNA preparations of Met^{27} larvae or adults failed to produce a predicted amplification product (unpublished), suggesting that even a minuscule level of transcript is not present. Based on PCR and sequence analysis of Met^{27} genomic DNA, we believe that this allele has a chromosomal break located between 42 and 128 bp (the locations of the forward primers used in the study) 5' upstream of the ATG start site of the *Met* ORF, presumably resulting in a failure of Met^{27} transcription.

What is the phenotype of Met^{27}? Homozygotes and Met^{27} /*deficiency* flies readily survive, ruling out a vital function for Met^+, although the flies are not as vigorous as wild-type (3). The most evident phenotype is a reduction of oogenesis, and subsequently oviposition, to about 20% of wild-type. Eggs laid by these females have good (85%) fertility, however. This phenotype is completely rescued in *w v* Met^{27}; *p[Met⁺]* transgenic flies bearing an ectopic copy Met^+ of Met^+, further demonstrating that the oogenic defect results from the Met^{27} mutation and not from another unrelated mutation in the background genome of Met^{27} females. This phenotype is consistent with previous demonstrations of a role for JH in *D. melanogaster* oogenesis (54-56)] as well as our detection of MET in ovary follicle cells. Usually, target-site resistance results from a slight alteration (due to a point mutation) in the target molecule (32, 35, 57); this work was the first demonstration of insecticide resistance resulting from a complete absence of a target-site resistance gene (3).

If MET is indeed a component of a JH receptor, then why has the null mutant such a mild phenotype? We have discussed several possibilities (58), but our favorite explanation is that flies are protecting a critical pathway with a redundant gene or mechanism. Genetic redundancy has been seen in a variety of organisms (59), so this result is not unusual. Redundancy has been seen with other PAS proteins as well; for example, *Arnt* and *Arnt2* in mice appear to be partially redundant (8). However, redundancy is not a foregone conclusion for PAS genes exhibiting high similarity of sequence and expression. The murine single-minded genes *SIM1* and *SIM2* are not interchangeable, as demonstrated with knock-out mutants of each gene (60).

7.   *MET-LIKE, A D. MELANOGASTER* PAS GENE SIMILAR TO *MET*

Several years ago an EST with about 70% homology to *Met* was posted to the Berkeley Drosophila Genome Project database. We used this EST in a library screen to isolate and sequence the entire cDNA, which we term *Met-like* (*Met-l*). Another laboratory has independently identified this gene by computer search of Flybase genomic sequences for bHLH genes and named it *germ cell expressed* (61). Amino acid sequence comparison shows high levels of identity in the bHLH (78%), PAS-A (68%), and PAS-B (86%) domains (Figure 3), much higher than found when other PAS genes, including *D. melanogaster* PAS genes, are compared to *Met*⁺. This high degree of homology extends beyond the end of the PAS-B domain and includes smaller islands of perfectly conserved sequence. Unlike the single intron found in *Met*⁺ from *D. melanogaster* and *D. virilis*, *Met-l* has 7 introns in the gene (unpublished and Flybase data). Presumably, *Met-l* and *Met* resulted from a duplication of an ancestral gene.

We do not have a mutant for *Met-l*, so the phenotype of a dysfunctional gene is unknown. MET-L may be serving either as a partner for MET in the putative heterodimer or more likely as the redundant MET substitute that can rescue the fly from more severe consequences of MET absence in *Met* mutants. Based on protein interaction studies and mutant rescue experiments, an analogous role for ARNT2 as a substitute or alternative partner for the similar homolog ARNT during interaction with AHR in mice has been proposed (62).

8.   *MET* FROM THE MOSQUITO, *ANOPHELES GAMBIAE*

The genome of the malarial mosquito, *Anopheles gambiae*, has been sequenced by Celera Genomics, Rockville, MD, and is currently available on GenBank. Recently, we carried out a BLAST search with MET sequence and identified an ORF (agCP5875) from *A. gambiae* that has sequence similarity to MET (Figure 3), although not to the extent seen for *D. virilis*. The overall structure of the gene resembles that for *Met*⁺ and most other PAS genes. The sequence of agCP5875 shows conserved amino acids to both MET and MET-L, although it doesn't appear to favor either gene.

BLAST inquires with various regions of *D. melanogaster Met* or *Met-l* did not reveal another gene in *A. gambiae* that showed nearly the level of similarity of agCP5875 to either *Met* or *Met-l*. Possibly, one exists in the mosquito genome that has yet to be reported to GenBank. Another possibility is that the *A. gambiae* gene resembles an ancestral gene presumed to have undergone duplication to yield *Met* and *Met-l* in the higher Diptera. If the latter is true, then the question of its function will be interesting. There are similarities in the endocrinology of lower and higher dipteran insects that have been examined. The mosquito *Aedes aegypti* has JH III as does higher Diptera (63), and JHA applied to mosquitoes results in toxicity and morphogenetic defects similar to those seen for higher Diptera (64, 65), presumably acting though a *Met*$^+$ mechanism. Although methoprene resistance has been reported in mosquito (the salt-marsh mosquito, *Aedes taeniorhynchus*) populations in Florida (25), it is not known if the resistance is *Met*-associated. Elucidation of the function of the *A. gambiae Met* homolog will be important for understanding the evolution of the *Met* and *Met-l* genes.

9.   IMMUNOLOCALIZATION OF MET

The tissue location of a PAS protein in an organism, while not revealing its precise function, can give clues about the general nature of its role. Several PAS proteins, particularly those given the designation of β-proteins, have a widespread tissue localization in organisms (1). These proteins function as a broad-spectrum heterodimeric partners with another class of PAS proteins that have been designated α-proteins. The latter have in turn been found in specific, presumably target tissues. Three examples of α-proteins found in *D. melanogaster* include the *trachealess* gene product, TRH, specific to tube-forming cells in the embryo (5, 66), *spineless* gene product, SS, found in antennal and tarsal cells, where it functions in specifying the identity of these cells (51), and *single-minded* gene product, SIM, found in the CNS midline cells of the embryo to function in specifying cell lineage (67, 68).

Similarly, the intracellular location, either in the nucleus or in the cytoplasm of target cells, can facilitate our understanding of a PAS protein. For example, AHR is a cytoplasmic protein that is complexed with HSP90/AIF chaperone proteins (69). If ligand diffuses into the cell, it can bind to AHR, displacing the chaperone proteins. The liganded AHR then enters the nucleus, forms a heterodimer with ARNT, and activates target gene transcription (1, 70).

To examine MET localization, rabbit polyclonal antibody was raised to MET protein expressed from a *Met*$^+$ cDNA in *E. coli*. Western analysis of *Met*$^+$ flies revealed a single band of 79 kD molecular weight (71), which is very similar to the size (78 kD) predicted from the cDNA ORF size (44).

Significantly, extracts from the null allele Met^{27} showed no 79 kD band when probed with MET antibody (71), thus confirming antibody specificity.

Our studies demonstrated the following (71): (i) First, MET protein localizes only in cell nuclei. We found no exceptions to this rule. (ii) Second, MET is located in specific tissues during all stages of *D. melanogaster* development. During embryogenesis MET is present from about the 256-cell stage through gastrulation. This time also corresponds of the appearance of the 3.3 kb transcript during the first 8-10 hours of embryonic development (44), but whether MET expressed during embryonic development results primarily from this transcript or from the 5.5 kb transcript is unknown. In larvae, MET occurs in larval fat body, salivary glands, and certain imaginal tissues, where it persists into the pupal stage. Finally, in adults MET is found in reproductive tissues of both sexes. In females it is found in ovarian follicle cells; in males, accessory glands and ejaculatory duct tissue. The localization in the larval fat body is consistent with earlier results that found JH binding protein in this tissue (34, 36), and the presence of MET in the adult tissues is consistent with roles for JH during reproduction (15). The presence of MET in embryonic tissue, although not unexpected in light of our finding of Met^{+} transcript during this stage (44), is not associated with a known function of JH, perhaps suggesting a novel role for JH during embryogenesis. Likewise, no role for JH is known for the larval salivary glands.

In summary, these results, with the exception of the embryo and larval salivary gland, have detected MET in tissues predicted from JH roles and further implicate MET in JH action. Whether JH has an unknown role during embryonic and larval salivary gland development or Met^{+} has a role unrelated to JH in these tissues is unknown

10. PATHOLOGY OF METHOPRENE TO *D. MELANOGASTER*

A better understanding of the role of MET during *D. melanogaster* development is predicated on an understanding of the pathology of methoprene applied to these insects. JH analog insecticides, including methoprene, kill *D. melanogaster* in a manner similar to that of exogenously applied JH III (11, 72). The sensitive period consists of an approximate 16-20-hour period centered on puparium formation at the onset of metamorphosis (72-74). Application of either JH III or JHA during the sensitive period introduces hormone to the larva or early pupa at a time in development when the endogenous JH titer is intentionally low or absent, resulting in a pathological condition. The lethal phase, however, occurs at the end of pupal development during the pharate adult stage several hours prior to eclosion; these pharate adults are relatively normal-appearing, but they either fail to eclose or eclose as weak adults and die soon thereafter. Morphogenetic defects, including aberrant sternite bristles and mal-rotated male external genitalia, are seen even at sublethal doses in many of the flies that eclose. A similar pathology has been seen following JHA treatment of other dipteran insects (75).

Effects of methoprene are also seen internally, and they probably account for the toxicity (76). In our experiments a dose of methoprene just sufficient to cause 100% lethality to susceptible *D. melanogaster* strains was either fed or topically applied to one of these strains during the JHA-sensitive period at the onset of metamorphosis. Pharate adults from these treated individuals were fixed, embedded in paraffin, sectioned, and examined for morphological defects. The defects found resembled many of those found in two mutants, *Broad-Complex* (*BR-C*), and *Deformed* (*Dfd*). Both of these genes are 20-HE-regulated, and expressed early during metamorphosis (77).

Our efforts have focused on understanding the relationship between JHA and BR-C. We found two types of defects upon examination of methoprene-treated pupae: (i) those similar to ones seen in *BR-C* mutants, and (ii) those unique to methoprene treatment. This array of defects is termed the "methoprene syndrome", is dose-dependent, and includes defects in the central nervous system, musculature, and adult salivary gland morphology (76). A chemical that phenocopies a mutant phenotype may be interpreted as interfering with the expression or function of the wild-type gene; thus, methoprene may be interfering, directly or indirectly, with BR-C expression or function.

Significantly, the methoprene syndrome was not seen in *Met* mutants (either the *Met¹* or *Met³* allele) treated with the same doses of methoprene as were susceptible flies (76). Since *Met* protects against the methoprene syndrome, then the wild-type allele is implicated in the pathology.

How can methoprene kill these insects in this manner? The pathology is not consistent with methoprene acting as a general toxin; instead, we believe that methoprene is interfering with the 20-HE cascade of gene regulation that directs metamorphosis and pupal development. BR-C appears to be a focus for this interference. *BR-C* encodes an alternatively spliced BTB-POZ-zinc finger transcription factor whose isoforms then transcriptionally regulate downstream target genes that direct metamorphic change (78-80). We have found that *BR-C* transcripts are not diminished in methoprene-treated pupae, suggesting that this JHA does not block transcription (Restifo and Wilson, unpublished). Indeed, application of pyriproxyfen, a potent JH analog, to pupae resulted in an abnormally extended appearance of *BR-C* transcripts, and overexpression of the Z1 BR-C isoform as a transgene mimicked methoprene effects (81). Presumably, the mal-expression of *BR-C* (and probably other genes) results in mis-regulation of downstream target genes, leading to the internal and external defects noted. An analogous situation in vertebrates has been described for toxic environmental chemicals disrupting gene expression that is regulated by the steroid hormone estrogen, thus leading to endocrine disruption (82). Additionally, a role for at least one PAS gene acting as a negative regulator of gene expression is known (83).

Overall, these results suggest an interaction between the 20-HE and JH signaling pathways and the involvement of MET. Interaction between other signaling pathways involving PAS genes have been found (82, 84), so there is precedence for this type of cross-talk between pathways.

11.   GENETIC INTERACTION BETWEEN *MET* AND BR-C MUTANTS

We have taken a separate, complementary approach to elucidate the interplay between *Met* and *BR-C*. This approach searches for a genetic interaction between *Met* and *BR-C* mutants. Since MET is involved in the action of JH and methoprene, we reasoned that flies made doubly mutant for *Met* and *BR-C* might express a more severe phenotype than expected from a summing of either single mutant. Such enhancement of a mutant phenotype has proven useful in other studies to demonstrate interactions between *D. melanogaster* genes or gene products (85).

BR-C consists of three genetically complementing isoforms that are generated by alternative splicing: *broad* (*br*), *reduced bristles on palpus* (*rbp*), and *lethal (2)BC* (86). Severe, presumably amorphic alleles, representing each isoform result in lethality at various stages in pupal development, and a null for all three undergoes larval development but fails to pupariate. Only lethal alleles are known for *l(2)BC*, but hypomorphic viable alleles are known for *br* (*br^1*) and *rbp* (*rbp^2*); the phenotype of flies carrying one of these hypomorphic alleles includes wing or palpus defects that are readily visible and gave rise to the gene names (79). When *br Met^{27}* and *rbp Met^{27}* double mutants were constructed and made homozygous, the loss-of-viability phenotype of each was more severe than expected from a simple summing of the component mutant gene phenotypes. For example, the *br^1* allele is homozygous viable (as is *Met^{27}*), but most (>90%) *br^1 Met^{27}/br^1 Met^{27}* flies die during pupal development. Likewise, the phenotype of *rbp^2 Met^{27}* homozygotes shows a similar viability shift to pupal lethality (unpublished). The presence of a *Met$^+$* transgene in the double mutant resulted in viable flies, demonstrating that *Met^{27}*, not another genetic variant in the background genome of the double mutant, is responsible for the loss-of-viability phenotype (Wilson and Restifo, unpublished).

The interaction between these mutant genes is evident also upon examination of a different phenotypic character in adults. Both br^1 or rbp^2 homozygous females show relatively normal oogenesis and oviposition, but occasional surviving adults that are double mutant with Met^{27} consistently have reduced oviposition to an extent approaching sterility. Both the viability and oogenesis phenotypes can be explained by an interaction of MET and BR-C as transcription factors to regulate a target gene(s) critical for pupal development and, in adults, oogenesis. Defective (or absent) MET and BR-C proteins fail to properly interact, resulting in misregulation of the target gene(s) and subsequent phenotypic consequences.

In summary, this genetic approach has corroborated our interpretation of the methoprene pathology studies indicating that MET is involved in gene regulation during pupal development. However, instead of a direct effect of MET on the transcription of the BR-C gene, we believe that the effect is indirect, with MET and BR-C proteins interacting to transcriptionally regulate a subset of target genes that direct late aspects of metamorphosis.

12. WORKING HYPOTHESIS OF THE MOLECULAR BIOLOGY OF *MET*

Results of the work described above plus knowledge from studies of other PAS genes have led to a proposal for MET function. It seems clear from several lines of indirect evidence that MET is a component of a JH receptor in flies. The findings that lead to this conclusion include *Met*-mediated resistance both to JH analog insecticides and to JH III; altered JH binding in cell extracts from *Met* mutants; cell-autonomous expression of *Met*, ruling out a circulating JH binding protein as defective in *Met*; the localization of MET to the nucleus; and the genetic interaction between *Met* and *BR-C*.

Unfortunately, we don't know what kind of receptor to help guide our search for the role of MET. The JH receptor may function as a heterodimer, as does the ecdysterone receptor. One of the subunits, probably the Met^+ encoded subunit, could bind JH with high affinity, as suggested by the binding data. The other subunit (or a totally different protein) could function as a low-affinity binder of JH; this result would account for the measurable, but weaker, JH binding results seen in *Met* mutants (34). MET-L could function as this weak binder. This weak binding is sufficient to enable the fly to carry out JH-dependent functions that are presumably critical during development.

A separate possibility is that the JH receptor is not a traditional heterodimeric hormone receptor but instead acts a co-regulator component of either the 20-HE receptor or one of several orphan receptors in *D. melanogaster* (87). In this role JH, acting though the co-regulator protein (or a partner), would alter the biological activity of the 20-HE response. MET may serve as the co-regulator protein, interacting with ECR/USP to affect its activity, presumably altering the expression of one or more target genes. During larval development when the JH titer is high in the insect, then a 20-HE-induced molt would be facilitated (and modulated) by ECR/USP/MET-JH regulation of genes involved in such a molt, but not in metamorphosis. When the JH titer is low or absent at the onset of metamorphosis, a 20-HE-induced molt would be facilitated by ECR/USP, regulating a somewhat different suite of genes, and culminating in a different, metamorphic molt. Alternatively, USP, not MET, might be the JH-binding component. USP has been shown to bind JH, although only at high JH concentration (88). In this scenario, MET interaction with USP would facilitate high affinity JH binding by USP, as suggested by the finding that *Met* mutants fail to show high-affinity JH binding (34). An analogous role for another PAS gene, the vertebrate *steroid hormone receptor coactivator-1* (*SRC-1*), in modulating steroid hormone action has been found, and *SRC-1* null mutants show hormone resistance (89).

PAS co-activator proteins contain one or more copies of a short motif, LXXLL, that is necessary and sufficient for their interaction with nuclear receptors (90, 91). MET in both *D. melanogaster* and *D. virilis* has one LXXLL copy at the beginning of the PAS-A region, but this motif is imperfect in MET-L. Since the motif is found in a conserved region (PAS-A), it is unclear if the PAS-A region is conserved and the LXXLL is simply hitchhiking, or if the LXXLL motif is a separate focus of selection in the protein. This motif is an obvious candidate for site-directed mutagenesis to establish its importance to MET function.

13.   WORK TO BE DONE

There are several directions in future work that should lead to a better understanding of the *Met*+ gene and its function in insects. First, proteins that interact with MET need to be identified. Given the history of the PAS family, another PAS protein is an obvious possibility, especially if MET functions as a "traditional" ligand receptor like AHR. However, if MET is acting in a co-activator role for the ecdysone receptor, then proteins other than PAS proteins involved in reception of this hormone may interact with MET. Likewise, our results with double mutants of *Met* and BR-C suggest a non-transcriptional interaction, perhaps a physical interaction between members of these two transcription families that results in the regulation of target genes during metamorphosis.

Second, the ability of MET or a MET/partner heterodimer to bind juvenile hormone needs to be evaluated. We have expressed *Met*[+] from a bacterial vector, carried out ^{3}H-JH III binding experiments, and failed to detect high-affinity binding by MET. However, this could be a false result due either to an improperly modified MET protein resulting from the bacterial expression or, more likely, to the absence of the presumed partner. By analogy, the 20-HE receptor ligand-binding component, ECR, does not bind hormone unless complexed with its partner, USP (20). The identity of the JH receptor and mechanism of action of this hormone has remained a difficult puzzle for insect endocrinologists (92), and its elucidation will add greatly to our understanding of the endocrinology of this hormone.

Third, the promiscuity of ligand interaction with MET, either directly or indirectly, needs to be examined. A surprising array of chemicals can act as JH agonists (23). Similar toxic and morphogenetic responses of *D. melanogaster* (and other dipteran insects) occur to these compounds (11, 30, 64), and *Met* mutants confer resistance to them. A single-gene mutation (*Met*) that results in resistance to a variety of JHA chemicals suggests that these chemicals are not acting through different gene products, as found for example for the cytochrome P450 family of genes (93). Since AHR also responds to a variety of chemicals, perhaps this ligand-binding PAS protein can serve as a model to understand MET. However, it is clear that the *ss* gene, not *Met*, is the *D. melanogaster* homolog to *Ah* (51), although invertebrate *Ah* homologs may have no ligand-binding ability (94, 95). Perhaps MET, although not highly similar to vertebrate AHR, shares a ligand-binding property.

Fourth, the function of *Met-l* needs to be understood. An obvious first-step in this process will involve the generation of a series of mutant alleles to evaluate the phenotype. Our failure to recover mutants of *Met-l* in typical screens for resistance to methoprene, when new *Met* alleles are readily recovered, suggests that either MET-L is a weak ligand binder that does not result in appreciable resistance when mutant, or that it does not bind ligand. In fact, this inability to confidently predict a mutant phenotype for *Met-l* makes the likelihood of success of a mutant screen questionable. Producing a mutation in the *Drosophila Met-l* gene by *in vitro* mutagenesis and homologous recombination is now possible. Any physical association of MET-L with MET may become evident in work identifying MET interacting proteins as described above, and, if found, this knowledge would help to define the role of MET-L in *D. melanogaster*.

Fifth, the evolutionary history of *Met$^+$*, both within Insecta and in related arthropods, needs to be explored as is being done by Hahn and coworkers for PAS genes in other phyla (96). We have found this gene to be conserved in dipteran insects, but other orders of insects may have very different *Met$^+$* genes, if any. JHAs are highly efficacious in dipteran insects, which in part have driven the examination of their effects and resistance characteristics in *D. melanogaster*. Insects from other orders, especially Lepidoptera, do not respond to JHAs nearly as dramatically, although it is clear that these insects use JH as an endocrine molecule. Perhaps the mild response is a reflection of a *Met$^+$* gene product with different characteristics than that in dipteran insects.

ACKNOWLEDGEMENTS

I wish to thank members of my laboratory for their work toward understanding the *Met$^+$* gene and its function. John Hogenesch first directed my attention to the *Met-1* EST. This work is supported by a grant from the National Science Foundation.

REFERENCES

1.   Gu, Y.-Z., J. B. Hogenesch, and C. A. Bradfield. 2000. The PAS superfamily: Sensors of environmental and developmental signals. *Annu. Rev. Pharmacol. Toxicol.* 40:519-561.

2.   Crews, S. T., and C. M. Fan. 1999. Remembrance of things PAS: regulation of development by bHLH-PAS proteins. *Curr. Opin. Genet. Dev.* 9:580-587.

3.   Wilson, T. G., and M. Ashok. 1998. Insecticide resistance resulting from an absence of target-site gene product. *Proc. Natl. Acad. Sci. USA* 95:14040-14044.

4.   Sonnenfeld, M., and J. R. Jacobs. 1994. Mesectodermal cell fate analysis in *Drosophila* midline mutants. *Mech. Dev.* 46:3-13.

5.   Isaac, D. D., and D. Andrew. 1996. Tubulogenesis in *Drosophila*: a requirement for the *tracheales* gene product. *Genes & Dev.* 10:103-117.

6.   Gehin, M., M. Mark, C. Dennefeld, A. Dierich, H. Gronemeyer, and P. Chambon. 2002. The function of TIF2/GRIP1 in mouse reproduction is distinct from those of SRC-1 and p/CIP. *Mol. Cell Biol.* 22:5923-5937.

7.   Michaud, J. L., T. Rosenquist, N. R. May, and C.-M. Fan. 1998. Development of neuroendocrine lineages requires the bHLH-PAS transcription factor SIM1. *Genes & Dev.* 12:3264-3275.

8.   Keith, B., D. M. Adelman, and M. C. Simon. 2001. Targeted mutation of the murine arylhydrocarbon receptor nuclear translocator 2 (*Arnt2*) gene reveals partial redundancy with *Arnt. Proc. Natl. Acad. Sci. USA* 98:6692-6697.

9. Maltepe, E., J. V. Schmidt, D. Baunoch, C. A. Bradfield, and M. C. Simon. 1997. Abnormal angiogenesis and responses to glucose and oxygen deprivation in mice lacking the protein ARNT. *Nature* 386:403-407.
10. Kozak, K. R., B. Abbott, and O. Hankinson. 1997. ARNT-deficient mice and placental differentiation. *Dev. Biol.* 191:297-305.
11. Wilson, T. G., and J. Fabian. 1986. A *Drosophila melanogaster* mutant resistant to a chemical analog of juvenile hormone. *Develop. Biol.* 118:190-201.
12. Sonnenfeld, M., M. Ward, G. Nystrom, J. Mosher, S. Stahl, and S. Crews. 1997. The *Drosophila tango* gene encodes a bHLH-PAS protein that is orthologous to mammalian Arnt and controls CNS midline and tracheal development. *Development* 124:457-.
13. Bacon, N. C. M., P. Wappner, J. F. O'Rourke, S. M. Bartlett, B. Shilo, C. W. Pugh, and P. J. Ratcliffe. 1998. Regulation of the *Drosophila* bHLH-PAS protein Sima by hypoxia: functional evidence for homology with mammalian HIF-1a. *Biochem. Biophys. Res. Comm.* 249:811-816.
14. Riddiford, L. M. 1994. Cellular and molecular actions of juvenile hormone I. General considerations and premetamorphic actions. *Adv. Insect Physiol.* 24:213-274.
15. Wyatt, G. R., and K. G. Davey. 1996. Cellular and molecular actions of juvenile hormone II. Roles of juvenile hormone in adult insects. *Adv. Insect Physiol.* 26:1-155.
16. Koelle, M. R., W. S. Talbot, W. A. Segraves, M. T. Bender, P. Cherbas, and D. S. Hogness. 1991. The *Drosophila* EcR gene encodes an ecdysone receptor, a new member of the steroid receptor superfamily. *Cell* 67:59-77.
17. Andres, A. J., and C. S. Thummel. 1992. Hormones, puffs, and flies: The molecular control of metamorphosis by ecdysone. *Trends Genet.* 8:132-138.
18. Willis, J. H. 1974. Morphogenetic action of juvenile hormones. *Annu. Rev. Entomol.* 19:97-115.
19. Henrich, V. C., T. J. Sliter, D. B. Lubahn, A. MacIntyre, and L. I. Gilbert. 1990. A steroid/thyroid hormone receptor superfamily member in *Drosophila melanogaster* that shares extensive sequence similarity with a mammalian homologue. *Nucl. Acids Res.* 18:4143-4148.
20. Yao, T.-P., B. M. Forman, Z. Jiang, L. Cherbas, J.-D. Chen, M. McKeown, P. Cherbas, and R. M. Evans. 1993. Functional ecdysone receptor is the product of *EcR* and *Ultraspiracle* genes. *Nature* 366:476-479.
21. Oro, A. E., M. McKeown, and R. M. Evans. 1992. The *Drosophila* retinoid X receptor homolog *ultraspiracle* functions in both female reproduction and eye morphogenesis. *Development* 115:449-462.
22. White, K. P., S. A. Rifkin, P. Hurban, and D. S. Hogness. 1999. Microarray analysis of *Drosophila* development during metamorphosis. *Science* 286:2179-2184.
23. Staal, G. B. 1975. Insect growth regulators with juvenile hormone activity. *Annu. Rev. Entomol.* 20:417-460.
24. Cerf, D. C., and G. P. Georghiou. 1974. Cross resistance to juvenile hormone analogues in insecticide-resistant strains of *Musca domestica*. *Pestic. Sci.* 5:759-767.
25. Dame, D. A., G. J. Wichterman, and J. A. Hornby. 1998. Mosquito (*Aedes taeniorhynchus*) resistance to methoprene in an isolated habitat. *J. Amer. Mosq. Control Assoc.* 14:200-203.

26. Schooley, D. A., F. C. Baker, L. W. Tsai, C. A. Miller, and G. C. Jamieson. 1984. Juvenile hormones 0,I and II exist only in Lepidoptera, p. 371-381. In M. Porchet (ed.), Biosynthesis, metabolism, and mode of action of invertebrate hormones. Springer-Verlag, Berlin.

27. Richard, D. S., S. W. Applebaum, T. J. Sliter, F. C. Baker, D. A. Schooley, C. C. Reuter, V. C. Henrich, and L. I. Gilbert. 1989. Juvenile hormone bisepoxide biosynthesis in vitro by the ring gland of *Drosophila melanogaster:* A putative juvenile hormone in the higher Diptera. *Proc. Natl. Acad. Sci. USA* 86:1421-1425.

28. Williams, C. M. 1967. Third-generation pesticides. *Sci. Am.* 217:13-17.

29. Wilson, T. G., and J. Fabian (ed.). 1987. Selection of methoprene-resistant mutants of *Drosophila melanogaster*, vol. 49. UCLA symposia on molecular and cellular biology, new series.

30. Riddiford, L. M., and M. Ashburner. 1991. Effects of juvenile hormone mimics on larval development and metamorphosis of *Drosophila melanogaster. Gen. Comp. Endocrinol.* 82:172-183.

31. Wilson, T. G. 2001. Resistance of *Drosophila* to toxins. *Annu. Rev. Entomol.* 46:545-571.

32. Taylor, M., and R. Feyereisen. 1996. Molecular biology and evolution of resistance to toxicants. *Mol. Biol. Evol.* 13:719-734.

33. Feyereisen, R. 1995. Molecular biology of insecticide resistance. *Toxicol. Lett.* 82:83-90.

34. Shemshedini, L., and T. G. Wilson. 1990. Resistance to juvenile hormone and an insect growth regulator in *Drosophila* is associated with an altered cytosolic juvenile hormone binding protein. *Proc. Natl. Acad. Sci. USA* 87:2072-2076.

35. ffrench-Constant, R. H. 1999. Target site mediated insecticide resistance: what questions remain? *Insect Biochem. Mol. Biol.* 29:397-403.

36. Shemshedini, L., M. Lanoue, and T. G. Wilson. 1990. Evidence for a juvenile hormone receptor involved in protein synthesis in *Drosophila melanogaster. J. Biol. Chem.* 265:1913-1918.

37. Poland, A., and E. Glover. 1980. 2,3,7,8-tetrachlorodibenzo-*p*-dioxin: Segregation of toxicity with the *Ah* locus. *Mol. Pharmacol.* 17:86-94.

38. Brattsten, L., C. W. J. Holyoke, J. R. Leeper, and K. F. Raffa. 1986. Insecticide resistance: challenge to pest management and basic research. *Science* 231:1255-1260.

39. Wilson, T. G., and J. W. Cain. 1997. Resistance to the insecticides lufenuron and propoxur in natural populations of *Drosophila melanogaster* (Diptera: Drosophilidae). *J. Econ. Entomol.* 90:1131-1136.

40. Gruntenko, N. E., T. G. Wilson, M. Monastirioti, and I. Y. Rauschenbach. 2000. Stress-reactivity and juvenile hormone degradation in *Drosophila melanogaster* strains having stress-related mutations. *Insect Biochem. Mol. Biol.* 30:775-783.

41. Bingham, P. M., M. G. Kidwell, and G. M. Rubin. 1981. Cloning of DNA sequences from the white locus of *D. melanogaster* by a novel and general method. *Cell* 25:693-704.

42. Turner, C., and T. G. Wilson. 1995. Molecular analysis of the *Methoprene-tolerant* gene region of *Drosophila melanogaster. Arch. Insect Biochem. Physiol.* 30:133-147.

43. Spradling, A. C. 1986. P element-mediated transformation, p. 175-197. In D. B. Roberts (ed.), Drosophila: a practical approach. IRL Press, Oxford.

44. Ashok, M., C. Turner, and T. G. Wilson. 1998. Insect juvenile hormone resistance gene homology with the bHLH-PAS family of transcriptional regulators. *Proc. Natl. Acad. Sci. USA* 95:2761-2766.

45. Ashok, M., C. Turner, and T. G. Wilson. 1998. Insect juvenile hormone resistance gene homology with the bHLH-PAS family of transcriptional regulators. *Proc. Natl. Acad. Sci. USA* 95:2761-6.

46. Huang, Z. J., I. Edery, and M. Rosbash. 1993. PAS is a dimerization domain common to *Drosophila* Period and several transcription factors. *Nature* 364:259-262.

47. Zelzer, E., P. Wappner, and B.-Z. Shilo. 1997. The PAS domain confers target gene specificity of *Drosophila* bHLH-PAS proteins. *Genes & Dev.* 11:2079-2089.

48. Burbach, K. M., A. Poland, and C. A. Bradfield. 1992. Cloning of the Ah-receptor cDNA reveals a distinctive ligand-activated transcription factor. *Proc. Natl. Acad. Sci. USA* 89:8185-8189.

49. Ponting, C. P., and L. Aravind. 1998. PAS: a multifunctional domain family comes to light. *Curr. Biol.* 7:R674-R677.

50. Beverley, S. M., and A. C. Wilson. 1984. Molecular evolution in Drosophila and the higher Diptera. II. A time scale for fly evolution. *J. Mol. Evol.* 21:1-13.

51. Duncan, D. M., E. A. Burgess, and I. Duncan. 1998. Control of distal antennal identity and tarsal development in *Drosophila* by *spineless-aristapedia*, a homolog of the mammalian dioxin receptor. *Genes & Dev* 12:1290-1303.

52. Nambu, J. R., J. O. Lewis, K. A. Wharton, Jr., and S. T. Crews. 1991. The *Drosophila* single-minded gene encodes a helix-loop-helix protein that acts as a master regulator of CNS midline development. *Cell* 67:1157-1167.

53. Wilson, T. G., and C. Turner (ed.). 1992. Molecular analysis of *Methoprene-tolerant*, a gene in *Drosophila* involved in resistance to juvenile hormone analog growth regulators, vol. 505. American Chemical Society, Washington,DC.

54. Soller, M., M. Bownes, and E. Kubli. 1999. Control of oocyte maturation in sexually mature *Drosophila* females. *Dev. Biol.* 208:337-351.

55. Wilson, T. G. 1982. A correlation between juvenile hormone deficiency and vitellogenic oocyte degeneration in Drosophila melanogaster. *Wilhelm Roux Arch. Entwicklungsmech. Org.* 191:257-263.

56. Jowett, T., and J. H. Postlethwait. 1980. The regulation of yolk polypeptide synthesis in *Drosophila* ovaries and fat body by 20-hydroxyecdysone and a juvenile hormone analog. *Develop. Biol.* 80:225-234.

57. Greenleaf, A. L., L. M. Borsett, P. F. Jiamachello, and D. E. Coulter. 1979. Alpha-amanitin-resistant *D. melanogaster* with an altered RNA polymerase. *Cell* 18:613-622.

58. Minkoff, C., III, and T. G. Wilson. 1992. The competitive ability and fitness components of the *Methoprene-tolerant* (*Met*) Drosophila mutant resistant to juvenile hormone analog insecticides. *Genetics* 131:91-97.

59. Brookfield, J. F. Y. 1997. Genetic redundancy. *Adv. Genet.* 36:137-155.

60.   Goshu, E., H. Jin, R. Fasnacht, M. Sepenski, J. L. Michaud, and C. M. Fan. 2002. *Sim2* mutants have developmental defects not overlapping with those of *Sim1* mutants. *Mol. Cell Biol.* 22:4147-4157.

61.   Moore, A. W., S. Barbel, L. Y. Jan, and Y. N. Jan. 2000. A genomewide survey of basic helix-loop-helix factors in Drosophila. *Proc. Natl. Acad. Sci. USA* 97:10436-10441.

62.   Hirose, K., M. Morita, M. Ema, J. Mimura, H. Hamada, H. Fujii, Y. Saijo, O. Gotoh, K. Sogawa, and Y. Fujii-Kuriyama. 1996. cDNA cloning and tissue-specific expression of a novel basic helix-loop-helix/PAS factor (Arnt2) with close sequence similarity to the aryl hydrocarbon receptor nuclear translocator (Arnt). *Mol. Cell Biol.* 16:1706-1713.

63.   Baker, F. C., H. H. Hagedorn, D. A. Schooley, and G. Wheelock. 1983. Mosquito juvenile hormone: identification and bioassay activity. *J. Insect Physiol.* 29:465-470.

64.   Klowden, M. J. 1997. Endocrine aspects of mosquito reproduction. *Arch. Insect Biochem. Physiol.* 35:491-512.

65.   O'Donnell, P. P., and M. J. Klowden. 1997. Methoprene affects the rotation of the male terminalia of *Aedes aegypti* mosquitoes. *J. Amer. Mosq. Control Assoc.* 13:1-4.

66.   Wilk, R., I. Wiezman, and B.-Z. Shilo. 1996. *trachealess* encodes a bHLH-PAS protein that is an inducer of tracheal cell fates in *Drosophila*. *Genes & Dev.* 10:93-102.

67.   Crews, S. T. 1998. Control of cell lineage-specific development and transcription by bHLH-PAS proteins. *Genes & Dev.* 12:607-620.

68.   Crews, S. T., J. B. Thomas, and C. S. Goodman. 1988. The Drosophila *single-minded* gene encodes a nuclear protein with sequence similarity to the *per* gene product. *Cell* 52:143-151.

69.   Ma, Q., and J. J. P. Whitlock. 1997. A novel cytoplasmic protein that interacts with the Ah receptor, contains tetratricopeptide repeat motifs, and augments the transcriptional response to 2,3,7,8-tetrachlorodibenzo-p-dioxin. *J. Biol. Chem.* 272:8878-8884.

70.   Wilson, C. L., and S. Safe. 1998. Mechanisms of ligand-induced aryl hydrocarbon receptor-mediated biochemical and toxic responses. *Toxicol. Path.* 26:657-671.

71.   Pursley, S., M. Ashok, and T. G. Wilson. 2000. Intracellular localization and tissue specificity of the *Methoprene-tolerant* (*Met*) gene product in *Drosophila melanogaster*. *Insect Biochem. Mol. Biol.* 30:839-845.

72.   Ashburner, M. 1970. Effects of juvenile hormone on adult differentiation of *Drosophila melanogaster*. *Nature* 227:187-189.

73.   Madhavan, K. 1973. Morphogenetic effects of juvenile hormone and juvenile hormone mimics on adult development of *Drosophila*. *J. Insect Physiol.* 19:441-453.

74.   Postlethwait, J. H. 1974. Juvenile hormone and the adult development of *Drosophila*. *Biol. Bull.* 147:119-135.

75.   Sehnal, F., and J. Zdarek. 1976. Action of juvenoids on the metamorphosis of cyclorrhaphous Diptera. *J. Insect Physiol.* 22:673-682.

76.   Restifo, L. L., and T. G. Wilson. 1998. A juvenile hormone agonist reveals distinct developmental pathways mediated by ecdysone-inducible *Broad Complex* transcription factors. *Develop. Genet.* 22:141-159.

77. Chao, A. T., and G. M. Guild. 1986. Molecular analysis of the ecdysterone-inducible 2B5 "early" puff in *Drosophila melanogaster*. *EMBO J.* 5:143-150.
78. Restifo, L. L., and K. White. 1991. Mutations in a steroid hormone-regulated gene disrupt the metamorphosis of the central nervous system in *Drosophila*. *Dev. Biol.* 148:174-194.
79. Kiss, I., A. H. Beaton, J. Tardiff, D. Fristrom, and J. W. Fristrom. 1988. Interactions and developmental effects of mutations in the *Broad-Complex* of *Drosophila melanogaster*. *Genetics* 118:247-259.
80. Crossgrove, K., C. A. Bayer, F. J.W., and G. M. Guild. 1996. The *Drosophila Broad-Complex* early gene directly regulates late gene transcription during the ecdysone-induced puffing cascade. *Dev. Biol.* 180:745-758.
81. Zhou, X., and L. M. Riddiford. 2002. Broad specifies pupal development and mediates the 'status quo' action of juvenile hormone on the pupal-adult transformation in *Drosophila* and *Manduca*. *Development* 129:2259-2269.
82. Safe, S., F. Wang, W. Porter, R. Duan, and A. McDougal. 1998. Ah receptor agonists as endocrine disruptors: antiestrogenic activity and mechanisms. *Toxicol. Lett.* 102-103:343-347.
83. Makino, Y., R. Cao, K. Svensson, G. Bertilsson, M. Asman, H. Tanaka, Y. Cao, A. Berkenstam, and L. Poellinger. 2001. Inhibitory PAS domain protein is a negative regulator of hypoxia-inducible gene expression. *Nature* 414:550-554.
84. Ma, Y., L. Certel, Y. Gao, E. Niemitz, J. Mosher, A. Mukherjee, M. Mutsuddi, N. Huseinovic, S. T. Crews, W. A. Johnson, et al. 2000. Functional interactions between Drosophila bHLH-PAS, Sox, and POU transcription factors regulate CNS midline expression of the slit gene. *J. Neurosci.* 20:4596-4605.
85. Chen, Y., M. J. Riese, M. A. Killinger, and F. M. Hoffmann. 1998. A genetic screen for modifiers of Drosophila decapentaplegic signaling identifies mutations in punt, Mothers against dpp, and the BMP-7 homologue, 60A. *Development* 125:1759-1768.
86. Belyaeva, E. S., M. G. Aizenzon, V. F. Semeshin, I. Kiss, K. Koczya, M. Baritcheva, T. D. Gorelova, and I. F. Zhimulev. 1980. Cytogenetic analysis of the 2B3-4-2B11 region of the X-chromosome of *Drosophila melanogaster*. I. Cytology of the region and mutant complementation groups. *Chromosoma* 81:281-306.
87. Segraves, W. A. 1994. Steroid receptors and other transcription factors in ecdysone response. *Rec. Prog. Horm. Res.* 49:167-194.
88. Jones, G., and P. A. Sharp. 1997. Ultraspiracle: An invertebrate nuclear receptor for juvenile hormones. *Proc. Natl. Acad. Sci. USA* 94:13499-13503.
89. Xu, J., Y. Qiu, F. J. DeMayo, S. Y. Tsai, M.-J. Tsai, and B. W. O'Malley. 1998. Partial hormone resistance in mice with disruption of the steroid receptor coactivator-1 (SRC-1) gene. *Science* 279:1922-1925.
90. Heery, D. M., E. Kalkhoven, S. Hoare, and M. G. Parker. 1997. A signature motif in transcriptional co-activators mediates binding to nuclear receptors. *Nature* 387:733-736.
91. Torchia, J., D. W. Rose, J. Inostroza, Y. Kamei, S. Westin, C. K. Glass, and M. G. Rosenfeld. 1997. The transcriptional co-activator p/CIP binds CBP and mediates nuclear-receptor function. *Nature* 387:677-684.

92. Jones, G. 1995. Molecular mechanisms of action of juvenile hormone. *Annu. Rev. Entomol.* 40:147-169.

93. Feyereisen, R. 1999. Insect P450 enzymes. *Annu. Rev. Entomol.* 44:507-533.

94. Emmons, R. B., D. D., P. A. Estes, P. Kiefel, J. T. Mosher, M. Sonnenfeld, M. P. Ward, I. Duncan, and S. T. Crews. 1999. The *Drosophila* spineless-aristapedia and tango bHLH-PAS proteins interact to control antennal and tarsal development. *Development* 126:3937-3945.

95. Butler, R. a., M. L. Kelley, W. H. Powell, M. E. Hahn, and R. J. Van Beneden. 2001. An aryl hydrocarbon receptor (AHR) homologue from the soft-shell clam, *Mya arenaria*: evidence that invertebrate AHR homologues lack 2,3,7,8-tetrachlorodibenzo-*p*-dioxin and β-naphthoflavone binding. *Gene* 278:223-234.

96. Hahn, M. E., S. I. Karchner, M. A. Shapiro, and S. A. Perera. 1997. Molecular evolution of two vertebrate aryl hydrocarbon (dioxin) receptors (AHR1 and AHR2) and the PAS family. *Proc. Natl. Acad. Sci. USA* 94:13743-13748.

Chapter 6

THE P160 FAMILY OF STEROID HORMONE RECEPTOR COACTIVATORS

Denise J. Montell
The Johns Hopkins University School of Medicine, Baltimore, MD 21205

1. INTRODUCTION

Among PAS domain proteins, the p160 coactivator class may be the most perplexing. Like many PAS domain proteins, members of the p160 family sense the presence of small, lipophilic molecules. And like many eukaryotic PAS domain proteins they regulate transcription. However, p160 coactivators have not been found to bind directly either to small molecules or to DNA; rather they bind to steroid hormone receptors in a ligand-dependent manner and serve as coactivators of transcription. As such, they form a bridge between the hormone receptors, general transcription machinery, and histone acetyl transferases (HATs), which regulate chromatin conformation. These proteins are essential to the organism for normal responses to steroid hormones and for energy homeostasis. Yet to this day, the precise function of the PAS domain within the P160 coactivators remains something of a mystery.

2.   DOMAIN ORGANIZATION, STRUCTURE, AND FUNCTION OF P160 COACTIVATORS

2.1   The p160 family

The family of p160 coactivators is composed of three members in mammals, and one protein each in *Drosophila melanogaster* and *Anopheles gambiae*. The domain organization of these proteins is shown schematically in Figure 1. The first identified member of this small family, SRC-1, was cloned by two-hybrid screening using the ligand-binding domain of the progesterone receptor (PR) as bait (1). Subsequent studies demonstrated that SRC-1 enhances transcriptional activation, not only by PR, but also by many other hormone receptors (reviewed in (2). The second family member was identified independently as transcription intermediary factor 2 (TIF2) (3) and as a glucocorticoid receptor-interacting protein (GRIP1) (4). Thus this protein is referred to here as SRC-2. The third was identified independently by several groups as a retinoic acid receptor interacting protein (RAC3), a CREB binding protein (CBP)-interacting protein (p/CIP), as a gene amplified in breast cancer (AIB1) (5), as a thyroid hormone receptor activating molecule (TRAM-1) (6), and in a yeast one hybrid screen as a retinoic acid receptor activating protein (ACTR) (6). This protein will be referred to as SRC-3. The family members share a common domain organization, which includes an N-terminal bHLH/PAS domain, a region containing multiple sequences referred to as LXXLL motifs, where L is leucine and X is any amino acid, and one or more polyglutamine tracts. The function of each domain will be discussed in subsequent sections.

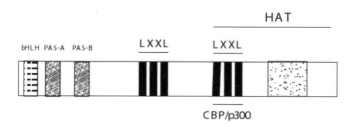

Figure 1. Schematic diagram of domain organization of P160 coactivators. The N-terminus contains the highly conserved bHLH and PAS A and B domains. The centrally located receptor-interacting domain and activation domain (AD) each contain three LXXLL motifs, while SRC-1 contains an additional, non-conserved motif at the C-terminus. The C-terminus contains a glutamine-rich domain. The specific domains for interaction with CBP/p300 and the histone acetyltransferase (HAT) domain are indicated.

Overall SRC-1 and SRC-2 are 40% identical and 56% similar, while SRC-1 and -2 are each about 36% identical and 50% similar to SRC-3. The p160 family members are best-known for their activities as ligand-dependent coactivators of transcription by the steroid hormone receptor superfamily (2, 7, 8). Therefore, to discuss the biochemical activities and *in vivo* functions of the p160 family, it is necessary to introduce the steroid hormone receptor superfamily.

2.2   The steroid hormone receptor family

There are two major types of steroid hormone receptors (Figure 2) (2, 9). Class I receptors, which include the estrogen receptor (ER), androgen receptor (AR), and glucocorticoid receptor (GR), are located in the cytoplasm in a complex with heat shock protein 90 (Hsp90), in the absence of ligand. Following ligand binding, these receptors dissociate from the heat shock proteins, dimerize, and translocate to the nucleus where they bind to palindromic response elements.   In contrast, Class II receptors, which include the thyroid hormone receptor (TR), retinoic acid receptor (RAR), and peroxisome proliferator activated receptors (PPARα, β, γ), form heterodimers with a common subunit known as the retinoid X receptor (RXR). These heterodimeric receptors bind specific, direct DNA repeats. In the absence of ligand, the latter receptors recruit repressor proteins and inhibit target gene expression.   Ligand binding causes a conformational change that displaces the co-repressor and leads to recruitment of a complex array of co-activator proteins. Chief among these are the p160 coactivators, which bind to both types of hormone receptors, exclusively in the presence of ligand.

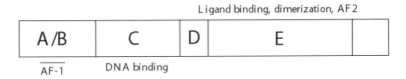

Figure 2. The variable NH2-terminal region contains the ligand-independent AF-1 transactivation domain and is labeled A/B. The conserved zinc finger DNA-binding domain corresponds to region C. Region D represents a variable linker domain. The E/F region is well-conserved and contains the ligand-binding domain as well as the dimerization interface. The ligand-dependent transactivation domain within the COOH-terminal portion of the ligand binding domain is the region that interacts with p160 coactivators.

Members of the steroid hormone receptor superfamily can be activated by a wide variety of small molecules, not only by steroids. These include glucocorticoids, thyroid hormone, retinoic acid, vitamin D, bile acids, fatty acids, and leukotrienes. Structural formulas for some of these compounds are shown in Figure 3. These ligands, acting through their receptors, serve a wide variety of critical functions, ranging from patterning the early embryo (10), to regulating energy homeostasis (11), to control of reproductive physiology.

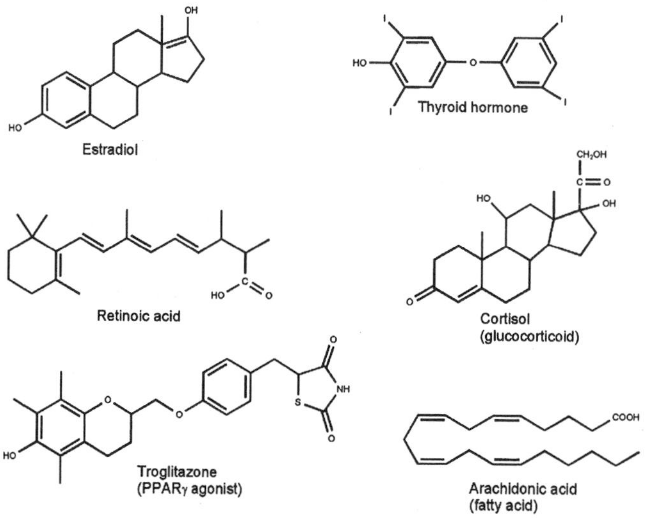

Figure 3. Structures of nuclear receptor ligands.

2.3   Invertebrate p160 proteins

The first member of the p160 coactivator family to be identified in an invertebrate is a coactivator for the ecdysone receptor in *Drosophila* (12, 13). The gene coding for the single *Drosophila* p160 family member is named *taiman* (*tai*), which means "too slow". This protein possesses all of the sequence characteristics of the other family members. It contains a bHLH-PAS domain, three LXXLL motifs, and multiple glutamine-rich regions. The protein is most similar to SRC-3, although the overall sequence identity is very low. BLAST searches only detect similarity in the bHLH domain whereas motif searches recognize both the bHLH and PAS domains. The bHLH domain is 47% identical and 69% similar to SRC-3/AIB1/RAC-3/pCIP/ACTR over 55 amino acids (BLAST E value 3e-04). The predicted TAI bHLH domain is 49% identical and 64% similar to SRC-2 over 51 amino acids (BLAST E value 0.058).

Despite the very low levels of amino acid sequence identity, it is clear that the *Drosophila* protein is a bona fide steroid hormone receptor coactivator, based on numerous *in vitro* and *in vivo* assays (12). Ecdysone is the only known steroid hormone in *Drosophila*. The TAI protein stimulates ecdysone-dependent transcription in cultured cells, colocalizes with the ecdysone receptor on chromosomes *in vivo* and binds to the ecdysone receptor *in vitro* in a hormone dependent manner. Furthermore, loss of function mutations in *tai* and in the genes coding for the ecdysone receptor subunits cause similar phenotypes.

The recently released genome sequence for the mosquito, *Anopheles gambiae*, contains a close homolog of the *Drosophila* TAI protein. However there is no obvious p160 family member in *C. elegans* or in yeast. The *C. elegans* genome encodes an unusually large number of proteins with homology to steroid hormone receptors (though no ligand has yet been identified), and so it is curious that there is no apparent p160 protein.

2.4 Biochemical functions of each domain

The vertebrate p160 family has been subjected to extensive structure/function analyses (reviewed in (2). The N-terminal bHLH domain is the most highly conserved domain both amongst the vertebrate family members and between vertebrate and invertebrate proteins. Yet this domain does not appear to bind DNA, or to be necessary for p160 interaction with steroid hormone receptors.

The critical region for interaction with the hormone receptors is the domain containing LXXLL motifs located near the center of the protein (6, 14-18). These motifs interact directly with the hormone receptors in a ligand-dependent manner. At least two LXXLL motifs are required for high affinity interactions between p160 proteins and hormone receptors. Most coactivators contain three or more such motifs and different motifs interact preferentially with each of the various hormone receptors. The amino acid sequences immediately flanking the LXXLL sequence contribute to the specificity of binding, and spacing between motifs is also important (16, 19). In SRC-1 there are additional LXXLL motifs at the C-terminus that are not present in the other family members.

Transcriptional activation by the p160 proteins is mediated in large part by interaction of these proteins with the more general activators p300 and CBP (20, 21), which coactivate not only for nuclear hormone receptors but for a wide variety of other transcription factors (22). The region of SRC-2 responsible for interaction with P300/CBP spans amino acids 108 to 1088 (23). Like p160 coactivators, p300 and CBP are multidomain proteins that contain LXXLL motifs; however they do not possess a bHLH or PAS domain. Both P300/CBP and p160 coactivators are histone acetyl transferase (HAT) enzymes. The HAT activity is found near the C-terminus, between residues 1088 and 1412 in SRC-2 (23). Acetylation of histones is thought to lead to alterations in chromatin conformation. Specifically, histone acetylation causes chromatin to open up and become more accessible to transcriptional activators. Thus one mechanism by which the p160 coactivators stimulate transcription appears to be by direct acetylation of histone proteins in the vicinity of hormone-bound receptors. HAT activity may also serve a negative feedback function, because it has been reported that acetylation of p300/CBP, possibly by p160, causes p300/CBP to dissociate from the complex. Such dissociation could account for the desensitization of certain hormone responsive promoters that occurs following prolonged stimulation.

The most mysterious domain in the p160 proteins is the bHLH-PAS domain. In many other proteins, as described elsewhere in this volume, these domains mediate DNA binding and dimerization, respectively. However no interaction has been detected between purified bHLH-PAS domain of p160 proteins and the full-length protein, making it unlikely that this domain mediates dimerization amongst family members. Nor does this protein seem to bind DNA directly. It is widely assumed that these domains perform some kind of protein-protein interaction function, and one study reported an interaction between the bHLH-PAS domain of SRC-2 and the muscle transcription factor MEF2 (24). This report provided the first evidence that the bHLH-PAS domain of the p160 coactivators can actually interact with another protein. A second study has reported an interaction between the bHLH-PAS domain of SRC-1 and the transcription factor TEF4 (transcriptional enhancer family) (25). SRC-1 was also able to potentiate the activity of TEF4 in HeLa cells. However the physiological significance of these protein interactions is not entirely clear.

Since the PAS domain of the aryl hydrocarbon receptor (AHR) binds directly to its ligand, the possibility exists that the PAS domains of p160 proteins bind small molecules in addition to, or instead of, binding other proteins. Regardless of exactly what binds or regulates the bHLH/PAS domains of these proteins, the larger question is what purpose is served by this domain. In some experiments p160 proteins lacking the bHLH/PAS domain are more active than the full length protein. This finding raises the possibility that the unbound bHLH/PAS domain may attenuate transcriptional activation and that binding of a ligand to that domain might relieve that attenuation. Such a mechanism would allow the p160 coactivator to serve as an integrator of signals. The coactivator might provide one level of transcriptional activation in response to a single signal, such as a steroid hormone, and a higher level of transcriptional output in the presence of two signals, for example the steroid hormone plus a ligand for the bHLH/PAS domain. Alternatively, if the bHLH/PAS domain binds a completely different class of transcription factor with different target genes than the steroid hormone receptor, then competition between the two p160 binding proteins could result in repression of one class of targets when the other class is highly activated. Further experiments will be required to distinguish between these possibilities.

2.5   Other protein partners

In addition to binding hormone receptors and p300/CBP, p160 proteins have been reported to bind HIF-1α (26), ARNT, AHR (27), NfκB (28), and AP1 (29). SRC proteins enhance transcriptional activation by these various factors as well. In some cases this has been analyzed only following overexpression in vitro while in other cases chromatin co-immunoprecipitation and other assays provide strong support for a functional interaction. Definitive evidence for a biologically significant interaction requires detailed analysis of animals lacking one or more of the SRC proteins, as described below.

3.   BIOLOGICAL FUNCTIONS

3.1   Mouse knockout studies

The extensive *in vitro* and cell culture studies on the steroid hormone receptor coactivators combined with the known physiological importance of hormone receptors and their ligands led to the prediction that mice lacking one or more of the p160 coactivators would exhibit defects similar to loss of one or more hormone receptors. It was difficult to predict how severe such defects would be, however. The three coactivators are widely expressed and there is significant overlap in their expression patterns. Therefore, it seemed possible that the three proteins would function redundantly and that elimination of any one gene would produce only mild consequences. On the other hand, given the diversity of transcription factors with which the p160 coactivators interact, it also seemed possible that the effect of knocking out one or more coactivators could be direr than loss of a single hormone receptor.

Mice have now been generated that lack each of the family members. The most striking finding to date is that SRC-1 and SRC-2 play non-overlapping, and in fact opposing, roles in energy homeostasis (30). SRC-2 knockout mice exhibit enhanced energy expenditure and reduced fat storage relative to wild-type mice, and they are protected from becoming obese when fed a high fat diet. On the other hand, SRC-1 knockout mice exhibit reduced energy expenditure, enhanced fat storage, and they are prone to obesity. These effects are likely to be due to the interaction that has been described between these coactivators and PPARγ. PPARγ is a member of the class II steroid hormone receptors. A number of endogenous and exogenous ligands are known for this receptor including fatty acids and their metabolites. In addition, pharmacological agents known as the thiazolidinediones are even more potent activators of this receptor.

Drugs in this family such as rosiglitazone are currently in use for the treatment of Type II, or insulin resistant, diabetes. These drugs were discovered and used before anything was known of their mechanism of action, and although it seems clear that these drugs act by binding PPARγ, the detailed physiology of their actions is not completely understood. This is because energy and glucose homeostasis involve complex interactions amongst different cells, tissues, and organs that respond to neural and hormonal influences. Therefore teasing apart the precise mechanism of action of any particular component can be daunting.

What is clear is that wild-type mice are prone to obesity when they are fed a high fat diet and SRC-2 knockout mice are protected from this effect, whereas SRC-1 knockout mice are more prone to obesity (30). At least some of these effects are due to the functions of SRC-1 and SRC-2 in both white and brown fat. White adipose tissue functions to store fat whereas brown adipose tissue plays an essential role in utilizing excess energy to generate heat in a process called adaptive thermogenesis. SRC-2 promotes adipogenesis and reduces energy expenditure, whereas SRC-1 enhances thermogenesis at the expense of fat stores. What is still somewhat unclear is exactly how SRC-1 and SRC-2, which are both coactivators for PPARγ, exert opposite physiological effects on energy balance. Part of the explanation is that SRC-1 is a more potent coactivator for PPARγ than SRC-2, especially in the presence of PGC-1 (PPARγ coactivator 1), which is another type of coactivator. SRC-2 does not interact well with PGC-1, and so the activity of PPARγ can be altered by changes in the relative amounts of SRC-1 and SRC-2, in the presence of a constant amount of PGC-1. Thus, in wild-type animals fed a regular diet, a balance between SRC-1 and SRC-2 results in balanced fat storage and energy expenditure. However when normal mice are fed a high fat diet, for some reason SRC-2 levels rise leading to increased fat storage, decreased energy expenditure, obesity and diabetes. SRC-2 knockout animals are protected from this effect, as are animals treated with high affinity PPARγ ligands such as rosiglitazone.

Although SRC-1 and SRC-2 play opposing roles in energy homeostasis, there is functional redundancy between SRC-1 and SRC-2 with respect to thyroid hormone function. SRC-1 mutant mice were initially characterized as being partially resistant to thyroid hormone (31). The mild nature of the phenotype was attributed to the compensatory upregulation of SRC-2 expression that was observed in these mice. Consistent with this notion, acute depletion of SRC-1 in adult rats or mice, by treatment with antisense oligonucleotides delivered via the ventricles of the brain, results in defects that are not apparent in the SRC-1 knockouts. The defects, which concern estrogen-dependent reproductive behavior, can also be evoked in the SRC-1 knockout mice by treating them with antisense oligonucleotides against SRC-2. SRC-3 antisense treatment does not cause this phenotype, consistent with the observation that SRC-3 is not expressed in the particular brain region thought to mediate this behavior (the VMN), whereas SRC-1 and SRC-2 are expressed there.

A direct test of the hypothesis that SRC-1 and SRC-2 are partially redundant in function and capable of compensation was carried-out by analyzing the phenotype of double knockout mice, at least with respect to thyroid hormone sensitivity. This study demonstrated that the double knockouts show a more severe defect than the SRC-1 single knockout. The SRC-2 single knockout does not show a phenotype on its own although, somewhat surprisingly the double heterozygote exhibits a phenotype that is similar to the SRC-1 homozygous mutant. Double mutants lacking both copies of both genes show thyroid hormone resistance equal to that of loss of the thyroid hormone receptor. However it is striking to note that the double mutant animals are viable and fertile, whereas estrogen receptor mutants are female sterile.

In addition to partial resistance to thyroid hormone, SRC-1 knockout mice showed reduced development of the mammary gland and uterus, testis and prostate in response to exogenous steroid treatment. These findings confirm the in vivo importance of SRC-1 in steroid hormone responsive tissues. Further investigation of the knockout mice revealed additional, subtle phenotypes. For example, the null mutant mice exhibit motor dysfunction and delayed development of Purkinje cells of the cerebellum (32). This is consistent with the relatively high level of expression of SRC-1 compared to SRC-2 or SRC-3 in this particular region of the brain. It is not yet clear which ligand(s) and receptor(s) function together with SRC-1 in Purkinje cell development. Analysis of double mutants would clarify whether the subtlety of this phenotype is due to compensatory up-regulation of SRC-2.

Mice lacking SRC-3 show growth retardation and delayed puberty, effects not seen in the SRC-1 or SRC-2 knockouts (33). Surprisingly the SRC-3 phenotypes appear to be secondary consequences of decreased estrogen production in these animals as systemic estrogen can rescue both phenotypes. The phenotype of a triple mutant lacking all three p160 coactivator proteins has not yet been reported.

3.2   Genetic studies in *Drosophila*

The existence of a single *Drosophila* gene that codes for a p160 family member provides a simpler model for genetic studies of coactivator function in vivo. The existence of a single *Drosophila* gene that codes for a p160 family member, *tai*, provides a simpler model for genetic studies of coactivator function in vivo (Bai et al., 2000). In contrast to the mammalian p160 knockouts, null mutations in the *tai* locus are homozygous lethal, although the precise reason for lethality has not been investigated. The only steroid hormone known in *Drosophila* is ecdysone and TAI is a coactivator for the ecdysone receptor, which is a heterodimer of a hormone binding subunit EcR and RXR homolog known as USP (34). It is interesting to note that ecdysone is the molting hormone and is responsible for metamorphosis, the insect transition from the juvenile form to the sexually mature animal. Therefore metamorphosis is the insect version of puberty and the role of ecdysone in this process may be analogous to the role of sex steroids in mammals.

The *tai* locus was not identified based on its role in metamorphosis however. The *tai* locus was first identified genetically in a screen for mutations that cause cell migration defects in mosaic clones, which are patches of homozygous mutant cells in an otherwise heterozygous organism (35). The specific migratory cell type under study in this screen was a small subset of follicle cells in the *Drosophila* ovary – referred to as border cells (reviewed in (13, 36). At a particular stage of ovarian development, this group of 6-10 cells separates from the remaining 650-900 cells of the follicular epithelium. The border cells invade the neighboring nurse cell cluster and migrate in between these cells to the anterior border of the oocyte. One function of the border cells is to form a pore in the eggshell that is necessary for sperm entry and therefore for fertility.

The ability to analyze newly induced mutations in mosaic clones circumvents the problem of pleiotropy and permits the analysis of the effect of cell-type specific loss of gene function. Loss of *tai* function specifically from the border cells causes a pronounced migration defect, whereas loss of *tai* from other follicle cells results in problems in their stage-specific behavior. All of these defects are due to defective ecdysone signaling. Ecdysone titers rise specifically at the stage in oogenesis when the border cells migrate. Ecdysone is required not only for border cell migration but also for oogenesis to progress (37-39). The function of the rise in ecdysone titer is apparently to coordinate multiple events that occur at this stage. So in *Drosophila*, as in mammals, a steroid hormone acts through its ligand-binding receptor and a p160 coactivator to regulate development and physiology of female reproductive tissues.

3.3   Functions in disease

The identification of AIB1 as a gene that is amplified or overexpressed in breast and ovarian cancer (5) is consistent with the known effects of steroid hormones in stimulating metastasis.   Drugs that inhibit the interaction of steroid hormone receptors with coactivators represent an important class of chemotherapeutics for hormone dependent cancers such as breast and prostate.   Selective estrogen receptor modulators like tamoxifen and raloxifene are clinically useful for the treatment and prevention of breast cancer (40).   In breast cancer cells, tamoxifen competes with estrogen for binding to the receptor, and causes a conformational change that favors association with co-repressors rather than co-activators.   However in endometrial cells, tamoxifen acts as an ER agonist rather than antagonist and can cause endometrial cancer.   One key difference between breast and endometrium, which is at least in part responsible for the tissue-specific responses to tamoxifen, is that endometrial cells express high levels of SRC-1 whereas breast cells do not.   Shang and Brown showed that expression of high levels of SRC-1 in breast cancer cells causes tamoxifen to act as an agonist rather than antagonist in this tissue (41).   Agonist activity is only observed for promoters to which ER binds indirectly.   When ER binds directly to ERE sequences in a hormone responsive promoter, then tamoxifen cannot activate transcription even in endometrial cells.   The basis of the difference is not yet established, but what is clear is that different SRC family members do have distinct biochemical properties that are biologically and clinically significant.

A somewhat more tenuous but nonetheless intriguing link between AIB1 and cancer emerged from a study of the relationship between cancer risk and AIB1 glutamine repeat length.   Women with mutations in BRCA1 or BRCA2 are at increased risk of breast cancer compared to women with wild-type alleles.   However the risk to a particular individual is impossible to predict and it is likely that individual genetic polymorphisms and backgrounds contribute significantly to risk.   Particularly, genetic differences at loci known to contribute to breast cancer progression are good candidates.   Rebbeck et al., tested whether variation in the glutamine repeat length in AIB1 contributed significantly to the probability that women with BRCA1 mutations would actually develop breast cancer (42).   The rationale for examining glutamine repeat length was that variations in trinucleotide repeats in a variety of genes (some of which encode glutamine and some of which occur in non-coding regions) are known to cause diseases such as Huntington's chorea and ATX (43). Rebbeck et al found that there was a statistically significant correlation between increased glutamine repeat length and the probability of developing breast cancer.

TIF2 has also been implicated in cancer, in this case a specific subtype of leukemia. In some patients with the M4/M5 subtype of acute myeloid leukemia (AML), a translocation between a gene known as the MOZ gene and the TIF2 gene has been found (42). This results in a fusion transcript, and presumably a fusion protein, which would contain the N-terminus of the MOZ protein and the C-terminus of TIF2, including the CBP interaction domain and the HAT activity, but excluding the bHLH/PAS domain and the LXXLL motifs. The biochemical function of the MOZ protein has not been extensively studied. However, a related protein in *Drosophila* is involved in dosage compensation, and a related yeast protein functions in transcriptional silencing. The MOZ protein also has HAT activity, which is presumably retained in the MOZ/TIF2 fusion. So it is likely that the MOZ protein plays a role in regulating chromatin states. It is interesting that the same subtype of leukemia can also result from a translocation that fuses the N-terminus of MOZ to CBP. The prediction then, is that the fusion proteins lead to aberrant transcriptional regulation, resulting in the overproliferation of myeloid blast cells (42).

4.   CLOSING THOUGHTS

Only eight years has passed since the identification of SRC-1 and yet in that short time intensive investigation of the p160 class of steroid hormone receptor coactivators has resulted in an abundance of information. The crystal structure has been solved of the LXXLL domain of SRC-1 complexed with the rozaglitazone-bound hormone receptor activation domain of PPARγ (17). The biochemical activities of the protein have been studied in detail. Mice lacking each of the coactivators have been generated, and phenotypic characterization of both single and double mutants is well-advanced. Approximately 350 papers have been published on this family of three proteins. And yet, some mysteries endure. It is as yet unclear what function the bHLH-PAS domain serves in these proteins, a mystery made deeper by the observation that this domain is the most highly conserved in the family. Technical advantages of *Drosophila* genetics may help in answering this particular question since it should be straightforward to generate transgenic animals expressing versions of the protein that lack this domain or bear subtle amino acid substitutions in it, in the context of a null mutant background. Another knot, this one for the mouse geneticists to untangle, is the complete phenotypic analysis of single, double and triple mutant animals. To what extent do these factors function uniquely or redundantly? Due to the subtlety of some of the single mutant phenotypes, it may take many years to fully document the defects in these animals. And then, to what extent do the defects seen in mice correlate with human development, physiology or disease? Are there mutations in the genes coding for SRC proteins in humans who have delayed puberty, growth defects, reproductive problems, neurological disorders, obesity, diabetes, etc? And finally, due to the sensitivity of this class of proteins to small lipophilic molecules, do they represent an ideal group of drug targets?

REFERENCES

1. Onate, S. A., S. Y. Tsai, M. J. Tsai, and B. W. O'Malley. 1995. Sequence and characterization of a coactivator for the steroid hormone receptor superfamily. *Science* 270:1354-7.
2. Leo, C., and J. D. Chen. 2000. The SRC family of nuclear receptor coactivators. *Gene* 245:1-11.
3. Voegel, J. J., M. J. Heine, C. Zechel, P. Chambon, and H. Gronemeyer. 1996. TIF2, a 160 kDa transcriptional mediator for the ligand-dependent activation function AF-2 of nuclear receptors. *EMBO J.* 15:3667-75.
4. Hong, H., K. Kohli, A. Trivedi, D. L. Johnson, and M. R. Stallcup. 1996. GRIP1, a novel mouse protein that serves as a transcriptional coactivator in yeast for the hormone binding domains of steroid receptors. *Proc. Natl. Acad. Sci. USA* 93:4948-52.
5. Anzick, S. L., J. Kononen, R. L. Walker, D. O. Azorsa, M. M. Tanner, X. Y. Guan, G. Sauter, O. P. Kallioniemi, J. M. Trent, and P. S. Meltzer. 1997. AIB1, a steroid receptor coactivator amplified in breast and ovarian cancer. *Science* 277:965-8.
6. Takeshita, A., G. R. Cardona, N. Koibuchi, C. S. Suen, and W. W. Chin. 1997. TRAM-1, A novel 160-kDa thyroid hormone receptor activator molecule, exhibits distinct properties from steroid receptor coactivator-1. *J. Biol. Chem.* 272:27629-34.
7. McKenna, N. J., and B. W. O'Malley. 2002. Combinatorial control of gene expression by nuclear receptors and coregulators. *Cell* 108:465-74.
8. Rosenfeld, M. G., and C. K. Glass. 2001. Coregulator codes of transcriptional regulation by nuclear receptors. *J. Biol. Chem.* 276:36865-8.
9. Aranda, A., and A. Pascual. 2001. Nuclear hormone receptors and gene expression. *Physiol. Rev.* 81:1269-304.
10. Marshall, H., A. Morrison, M. Studer, H. Popperl, and R. Krumlauf. 1996. Retinoids and Hox genes. *FASEB J.* 10:969-78.
11. Kliewer, S. A., H. E. Xu, M. H. Lambert, and T. M. Willson. 2001. Peroxisome proliferator-activated receptors: from genes to physiology. *Recent Prog. Horm. Res.* 56:239-63.
12. Bai, J., Y. Uehara, and D. J. Montell. 2000. Regulation of Invasive Cell Behavior by Taiman, a Drosophila Protein Related to AIB1, a Steroid Receptor Coactivator Amplified in Breast Cancer. *Cell* 103:1047-58.

13.   Montell, D. J. 2001. Command and control: regulatory pathways controlling invasive behavior of the border cells. *Mech. Dev.* 105:19-25.

14.   Heery, D. M., E. Kalkhoven, S. Hoare, and M. G. Parker. 1997. A signature motif in transcriptional co-activators mediates binding to nuclear receptors. *Nature* 387:733-6.

15.   Hong, H., B. D. Darimont, H. Ma, L. Yang, K. R. Yamamoto, and M. R. Stallcup. 1999. An additional region of coactivator GRIP1 required for interaction with the hormone-binding domains of a subset of nuclear receptors. *J. Biol. Chem.* 274:3496-502.

16.   McInerney, E. M., D. W. Rose, S. E. Flynn, S. Westin, T. M. Mullen, A. Krones, J. Inostroza, J. Torchia, R. T. Nolte, N. Assa-Munt, et al. 1998. Determinants of coactivator LXXLL motif specificity in nuclear receptor transcriptional activation. *Genes Dev.* 12:3357-68.

17.   Nolte, R. T., G. B. Wisely, S. Westin, J. E. Cobb, M. H. Lambert, R. Kurokawa, M. G. Rosenfeld, T. M. Willson, C. K. Glass, and M. V. Milburn. 1998. Ligand binding and co-activator assembly of the peroxisome proliferator-activated receptor-gamma. *Nature* 395:137-43.

18.   Westin, S., R. Kurokawa, R. T. Nolte, G. B. Wisely, E. M. McInerney, D. W. Rose, M. V. Milburn, M. G. Rosenfeld, and C. K. Glass. 1998. Interactions controlling the assembly of nuclear-receptor heterodimers and co-activators. *Nature* 395:199-202.

19.   Needham, M., S. Raines, J. McPheat, C. Stacey, J. Ellston, S. Hoare, and M. Parker. 2000. Differential interaction of steroid hormone receptors with LXXLL motifs in SRC-1a depends on residues flanking the motif. *J. Steroid Biochem. Mol. Biol.* 72:35-46.

20.   Kamei, Y., L. Xu, T. Heinzel, J. Torchia, R. Kurokawa, B. Gloss, S. C. Lin, R. A. Heyman, D. W. Rose, C. K. Glass, et al. 1996. A CBP integrator complex mediates transcriptional activation and AP-1 inhibition by nuclear receptors. *Cell* 85:403-14.

21.   Torchia, J., D. W. Rose, J. Inostroza, Y. Kamei, S. Westin, C. K. Glass, and M. G. Rosenfeld. 1997. The transcriptional co-activator p/CIP binds CBP and mediates nuclear-receptor function. Nature 387:677-84.

22.   Janknecht, R. 2002. The versatile functions of the transcriptional coactivators p300 and CBP and their roles in disease. *Histol. Histopathol.* 17:657-68.

23.   Chen, H., R. J. Lin, R. L. Schiltz, D. Chakravarti, A. Nash, L. Nagy, M. L. Privalsky, Y. Nakatani, and R. M. Evans. 1997. Nuclear receptor coactivator ACTR is a novel histone acetyltransferase and forms a multimeric activation complex with P/CAF and CBP/p300. *Cell* 90:569-80.

24.   Chen, S. L., D. H. Dowhan, B. M. Hosking, and G. E. Muscat. 2000. The steroid receptor coactivator, GRIP-1, is necessary for MEF-2C- dependent gene expression and skeletal muscle differentiation. *Genes Dev.* 14:1209-28.

25.   Belandia, B., and M. G. Parker. 2000. Functional interaction between the p160 coactivator proteins and the transcriptional enhancer factor family of transcription factors. *J. Biol. Chem.* 275:30801-5.

26.   Carrero, P., K. Okamoto, P. Coumailleau, S. O'Brien, H. Tanaka, and L. Poellinger. 2000. Redox-regulated recruitment of the transcriptional coactivators CREB-binding protein and SRC-1 to hypoxia-inducible factor 1alpha. *Mol. Cell. Biol.* 20:402-15.

27.   Beischlag, T. V., S. Wang, D. W. Rose, J. Torchia, S. Reisz-Porszasz, K. Muhammad, W. E. Nelson, M. R. Probst, M. G. Rosenfeld, and O. Hankinson. 2002. Recruitment of the NCoA/SRC-1/p160 family of transcriptional coactivators by the aryl hydrocarbon receptor/aryl hydrocarbon receptor nuclear translocator complex. *Mol. Cell. Biol.* 22:4319-33.

28. Na, S. Y., S. K. Lee, S. J. Han, H. S. Choi, S. Y. Im, and J. W. Lee. 1998. Steroid receptor coactivator-1 interacts with the p50 subunit and coactivates nuclear factor kappaB-mediated transactivations. *J. Biol. Chem.* 273:10831-4.

29. Lee, S. K., H. J. Kim, S. Y. Na, T. S. Kim, H. S. Choi, S. Y. Im, and J. W. Lee. 1998. Steroid receptor coactivator-1 coactivates activating protein-1-mediated transactivations through interaction with the c-Jun and c-Fos subunits. *J. Biol. Chem.* 273:16651-4.

30. Picard, F., M. Gehin, J. Annicotte, S. Rocchi, M. F. Champy, B. W. O'Malley, P. Chambon, and J. Auwerx. 2002. SRC-1 and TIF2 control energy balance between white and brown adipose tissues. *Cell* 111:931-41.

31. Xu, J., Y. Qiu, F. J. DeMayo, S. Y. Tsai, M. J. Tsai, and B. W. O'Malley. 1998. Partial hormone resistance in mice with disruption of the steroid receptor coactivator-1 (SRC-1) gene. *Science* 279:1922-5.

32. Nishihara, E., H. Yoshida-Komiya, C. S. Chan, L. Liao, R. L. Davis, B. W. O'Malley, and J. Xu. 2003. SRC-1 null mice exhibit moderate motor dysfunction and delayed development of cerebellar Purkinje cells. *J. Neurosci.* 23:213-22.

33. Xu, J., L. Liao, G. Ning, H. Yoshida-Komiya, C. Deng, and B. W. O'Malley. 2000. The steroid receptor coactivator SRC-3 (p/CIP/RAC3/AIB1/ACTR/TRAM-1) is required for normal growth, puberty, female reproductive function, and mammary gland development. *Proc. Natl. Acad. Sci. USA* 97:6379-84.

34. Yao, T. P., B. M. Forman, Z. Jiang, L. Cherbas, J. D. Chen, M. McKeown, P. Cherbas, and R. M. Evans. 1993. Functional ecdysone receptor is the product of EcR and Ultraspiracle genes. *Nature* 366:476-9.

35. Xu, T., and S. D. Harrison. 1994. Mosaic analysis using FLP recombinase. *Methods in Cell Biology* 44:655-681.

36. Montell, D. J. 2003. Border-cell migration: the race is on. *Nat. Rev. Mol. Cell Biol.* 4:13-24.

37. Bender, M., F. B. Imam, W. S. Talbot, B. Ganetzky, and D. S. Hogness. 1997. Drosophila ecdysone receptor mutations reveal functional differences among receptor isoforms. *Cell* 91:777-88.

38. Buszczak, M., M. R. Freeman, J. R. Carlson, M. Bender, L. Cooley, and W. A. Segraves. 1999. Ecdysone response genes govern egg chamber development during mid- oogenesis in Drosophila. *Development* 126:4581-9.

39. Carney, G. E., and M. Bender. 2000. The Drosophila ecdysone receptor (EcR) gene is required maternally for normal oogenesis. *Genetics* 154:1203-11.

40. Fisher, B., J. P. Constantino, D. L. Wickerham, C. K. Redmond, M. Kavanah, W. M. Cronin, V. Vogel, A. Robidoux, N. Dimitrov, J. Atkins, et al. 1998. Tamoxifen for prevention of breast cancer: report of the National Surgical Adjuvant Breast and Bowel Project P-1 Study. *J. Natl. Cancer Inst.* 90:1371-88.

41. Shang, Y., and M. Brown. 2002. Molecular determinants for the tissue specificity of SERMs. Science 295:2465-8.

42. Carapeti, M., R. C. Aguiar, A. E. Watmore, J. M. Goldman, and N. C. Cross. 1999. Consistent fusion of MOZ and TIF2 in AML with inv(8)(p11q13). *Cancer Genet. Cytogenet.* 113:70-2.

43. Zoghbi, H. Y., and H. T. Orr. 2000. Glutamine repeats and neurodegeneration. *Annu. Rev. Neurosci.* 23:217-47.

Chapter 7

THE AH RECEPTOR

Guang Yao, Eric B. Harstad, and Christopher A. Bradfield
McArdle Laboratory for Cancer Research, Madison, WI 53706

1.   BACKGROUND

Adaptive Metabolism: As early as the 1950s, it was observed that exposure to certain polycyclic aromatic hydrocarbons (PAHs) influenced the metabolism of structurally related xenobiotic chemicals.   For example, administration of 3-methylcholanthrene (3MC) to rats led to a marked increase in hepatic microsomal 3-methyl-4-monomethylazobenzene demethylase and 4-dimethylaminoazobenzene reductase activities (1). The observation that the inducing PAHs were commonly substrates of the upregulated enzymes gave rise to the idea that this was an adaptive metabolic response.   This early work also gave rise to important nomenclature that still colors this field. Most important in this regard is the term <u>a</u>ryl <u>h</u>ydrocarbon <u>h</u>ydroxylase or AHH (2). In the early literature, this term referred to the enzymatic activities of a number of cytochrome P450-dependent monooxygenases that convert benzo[a]pyrene to hydroxylated metabolites (reviewed in (3, 4). Collectively, data from this early period demonstrated the existence of a mechanism for the adaptive metabolism of PAHs.   This pathway appeared very similar to the adaptive metabolic pathways that had been previously described in simpler organisms and thus is commonly referred to as induction (5).

Dioxin: Other xenobiotics were also shown to induce AHH activity. Perhaps the most extensively studied inducers are halogenated dioxin congeners such as, 2,3,7,8-tetrachlorodibenzo-p-dioxin ("dioxin").   The early interest in dioxin was related to its dramatic toxicity and its widespread distribution in the environment (6). Because of its remarkable potency and biological stability, dioxin has proven to be invaluable in elucidating the mechanism AHH induction. For the purposes of this chapter, we will limit the discussion of dioxin primarily to its utility as a tool to study the aryl hydrocarbon receptor.

Ah Locus: Using both PAHs and dioxins as model compounds, it was discovered that the inductive response was strain dependent (7). This observation defined a genetic component in the regulation of AHH activity and influenced the research of the 1970s (8-12). The locus responsible for the inductive phenotype segregated as a single gene in a simple autosomal dominant fashion and was named *Ah*, for *aryl hydrocarbon responsiveness* (8, 9, 12, 13). The *Ah* allele in PAH-responsive strains was designated Ah^b, derived from the C57BL strain, whereas the allele in the less-responsive DBA strain was designated Ah^d (13).

The Ah locus encodes the Ah receptor: Classical pharmacological approaches demonstrated that the *Ah* locus encoded a receptor protein for dioxins and PAHs. The observation that dioxin was about 3 orders of magnitude more potent as an inducer of AHH, as compared to PAHs, led to the use of ^3H and ^{14}C labeled dioxins as radioligands for the putative receptor (14). Using such radioligands, saturable and high-affinity binding sites were identified from hepatic cytosol, indicating the presence of a dioxin-binding receptor (15). Competitive binding studies showed that the relative binding affinity of a congener for these sites was directly correlated with its potency to induce AHH activity and its ability to elicit various toxic endpoints (16-18). These studies definitively linked the biological responses to dioxins and PAHs to a soluble binding site that has now become widely known as the Ah receptor or AHR.

A genetic analysis of the *Ah* locus provided additional evidence that the AHR was a signal transduction molecule at the center of dioxin toxicity and the induction of AHH. When these binding sites were examined across mouse strains, it was observed that high-affinity sites could be readily identified in Ah^b strains, whereas the less responsive Ah^d strains displayed binding sites with a much lower affinity (15, 19). The strain distribution and the co-segregation of the high and low affinity binding sites with the Ah^b and Ah^d responsiveness, provided compelling experimental evidence that the *Ah* locus encoded a receptor protein for dioxins and that this receptor mediated the induction of AHH (15).

Identification of the AHR protein: The AHR protein initially proved difficult to purify due to its low concentrations in cell extracts, marked heterogeneity upon chromatography, and high instability upon purification in nondenaturing conditions (20-22). Purification of the AHR only became possible after the synthesis of a photoaffinity ligand 2-azido-3-[^{125}I]iodo-7,8-dibromodibenzo-*p*-dioxin, which allowed covalent labeling of the AHR (20). Use of this ligand greatly simplified the quantitation of the receptor, and permitted the identification of the AHR through denaturing purification (23). Once the receptor was purified to homogeneity, picomole quantities were obtained and the N-terminal 27 amino acid sequence was determined. Using western blot analysis, antibodies raised against the N-terminal amino acid sequence were shown to cross react with AHR orthologs from a variety of animal species and could immunoprecipitate the ligand-bound AHR from cytosol (24). These studies led us to the understanding that the AHR protein is found in a number of cell types, as well as in a number of different species. Moreover, they provided an early biochemical look at the protein and indicated that the denatured AHR protein from common laboratory mice had an apparent molecular weight of between 95 – 104 kDa. Receptors found across species could differ in their apparent molecular weight from approximately 104-106 kDa in rats, up to 130kDa in deer mice (*Peromyscus maniculatus*) and even 143 kDa in rainbow trout (*Salmo gardineri*) (24, 25). This clearly illustrated the marked heterogeneity in molecular weight of the AHR protein across species, yet the common antigenicity of the *N*-terminal region. In mice, the disparity of molecular weight was determined to be due to a hypervariable *C*-terminal region (illustrated in Figure 1, discussed later in this chapter).

2. CHARACTERIZATION OF THE AHR AND ARNT

Signaling through the AHR pathway has served as an important model system of the PAS superfamily. This research foreshadowed the general importance of PAS protein heterodimerization in biology and allowed the early definition of those sequences that play functional roles in this interesting domain. In this regard, the Ah receptor nuclear translocator, or "ARNT," was the first vertebrate PAS protein to be recognized, and the AHR:ARNT complex was the first PAS protein heterodimer shown to be biologically relevant. In this section, we will review the important features of these two interesting PAS proteins and provide the evidence that helped define important functional domains.

Cloning of the Ah Receptor: The cloning of the AHR cDNA allowed a description of how sequence motifs relate to receptor function. Cloning of the AHR cDNA was accomplished using degenerate oligonucleotide probes that had been deduced from the protein's N-terminal amino acid sequence. Use of these probes led to the identification of a cDNA from the C57BL/6J mouse that harbored an open reading frame of 805 amino acids with a predicted molecular weight of 95 kDa (26, 27). The identity of the cloned AHR cDNA was established by (i) the agreement of its deduced molecular weight with that observed for the purified protein, (ii) a concordance between this mRNA's expression and functional receptor expression in a panel of hepatoma cell lines with mutations in AHR signaling and (iii) functional expression of a receptor protein from the cDNA (26-28). Sequence analysis of the AHR cDNA revealed a N-terminal bHLH domain, a C-terminal glutamine-rich region, and a central region with significant homology to the newly recognized PAS domain (Figure 1) (26, 29).

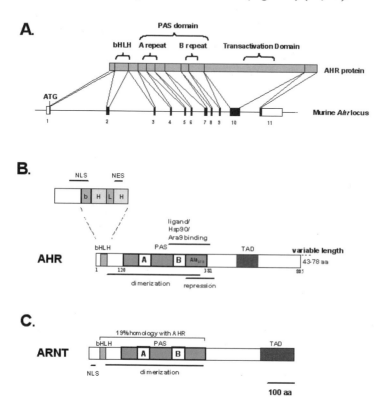

Figure 1. Domain maps of murine AHR and ARNT. A. The murine Ah locus and the AHR protein. The 11 exons are indicated by the solid boxes that correspond to the coding region of the AHR, and the 5' and 3' untranslated regions are marked by the open boxes. B. Domain map of the murine AHR protein, including identification of the basic-helix-loop-helix domain (bHLH), PAS domain, and transcriptional activation domain (TAD). In addition, the regions that have been shown to mediate nuclear translocation (NLS), nuclear export (NES), dimerization, ligand/Hsp90/Ara9 binding and repression of AHR activity are labeled. Amino acid residues that mark domain boundaries are numbered according to their positions within the murine AHR protein. Ala375 is marked for its importance for high-affinity ligand binding. The C-terminal end "variable length" represents the length of different Ah alleles in various mouse strains. C. Domain map of the murine ARNT protein. The bHLH, PAS, TAD domains, the regions that mediate the dimerization, nuclear localization (NLS), and its percentage similarity with the AHR are marked.

ARNT: The cloning of the Ah receptor nuclear translocator, or "ARNT," was accomplished about one year before the cloning of the AHR (30). The ARNT cDNA was cloned from a genetic screen designed to identify gene products that were able to rescue mutants derived from the Hepa1c1c7 cell line that displayed defects in AHR signaling (31, 32). One particular mutant line displayed a phenotype where the AHR was expressed at normal levels, but remained in the cytosolic fraction of cell homogenates after exposure to AHR agonists. To clone the gene responsible for this phenotype, a rescue approach was employed using transformation with human genomic DNA (30). Once a rescue was achieved by this method, human specific *Alu* probes and chromosome walking yielded a genomic fragment that rescued the mutant phenotype. These genomic fragments were used as a probe in cDNA library screening and led to the identification of the corresponding cDNA (30). Introduction of this cDNA into the mutant cell line restored the AHR signaling in response to agonists and led to the agonist-dependent presence of the AHR in the nuclear fraction of isolated cells. Thus, the protein was named the Ah receptor nuclear translocator, or ARNT (30). Sequence analysis of the cDNA indicated that it encoded an open reading frame of 87 kDa protein and that the protein harbored a bHLH domain at the amino terminus, and an internal PAS homology domain. Importantly, the homology to homologous regions in the *Drosophila* SIM and PER proteins led to the common nomenclature now used to refer to this motif, "PER, ARNT, SIM homology domains" or simply "PAS" domains (Figure 1).

The AHR:ARNT heterodimer: The observation that the AHR and ARNT were members of the emerging PAS superfamily was a surprise to a research community that had begun to think of the AHR with respect to its similarities to the steroid-thyroxine receptor superfamily (33). The fact that both the AHR and ARNT contained bHLH domains and that their structural symmetry extended into the PAS domains led to the idea that these two proteins were heterodimeric partners. This concept was initially demonstrated through *in vitro* expression studies, which showed that AHR:ARNT coexpression generated a complex that bound, with high affinity, to specific DNA sequences known as dioxin response elements, or "DREs" (34, 35). The follow-up observation that DRE binding complexes isolated from cell fractions contained both the AHR and ARNT confirmed the biological relevance of this complex (36). From parallel localization studies, a signaling model took shape, where in the absence of ligand, the AHR resides in the cytosol and ARNT resides in the nucleus (37). Upon ligand binding, the AHR translocates to the nucleus, dimerizes with ARNT, and forms a competent DRE binding transcription factor that regulates the expression of target genes such as those involved in AHH activity (38).

The PAS domain: The PAS domain has now emerged as the signature of a superfamily of proteins with roles in environmental adaptation and development (39). The PAS domain typically encompasses 250-300 amino acids and contains a pair of highly degenerate 50 amino acid motifs, often termed the A and B repeats or PAS repeats (30, 40, 41). The PAS domain of the mouse AHR spans amino acid residues 120-381. In the ARNT protein, this domain spans residues 173-458 (30). The analysis of this region in the AHR and ARNT has provided evidence that the domain harbors a number of discrete functional subdomains and that PAS domains found in different proteins often display distinct biological properties.

The PAS domains of the AHR and ARNT harbor subdomains that correspond to a number of important biological properties. A large body of indirect evidence suggests that these PAS subdomains play a role in mediating AHR:ARNT dimerization. Early on it was observed that regions within the PAS domains of both AHR and ARNT were essential for heterodimerization (28, 42, 43). In addition, the PAS domain has been shown to confer pairing specificity between different PAS proteins (43-47). There is also emerging evidence to indicate that the PAS domain plays a direct role in DNA recognition. Recently, a point mutation within the PAS domain of the AHR ($Cys^{216} \rightarrow Trp^{216}$) was found to abolish the DNA-binding capacity of the AHR:ARNT complex, but did not appear to impair the AHR's ligand binding ability or dimerization with ARNT. This observation suggests that the PAS domain might be involved in sequestering the DNA binding domains of the AHR or ARNT (48).

The PAS domain of the AHR also functions as the surface for both ligand binding and interaction with chaperones (28, 49, 50). In contrast, the PAS domain of ARNT does not bind any known ligands and has not yet been shown to interact with cellular chaperones in a stable manner. The minimal ligand-binding region of the AHR was determined to be located between residues 232 and 402 by microsequencing the photoaffinity ligand labeled region and also by deletion analysis of the cDNA in expression studies (26, 28, 51, 52). The ligand binding domain of the AHR was shown to overlap with a domain that formed stable interactions with the cellular chaperone Hsp90 (52). The AHR:Hsp90 interaction requires domains within and beyond the PAS domains. Interaction sites have been mapped to a region near the PAS-B repeat, as well as sequences within the bHLH domain (53, 54). In a manner similar to what has been observed for steroid receptors, the ligand binding domain and Hsp90 interaction domain overlap (Figure 2).

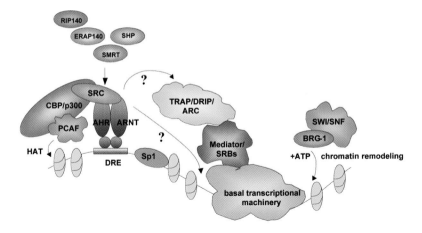

Figure 2. Model of AHR transcriptional activation. (Figure adapted from Lee and Kraus, 2001) The AHR:ARNT heterodimer binds to the dioxin responsive element (DRE) and recruits the SRC/CBP/PCAF complex. This results in chromatin acetylation to generate a more open chromatin conformation through its intrinsic histone acetyltransferase (HAT) activity. The AHR:ARNT also recruits the ATP-dependant SWI/SNF chromatin remodeling complex that "twists and writhes" the nucleosomal DNA and exposes the proximal promoter region to facilitate the binding of the basal transcriptional machinery. The AHR:ARNT appears to recruit the basal transcriptional apparatus through direct interaction or via TRAP/DRIP/ARC/SRBs complex. Transcriptional is potentially regulated through interaction with coactivators (RIP140, ERAP140, and Sp1) and corepressors (SMRT and SHP). See also Table 2.

The Hsp90/ligand binding domain was originally identified as a domain that repressed AHR signaling (28). This designation was made through the observation that the removal of this region increased the ligand-independent dimerization and DNA-binding capacity of the AHR:ARNT complex in gel mobility shift assays (28, 55, 56). Cell culture experiments support this analysis as the deletion of the amino acids between residues 340 and 421 result in a receptor that is constitutively active and ligand independent (55, 57, 58). The fact that this repressor domain overlaps with the Hsp90 binding region conforms with a model where Hsp90 represses the capacity of the AHR to dimerize with ARNT.

bHLH domain: Adjacent to the PAS domain of the AHR and ARNT is a basic-helix-loop-helix (bHLH) domain. The basic region corresponds to amino acid residues 27-39 in the mouse AHR and 86-102 in mouse ARNT. Assuming functional similarity to other bHLH domains, we can predict that the basic region mediates DNA binding while the HLH region supports dimerization (59, 60). The predicted roles of these subdomains are supported by the observation that mutation of the bHLH domain in either protein results in the abolition of the DNA binding and transactivation capacity of the AHR:ARNT complex (28, 42, 61). More recently, site-directed mutagenesis determined that the DNA recognition residues of ARNT closely resembled the consensus of other bHLH proteins like Myc and Max, whereas the AHR's DNA recognition sequence was unique to this family (62-65). These experiments have led to the conclusion that ARNT binds to the GTG half-site of the DRE, whereas the AHR binds to a TNGC half-site that deviates from the canonical E-box sequence bound by most bHLH dimers (65, 66). Deletion studies also demonstrated that the bHLH regions of the AHR and ARNT play an important role in dimerization (29, 42, 61). Dimerization of the AHR and ARNT likely facilitates interactions between respective basic regions and allows the production of a composite DNA-binding domain, as is common in the case of other bHLH protein dimers (67, 68).

Nuclear Localization Domains: Recent evidence has demonstrated that the bHLH domain of the AHR also harbors a nuclear localization sequence (NLS) and a nuclear export sequence (NES) (Ikuta et al., 1998, Pollenz and Barbour, 2000). Using microinjection of a recombinant GST-AHR-GFP fusion protein, a minimal bipartite NLS was identified at amino acid residues 13-39 of the human AHR. A NES was also found at residues 55-75, indicating that the AHR acts as a nucleocytoplasmic shuttling protein (69). In contrast, the human ARNT harbors only an NLS at amino acid residues 39-61. The fact that no NES has been found in ARNT is consistent with the observation that ARNT is almost exclusively found in the nuclear compartment of cells (37, 70).

The hypervariable C-terminal half: The C-terminal half of the AHR is poorly conserved across species. The transcriptional activation domains (TADs) of both the AHR and ARNT are located in this C-terminal region (42, 53, 55, 57, 71). The fact that both the AHR and ARNT contain a TAD suggests that they cooperate in the transactivation of target genes or that each has independent activity at distinct promoters.

Some investigators have divided the AHR's TAD into three distinct subdomains, including a glutamine/aspartic acid-rich region (amino acid residues 490-593), a glutamine/proline-rich region (residues 590-718), and a glutamine/proline/serine-rich region (residues 719-805) (57). Each subdomain was capable of activating transcription by itself, but when put together they acted synergistically to strongly activate transcription. This suggests the capacity of the AHR to interact with a variety of transcription coactivators and activate transcription from a variety of promoters (57, 72). However, the glutamine-rich domain appears to be critical for hAhR-mediated transactivation potential (73). It has also been observed that the strength of the AHR's TAD varies depending on the cell type. This observation has been used to support the idea that these domains interact with cell-specific coactivators (53). In contrast to AHR, the TAD in ARNT has been divided into two subdomains (amino acid residues 757-767 and 777-789) which share some sequence similarity to each other (71). It was shown that the acidic and bulky hydrophobic residues within ARNT's TAD, but not the glutamine-rich region, were crucial for ARNT's transcriptional activation activity.

3.   COMPONENTS OF THE AHR PATHWAY

Great progress has been made in our understanding of the AHR:ARNT signaling pathway. In addition to the core PAS components, a number of additional chaperones and coactivators are now known to play important roles (Figure 2). The chaperones appear to play an important role in maintaining properly folded receptor and preventing its spontaneous dimerization with ARNT. In addition to chaperones, a host of transcriptional coactivators (Table 2) appear to influence the ability of the AHR:ARNT heterodimer to activate transcription of target genes (Figure 3). Considering that the regulation of AHR signaling primarily resides with the associated chaperone and coactivator proteins, this next section will focus on each protein's role in AHR signaling.

Hsp90: The AHR forms a stable complex with Hsp90 in the cytosol (49, 50, 74). The Hsp90 protein is a well characterized chaperone that participates in the signal transduction by a variety of nuclear receptors, growth factor receptors and protein kinases (75). Evidence that the AHR:Hsp90 association is biologically relevant is based upon a large body of experimental evidence. *In vivo* data from yeast models have shown that the ligand-dependent activation of the AHR is completely blocked in those strains that are Hsp90-deficient (56, 76). *In vitro* data has suggested a dual-role for Hsp90 in the AHR pathway. First, Hsp90 association is correlated with high affinity ligand binding by the AHR (74). This suggests that Hsp90 assists AHR's folding into a conformation that is capable of ligand binding. Second, Hsp90 dissociation is correlated with increased DNA-binding activity by the AHR:ARNT complex (77). This observation supports the idea that Hsp90 prevents unliganded AHR from dimerizing with ARNT and thus, forming a DNA-binding complex in the absence of ligand. Another piece of evidence to support such a role for Hsp90 is the observation that deletion of the Hsp90 binding domain from the AHR yields a receptor which constitutively dimerizes with ARNT and binds to DNA (52, 54, 88). Third, functional Hsp90 is critical to the stability of the AHR in the cell (89).

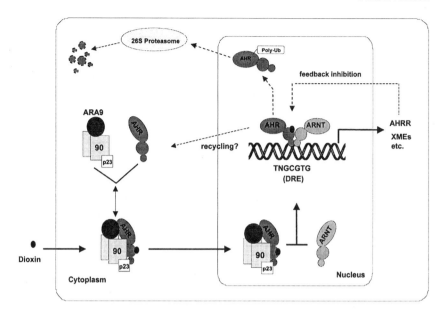

Figure 3. Model of the AHR signaling pathway. The unliganded AHR is sequestered in the cytoplasm as part of a multimeric protein complex composed of Hsp90, a cochaperone p23, and an immunophilin-like protein, known as ARA9. Upon ligand binding, the AHR translocates into nucleus, and exchanges associated chaperones for its nuclear partner ARNT. This AHR:ARNT heterodimer recognizes and binds to specific DRE sequences, leading to the transcriptional activation of genes such as XMEs and AHRR. The AHR signal transduction pathway is attenuated by AHRR and ubiquitin (Poly-Ub) and proteasome pathways.

ARA9: A second AHR-associated chaperone has been identified by a number of laboratories and thus goes by three names, AHR-associated protein 9 (ARA9), AHR-interacting protein 1 (AIP1), or Hepatitis B virus X-associated protein 2 (XAP2) (90-92). For the remainder of this review, we will refer to this protein as ARA9. The association of ARA9 with the AHR has been demonstrated in yeast two-hybrid assays designed to detect proteins that formed stable complexes with the AHR. The ARA9:AHR association was also demonstrated by isolation of the complex from the cytosolic fractions of cells (92-94). More recent biochemical analysis suggests that ARA9 directly interacts with Hsp90, with domain mapping studies suggesting a ternary AHR:Hsp90:ARA9 complex structure where any single subunit is bound to the other two subunits in a tetrameric complex (95-98).

Table 1. Transcriptional Cofactors.

Complexes	Subunits	Proposed Mechanisms	References
HAT	SRC/NcoA/p160* (SRC-1/NcoA-1, NcoA-2/TIF-2/ GRIP-1, pCIP/ACTR/AIB1/RAC3) p300/CBP	AHR·ARNT recruits HAT coactivators to DRE region and facilitate acetylation of chromatin and thus weaken the contacts between DNA and histone.	(78-81)
SWI/SNF	ATPase subunit BRG-1	ATP-dependant chromatin remodeling activity is involved in AHR transactivation to expose the proximal promoter region to components of the basal transcriptional apparatus.	(82, 83)
Basal transcriptional machinery	TBP, TFIIF, TFIIB	AHR·ARNT may recruit basal transcriptional apparatus by direct interacting with certain subunits in the basal machinery (*in vivo* evidence is yet to be shown).	(72, 84)
Other cofactors	Coactivators: - RIP140 - ERAP140 - Sp1 Corepressors: - SHP - SMRT	The fact that many of these cofactors also participate in the transactivation mediated by other nuclear receptors such as estrogen receptor indicates that AHR may compete with other nuclear receptors for a common coactivator/corepressor pool.	(78, 85-87)

A number of lines of evidence demonstrate the *in vivo* relevance of the ARA9:AHR interaction. For example, it has been shown that ARA9 is able to enhance AHR signaling in both mammalian cells and yeast model systems (91, 92, 95). Subsequent experiments demonstrated that the mechanism of this enhancement is related to ARA9's ability to increase the amount of properly folded AHR in the cytoplasm (99-102). Although the mechanism underlying cytosolic stabilization of the AHR is unknown, it has been shown that ARA9 increases AHR protein levels by protecting the ligand-free AHR against ubiquitination (100, 102). ARA9 presence in the AhR complex also appears to inhibit importin beta binding to the AhR and thus block nucleocytoplasmic shuttling (103).

Structural analysis demonstrated that ARA9 shares significant homology to the GR-associated immunophilin FK506 binding protein 52 kDa or simply FKBP52. The ARA9 protein has two FK506-binding domains (FKBP domain) in its amino terminus and three tetratricopeptide repeats (TPRs) in the carboxyl terminus (104). The TPRs together with other C-terminal domains of ARA9 are required for AHR contacts with Hsp90 (97, 105, 106). The FKBP domain within the amino terminus of ARA9 is somewhat unusual in that it does not appear to bind molecules like FK506 and has not been shown to harbor peptidyl-prolyl-isomerase activity that is commonly associated with such domains. Instead, this domain was shown to be essential in the stabilization of the AHR:Hsp90:ARA9 ternary complex, and important in the regulation of the intracellular localization of the AHR (95, 105).

p23: The Hsp90 cochaperone known as p23, has also been found to be associated with AHR in coimmunoprecipitation assays. It has been suggested that p23 plays a role in regulating ligand responsiveness and nuclear translocation via stabilizing the AHR-Hsp90 interaction (107, 108). The involvement of p23 in AHR signaling *in vivo* is suggested by the observation that, in a yeast model system, the deletion of the yeast p23 homolog, SBA1, decreases AHR signaling by 40% (109). This incomplete down-regulation indicates the possible involvement of other Hsp90-associated cochaperones that are functionally redundant with p23. Collectively, a chaperone system consisting of at least Hsp90, ARA9, and p23 is associated with cytosolic AHR and plays an essential role in mediating its folding, ligand binding, and subsequent nuclear translocation. Other participants in this chaperone complex are likely to be revealed in the years to come.

4.   DIOXIN RESPONSIVE ELEMENTS

Dioxin-responsive elements (DREs) were first identified through the examination of the upstream sequences of the AHR-responsive gene, cytochrome P450-1A1, or "CYP1A1." The initial analyses of the CYP1A1 promoter revealed multiple regulatory elements upstream of the transcriptional start site (110, 111). Deletion analysis of the CYP1A1 promoter identified cis-acting regulatory regions and localized the most powerful dioxin-responsive elements between -1200 to -800 bp upstream (110, 112). These elements were shown to act as classical enhancers as they were able to activate transcription in a manner that was independent of their orientation and distance from the promoter (110, 112-114). These responsive elements were independently named "dioxin-responsive elements" or DREs, "xenobiotic-responsive elements", or XREs, and AH-responsive elements" or AHREs (115-118). These terms all represent the same AHR:ARNT binding site, so for this review, we will use the term "DRE."

The exact nucleotide sequence of the DRE was identified using several independent approaches (119-128). As a result of these experiments, a non-palindromic core consensus sequence (5'-TNGCGTGA/C-3') was identified that conferred the minimal capacity for DNA binding and transcriptional enhancement of the ligand-bound AHR (65). Additionally, the immediate flanking nucleotides of this core consensus sequence affected AHR activation of the DRE primarily through alterations in binding affinity (65, 125, 126). Thus, beyond the specific binding of the AHR:ARNT complex to DNA, the flanking DNA appears to provide an additional level of specificity for accessory proteins.

In addition to CYP1A1, functional DREs were identified in the upstream regulatory regions of other AHR responsive genes that collectively have been designated the "Ah battery" (reviewed in (129). Although this list of genes is continually expanding, the Ah battery primarily consists of enzymes thought to be involved in hepatic metabolism and clearance of xenobiotics. These gene products include the CYP1A1, CYP1A2 and CYP1B1 monooxygenases, a quinone reductase, NQO1, an aldehyde dehydrogenase, ALDH3c, and the phase II conjugating enzymes, UGT1A6 and GST-Ya (129-132). The conservation of DRE sequences upstream of these promoters is a significant component underlying the idea that the AHR functions as a xenobiotic sensor and regulator of xenobiotic metabolism.

5. AHR CROSS-TALK WITH OTHER SIGNALING PATHWAYS

Cross-talk between AHR signaling and other cellular pathways may occur as the result of competition for common binding partners or DNA-binding sites. The potential for cross-talk between the AHR signaling pathway and hypoxia-inducible factor, "HIF-1α," pathway has been extensively studied. Such experiments became obvious once it was learned that both the AHR and HIF-1α share ARNT as a heteromeric dimerization partner (34, 38, 133). While the AHR:ARNT dimer drives the expression of DRE-regulated genes, the HIF-1α:ARNT complex binds to hypoxia responsive elements (HREs) and activates the transcription of a battery of hypoxia responsive genes (*e.g. Epo* and *Vegf*) (133). Thus, a number of laboratories tested a model where AHR and HIF-1α compete for ARNT dimerization. The first experimental support for this idea came from the observation that the activation of the HIF-1α pathway inhibited the AHR-dependant CYP1A1 up-regulation in both human and mouse cell lines (134-136). Interestingly, in that system AHR activation did not interfere with HIF-1α signaling. It was proposed that this uni-directional interference was due to a higher affinity of HIF-1α for ARNT than the AHR, thus out-competing AHR for ARNT binding. In our own lab, we observed that the *Epo* gene has five weak DREs within its promoter region (135). We proposed that AHR activation had two effects on the *Epo* promoter. First, it activates the promoter through the proximal DRE, and second, it competes with ARNT for the HRE site. Others have provided evidence that ARNT may not be limiting in a number of cell types (137). Such an observation suggests that cross-talk between the AHR and HIF-1α pathways may be much more complicated than simple competition for ARNT. Although cross-talk between these two PAS pathways does indeed occur *in vitro*, it appears to be highly dependent upon cell-type and experimental conditions, and as such it is difficult to assign a biological significance at this time.

Estrogen Receptor: The AHR has also been shown to display cross-talk with the estrogen receptor (ER) pathway (reviewed in (138). Activation of the AHR by dioxin has been shown to repress multiple estrogen-induced responses (138). The underlying mechanisms for this cross-talk may be numerous. The most likely mechanism proposed to date involves the binding of the AHR:ARNT heterodimer to imperfect DREs, also termed "inhibitory DREs" or simply "iDREs," found upstream of the promoters of certain ER responsive genes such as c-fos and pS2 (139, 140). Inhibitory DREs are "DRE-like" sequences that lie adjacent to or overlap estrogen responsive elements (EREs). The AHR:ARNT heterodimer has been reported to be capable of binding to iDREs found upstream of many estrogen responsive genes. It has been proposed that the binding of the AHR:ARNT complex to iDREs might inhibit the interaction of ER or its coactivators with their adjacent binding sites.

The AHR has also been reported to be capable of directly interacting with nuclear factors, including NF-κB, Rb, pp60^{src}, and even ERα, although the physiological significance of these interactions remains unclear. Klinge, et al., reported that AHR may directly interact with ERα in a ligand-specific manner *in vitro* (141). The physical interaction between the AHR and NF-κB leads to the mutual repression of both pathways (142). The association of dioxin-activated AHR with Rb has been shown to induce cell cycle arrest and enhance AHR-dependant CYP1A1 expression (143, 144). Together with AHR-associated pp60^{src} kinase activity, all these DRE-independent aspects of the AHR function illustrate the complexity of the AHR signaling pathway. Yet, it is important to note that, like the data for interaction with the HIF-1α pathway, it has yet to be conclusively demonstrated that any of the proposed cross-talk pathways have significance *in vivo*. Such data may emerge in the years to come.

6. REGULATION OF THE AHR PATHWAY

For many agonists, signaling through the AHR is quickly attenuated. The mechanism by which AHR signaling is downregulated appears to occur by at least two mechanisms. One mechanism involves a bHLH-PAS protein known as the AHR repressor, or "AHRR," although the significance of this mechanism may be limited by tissue-specific expression (145, 146). The AHRR protein is structurally very similar to the AHR and contains both a bHLH and PAS-A domain. Yet, it does not possess a domain corresponding to the PAS-B domain. In the simplest terms, the AHRR can be thought of as a "PAS-B deletion" of the AHR (145). In keeping with this analogy, the AHRR is not repressed by Hsp90 and does not require an agonist to dimerize with ARNT. The basic region of AHRR also has high sequence homology to that found in the AHR's, as a result, the AHRR:ARNT heterodimer has a high affinity for DREs (145). An interesting difference between the AHR and AHRR, is that the carboxy-terminal region of the AHRR has activity as a transcriptional repressor (145). As a result, AHRR:ARNT heterodimers repress transcription of the *Ah* battery of genes and thus can be thought of as antagonists of the AHR:ARNT complex. One additional important observation about AHRR is that it is a DRE-driven gene and is rapidly upregulated through AHR's activation by agonist. Thus, a model emerges where ligand-activated AHR upregulates the transcription of AHRR, and AHRR subsequently inhibits AHR signaling by competing for ARNT dimerization, competition for DREs, and repression of *Ah* gene battery transcription. Thus, the AHRR appears to lie at the center of an important negative feedback loop for the AHR signal transduction pathway (145).

The ubiquitination/proteasome pathway was identified as a second mechanism for attenuating AHR signaling. Upon entering the nucleus, the ligand-activated AHR is rapidly exported and degraded in a number of cell types (147, 148). In contrast, ARNT protein levels do not seem to be down-regulated in response to ligand. The evidence that AHR degradation involves the ubiquitin/proteasome pathway comes from the observation that its degradation can be blocked by the proteasome inhibitor MG132 or a ubiquitin mutant UbK48R (147, 148).

The significance of AHR attenuation by these two mechanisms is as yet unclear. From a teleological perspective, multiple mechanisms for receptor attenuation would suggest that deleterious effects might arise from prolonged activation of this pathway. This becomes an interesting concept when one considers the known toxic effects of halogenated dioxins. These compounds are high affinity agonists that are poorly metabolized. Taken in sum, these observations support the idea that attenuation of the AHR response plays an important role in preventing the toxic consequences that arise from prolonged or hyper-activation of certain target genes.

7.   AHR POLYMORPHISMS

Identification of Ah alleles: The initial description of the *Ah* locus used the designations, Ah^b and Ah^d, to describe the alleles that code for receptors with high and low affinity for ligands. In later experiments using the photoaffinity ligand (^{125}I)-2-azido-3-iodo-7,8-dibromodibenzo-*p*-dioxin, it was found that the high affinity Ah^b allele actually represented three distinct alleles encoding receptor forms with molecular weights of 95 kDa, 104 kDa, and 105 kDa (20, 149, 150). These alleles were designated as Ah^{b-1}, Ah^{b-2}, and Ah^{b-3}, respectively. The Ah^{b-1} and Ah^{b-2} are found in common laboratory strains (e.g. *Mus musculus* and *domesticus*), whereas the Ah^{b-3} allele is found in a number of strains established from feral mice (Table 2). The Ah^d allele encodes a 104 kDa receptor that has a 10-fold lower affinity for agonists compared to the Ah^b alleles. Because of this low affinity, the Ah^d mice are much less susceptible to the toxic effects of dioxin or to the inductive effects of all agonists (19, 151).

Table 2. Murine Ahr alleles.

Ah^d	Ah^{b-1}	Ah^{b-2}	Ah^{b-3}
Low affinity MW = 104 kDa 848 aa.	High affinity MW = 95 kDa 805 aa.	High affinity MW = 104 kDa 848 aa.	High affinity MW = 105 kDa 883 aa.
Strains[*]: - DBA - AKR - 129 - LP - SJL - ST	Strains: - C58 - C57BL - C57BR	Strains: - A - BALB - C3H - CBA - SEA - CE	Strains: - *Mus spretus* - *Mus parahi* - *Mus hortulanus* - *Mus cookii* - *Mus caroli* - *Mus molossinus*

The murine polymorphisms have been characterized by cloning and sequencing the cDNAs of all four known murine *Ah* alleles (51, 152, 153). In keeping with the molecular masses estimated from photoaffinity labeling studies, the Ah^{b-1}, Ah^{b-2}, Ah^{b-3}, and Ah^d cDNAs were found to encode proteins of 805, 848, 883, and 848 amino acids, respectively (51). Among the four alleles, eight single amino acid polymorphisms were identified in the initial common 805 residues. These alleles also differed by various additional sequences at the carboxyl end resulting from different use of stop codons. In fact, the variable length of the C-terminal end largely explains the high degree of observed molecular weight polymorphism across *Ah* alleles (51).

A polymorphism in the ligand-binding domain of *Ah^d* mice explains their decreased responsiveness to PAHs. A point mutation (Ala^{375}→Val^{375}) within the PAS domain of the *Ah^d* was identified that resulted in low affinity binding of ligands (51, 152, 153). This observation was consistent with the localization of this residue within the region mapped as the ligand binding domain (26, 28, 51). The additional 43 amino acids at the carboxyl terminus of the *Ah^d* receptor also appeared to be an influencing factor in the decreased sensitivity to ligands (152). Yet, this relationship is not fully understood at the present time.

Other genomic polymorphisms: Most recently, a high-throughput sequence comparison was performed using 13 AHR mouse lines, comprised of 8 laboratory strains, 2 *Mus musculus* subspecies and 3 additional *Mus* species (154). As expected, the exonic regions were shown to be less polymorphic as compared to the intronic regions. A total of 111 non-synonymous nucleotide polymorphisms were identified in the AHR protein-coding region which expanded the known amino acid changes from the previously identified 10 to 42. Notably, no amino acid changes were identified in the bHLH domains of all the thirteen mouse lines studied. The PAS domains were also found highly conserved, whereas the C-terminal ends were hyperpolymorphic. Interestingly, the analysis of "evolutionary pressure" (non-synonymous:synonymous substitutions) revealed that the function of the AHR as a whole is highly constrained. However, ligand binding is not constrained to the high-affinity alleles. It is worth noting that the survival of the low affinity *Ah^d* allele under evolutionary pressure may be biologically meaningful. The fact that the *Ah^d* mice do not display obvious evolutionary and physiological handicaps is compelling evidence that acting as a xenobiotic sensor may be secondary to a ligand-independent innate function for AHR.

8. AHR EVOLUTION/ORTHOLOGS

Phylogenetic analyses demonstrated that the AHR is an ancient protein that existed early in vertebrate evolution (450-510 million years ago) (155). Today, the AHR gene can be found in living representatives of early vertebrates, including bony, cartilaginous, and jawless fish (155). The AHR has been found in all living vertebrate groups: including mammals, amphibians, reptiles, and birds (155-157). These species are evolutionarily diverse and live in different chemical environments. The conservation of the

AHR in these species provides an additional argument to suggest an endogenous role for the AHR beyond the singular metabolic response to environmental chemicals.

Vertebrate AHRs: The majority of vertebrate AHR analysis has centered on the mouse, rat, and human species. Sequence comparison among these species identified differential conservation patterns for each individual domain within the AHR protein. The bHLH and PAS domains of the AHR were found highly conserved, whereas the C-terminal region exhibited hypervariability. Rat AHR shares 100% amino acid identity with mouse AHR in the bHLH region and 96% in the PAS domain, but only 79% identity at the C-terminus (158). The cloned human AHR cDNA encodes a protein of 848 amino acids that has 73% overall sequence similarity with mouse AHR (29, 152). The carboxyl termini of human and mouse AHRs exhibit hypervariability (< 60% identity), whereas the amino terminal halves of these two are highly conserved, with 100% identity in the basic region, 98% in the HLH domain, and 87% in the PAS domain (29). The human AHR contains Val at amino acid residue 381 equivalent to Val^{375} in the mouse Ah^d allele. This residue contributes to the low ligand-binding affinity of human AHR (152). Thus, the low-affinity Ah^d mouse strains may serve as the most appropriate model when attempting to understand the importance of this signaling pathway on the human condition.

Invertebrate AHRs: Homologs of the AHR have also been identified in invertebrates such as *C. elegans* and *Drosophila*. Given the fact that these invertebrate homologs have not yet been shown to bind any AHR ligands, it is not yet clear if these homologs are *bona fide* orthologs that reflect a common ancestry at this locus, or if they are simply examples of divergent evolution that merely exhibit sequence homology to vertebrate AHR (159, 160). In this regard, the *C. elegans* AHR (CEC41G7.5, or "CeAHR-1") possess both bHLH and PAS domains. While the bHLH and PAS-A domains are fairly well conserved with respect to vertebrate AHRs, the PAS-B domain of CeAHR-1 is poorly conserved and does not appear to bear the mammalian AHR-like ligand-binding property (155, 159). However, CeAHR-1 was found to form a heterodimer with a *C. elegans* ARNT homolog known as AHA-1 and bind to mammalian DREs (159). In *Drosophila*, the putative AHR and ARNT homologs Ss and Tgo, respectively, have been shown to form a heterodimer and exhibit a mammalian AHR:ARNT-like DNA-binding specificity. Both Ss and Tgo play a central role in defining the distal regions of both the antenna and leg of *Drosophila* (161, 162). Although the relationship of the putative invertebrate orthologs are not yet clear, it is possible they reflect multiple functions of AHR that extend beyond its role as a xenobiotic sensor.

9. PHYSIOLOGICAL FUNCTION OF AHR

Although it is most common to study the AHR with respect to its role in the adaptive metabolic responses to PAHs or the toxicity of dioxins, the physiological significance of this signaling pathway remains somewhat enigmatic. The focus on the relationship between the AHR and toxicology is largely the result of the historical circumstances that led to the receptors discovery (reviewed in (38, 130, 131, 163). Given that the AHR clearly mediates adaptive and toxic responses to environmental chemicals, the biological relevance of such study is of clear importance (Figure 4). Yet, considering that the AHR arose over 450 million years ago, and given that it is expressed in species found in markedly differing chemical environments, it is rather unlikely that the products of modern industrial society, such as the PAHs and dioxins, have provided the selective pressure for the conservation of the AHR throughout evolution (130). More clearly stated, the putative endogenous signal that activates the AHR remains to be identified.

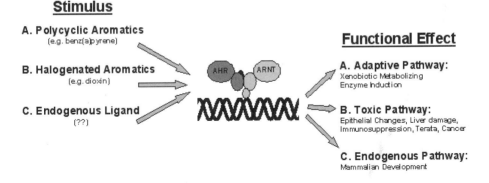

Figure 4. AHR-mediated effects. The AHR appears to mediate varying metabolic, toxic, and developmental effects, depending upon the stimulus. The induction of XMEs is conventionally thought to be an adaptive response to environmental AHR ligands. Toxicity results from chemicals which are either not detoxified by AHR-mediated XME induction or activate the AHR at incorrect times during fetal development. Lastly, the endogenous ligand remains unknown. Through this pathway, the AHR appears to mediate vascular development.

An interest in understanding the most basic aspects of AHR biology led to investigations into this protein's role in normal murine development. Using classic gene targeting strategies, several laboratories developed mice deficient in AHR (AHR-null) (164-166). As anticipated, these mice were refractory to agonist-induced expression of xenobiotic-metabolizing enzymes and to the toxicities of dioxin (164-173). Perhaps of greater interest was the observation that naïve AHR-null mice displayed a spectrum of developmental abnormalities (164-167, 174-176). Although the reported phenotypes of AHR-null mice were quite varied, the observation of decreased liver size throughout life and slower growth rates is highly consistent across laboratories. Taken in sum, these experiments provided compelling evidence that the AHR was necessary for normal development and imply that an endogenous ligand/signal for this receptor is important in normal physiology.

Two models have been proposed to explain the smaller livers observed in AHR-null mice. One line of evidence suggests that smaller livers are the result of an increased rate of hepatocyte apoptosis in null animals (177). Our own lab has provided evidence that the smaller liver size is the direct result of decreased hepatic perfusion (175). In this regard, it has been observed that AHR-null mice display a persistent vascular shunt known as the ductus venosus (DV) (Figure 5). The DV is a fetal vein that normally resolves post parturition (178). In the absence of resolution, the DV effectively functions as a portosystemic bypass of the liver, connecting the portal vein to the inferior vena cava (179). A patent DV, as found in AHR-null mice, diverts approximately 50% of hepatic blood flow around the liver parenchyma (175, 180). In addition to the persistent DV, AHR-null mice also display a number of other subtle vascular phenotypes, such as persistence of the hyaloid artery in the eye (175). Collectively, these results indicate a role for the AHR in vascular development, particularly in the resolution of fetal vascular structures.

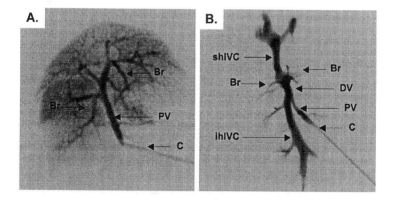

Figure 5. Radiographic images of livers from wild-type and AHR-null mice. A. The portal vein (PV) was perfused and injected with contrast dye via a cannula (C) inserted into the PV. The contrast dye fills the PV and the intrahepatic branches (Br) of the PV of wild-type mice. B. In AHR-null mice ($AHR^{\Delta 2/\Delta 2}$), the presence of a patent ductus venosus (DV) allows the contrast dye to bypass the liver and fill the suprahepatic (shIVC) and infrahepatic inferior vena cava (ihIVC).

The discovery that the AHR is required for normal hepatic development is the strongest indication, apart from the evolutionary argument noted above, to support the idea that this protein holds an important function outside the realm of toxicology. Yet, efforts to identify a receptor agonist or signal that relates to these developmental phenotypes are still lacking. On the other hand, a number of physiological compounds are emerging as potential agonists of the AHR (reviewed by (181)). Included among these candidates are metabolically derived indole derivatives of tryptophan, heme degradation products, such as bilirubin and biliverdin, and arachidonic acid metabolites, (182-189). Still others suggest that AHR is activated in the absence of ligand, perhaps by direct phosphorylation (190-192). Whatever the activating stimulus for AHR, it remains clear that this protein has a physiological function *in vivo* beyond mediating the adaptive and toxic responses to PAHs. The exact nature of these signals is yet to be determined.

10.   SUMMARY

The AHR is an ancient protein that has been highly conserved during vertebrate evolution. This receptor was originally discovered as the common mediator of the adaptive response to environmental xenobiotics as well as the toxic effects of dioxins. More recently, the AHR has been shown to play an important physiological role in liver and possibly vascular development through an as yet unidentified endogenous pathway. Great strides have been made in recent years to understand the complicated molecular interactions of the AHR. As such, many putative ligands have been suggested as the stimulus for the AHR. Additionally, the AHR has been shown to interact with many other signal transduction pathways. However, these research areas need to be more fully investigated before the true biological implications are known. In the near future, we hope that many important questions will be answered. For example, to what extent does the mechanism of the adaptive signaling pathway mirror that of the toxic and endogenous pathways? Is there a "toxic gene battery" that is distinct from the adaptive battery? Is there a true endogenous ligand for the AHR? What was the selective pressure for the evolution of the AHR? Future research will surely elucidate convincing answers to these and further questions regarding the AHR.

ACKNOWLEDGEMENTS

The authors would like to thank Drs. Jacqueline Walisser, Gary Perdew, Michael Denison, and Mark Hahn for their critical review during the preparation of this chapter. This work was supported by The National Institutes of Health (grants R37-ES05703, R01-ES06883, P01-CA22484, P30-CA14520, T32-CA09681).

REFERENCES

1.   Conney, A. H., E. C. Miller, and J. A. Miller. 1956. The metabolism of methylated aminoazo dyes. V. Evidence for induction of enzyme synthesis in the rat by 3-methylcholanthrene. *Cancer. Res.* 16:450-459.

2.   Conney, A. H., J. R. Gillette, J. K. Inscoe, E. R. Trams, and H. S. Posner. 1959. Induced synthesis of liver microsomal enzymes which metabolize foreign compounds. *Science* 130:1478-1479.

3.   Whitlock, J. P., Jr. 1999. Induction of cytochrome P4501A1. *Annu. Rev. Pharmacol. Toxicol.* 39:103-25.

4.   Guengerich, F. P., A. Parikh, C. H. Yun, D. Kim, K. Nakamura, L. M. Notley, and E. M. Gillam. 2000. What makes P450s work? Searches for answers with known and new P450s. *Drug Metab. Rev.* 32:267-81.

5.   Jacob, F., and J. Monod. 1961. Genetic regulatory mechanisms in the synthesis of proteins. *J. Mol. Biol.* 3:318-356.

6.   Poland, A., and A. Kende. 1976. 2,3,7,8-Tetrachlorodibenzo-p-dioxin: environmental contaminant and molecular probe. *Fed. Proc.* 35:2404-11.

7.   Nebert, D. W., and H. V. Gelboin. 1969. The *in vivo* and *in vitro* induction of aryl hydrocarbon hydroxylase in mammalian cells of different species, tissues, strains, and development and hormonal states. *Arch. Biochem. Biophys.* 134:76-89.

8.   Gielen, J. E., F. M. Goujon, and D. W. Nebert. 1972. Genetic regulation of aryl hydrocarbon hydroxylase induction. *J. Biol. Chem.* 247:1125-1137.

9.   Nebert, D. W., F. M. Goujon, and J. E. Gielen. 1972. Aryl hydrocarbon hydroxylase induction by polycyclic hydrocarbons: Simple autosomal dominant trait in the mouse. *Nat. New Biol.* 236:107.

10.   Thomas, P. E., R. E. Kouri, and J. J. Hutton. 1972. The genetics of aryl hydrocarbon hydroxylase induction in mice: A single gene difference between C57BL/6J and DBA/2J. *Biochem. Genetics* 6:157-168.

11.   Thomas, P. E., J. J. Hutton, and B. A. Taylor. 1973. Genetic relationship between aryl hydrocarbon hydroxylase inducibility and chemical carcinogen induced skin ulceration in mice. *Genetics* 74:655-659.

12.   Thomas, P. E., and J. J. Hutton. 1973. Genetics of aryl hydrocarbon hydroxylase induction in mice: Additive inheritance in crosses between C3H/HeJ and DBA/2J. *Biochem. Genet.* 8:249-257.

13.   Green, M. C. 1973. Nomenclature of genetically determined biochemical variants in mice. *Biochem. Genet.* 9:369-374.

14.   Poland, A., and E. Glover. 1974. Comparison of 2,3,7,8-tetrachlorodibenzo-p-dioxin, a potent inducer of aryl hydrocarbon hydroxylase, with 3-methylcholanthrene. *Mol. Pharmacol.* 10:349-59.

15.   Poland, A., E. Glover, and A. S. Kende. 1976. Stereospecific, high affinity binding of 2,3,7,8-tetrachlorodibenzo-*p*-dioxin by hepatic cytosol. *J. Biol. Chem.* 251:4936-4946.

16.   Poland, A., and E. Glover. 1977. Chlorinated biphenyl induction of aryl hydrocarbon hydroxylase activity: a study of the structure-activity relationship. *Mol. Pharmacol.* 13:924-38.

17.   Goldstein, J. A. 1979. The structure-activity relationship of halogenated biphenyls as enzyme inducers. *Ann. N.Y. Acad. Sci.* 320:164-171.

18.   Parkinson, A., L. Robertson, L. Safe, and S. Safe. 1981. Polychlroinated biphenyls as inducers of hepatic microsomal enzymes: structure-activity rules. *Chem. Biol. Interact.* 30:271.

19.   Okey, A. B., L. M. Vella, and P. A. Harper. 1989. Detection and characterization of a low affinity form of cytosolic Ah receptor in livers of mice nonresponsive to induction of cytochrome P1-450 by 3-methylcholanthrene. *Mol. Pharmacol.* 35:823-830.

20.   Poland, A., E. Glover, F. H. Ebetino, and A. S. Kende. 1986. Photoaffinity labeling of the Ah receptor. *J. Biol. Chem.* 261:6352-65.

21.   Perdew, G. H., and A. Poland. 1988. Purification of the Ah receptor from C57BL/6J mouse liver. *J. Biol. Chem.* 263:9848-52.

22.   Perdew, G. H., and C. E. Hollenback. 1990. Analysis of photoaffinity-labeled aryl hydrocarbon receptor heterogeneity by two-dimensional gel electrophoresis. *Biochemistry* 29:6210-4.

23.   Bradfield, C. A., E. Glover, and A. Poland. 1991. Purification and N-terminal amino acid sequence of the Ah receptor from the C57BL/6J mouse. *Mol. Pharmacol.* 39:13-9.

24.   Poland, A., E. Glover, and C. A. Bradfield. 1991. Characterization of polyclonal antibodies to the Ah receptor prepared by immunization with a synthetic peptide hapten [published erratum appears in *Mol. Pharmacol.* 1991, 4:435]. *Mol. Pharmacol.* 39:20-6.

25.   Swanson, H. I., and G. H. Perdew. 1991. Detection of the Ah receptor in rainbow trout: use of 2-azido-3-[125I]iodo-7,8-dibromodibenzo-p-dioxin in cell culture. *Toxicol. Lett.* 58:85-95.

26.   Burbach, K. M., A. Poland, and C. A. Bradfield. 1992. Cloning of the Ah receptor cDNA reveals a distinctive ligand-activated transcription factor. *Proc. Natl. Acad. Sci. USA* 89:8185-8189.

27.   Ema, M., K. Sogawa, N. Watanabe, Y. Chujoh, N. Matsushita, O. Gotoh, Y. Funae, and Y. Fujii-Kuriyama. 1992. cDNA cloning and structure of mouse putative Ah receptor. *Biochem. Biophys. Res. Comm.* 184:246-253.

28.   Dolwick, K. M., H. I. Swanson, and C. A. Bradfield. 1993. *In vitro* analysis of Ah receptor domains involved in ligand-activated DNA recognition. *Proc. Natl. Acad. Sci. USA* 90:8566-8570.

29.   Dolwick, K. M., J. V. Schmidt, L. A. Carver, H. I. Swanson, and C. A. Bradfield. 1993. Cloning and expression of a human Ah receptor cDNA. *Mol. Pharmacol.* 44:911-917.

30.   Hoffman, E. C., H. Reyes, F. F. Chu, F. Sander, L. H. Conley, B. A. Brooks, and O. Hankinson. 1991. Cloning of a factor required for activity of the Ah (dioxin) receptor. *Science* 252:954-958.

31. Hankinson, O. 1983. Dominant and recessive aryl hydrocarbon hydroxylase-deficient mutants of mouse hepatoma line, Hepa-1, and assignment of recessive mutants to three complementation groups. *Somatic Cell Genet* 9:497-514.

32. Watson, A. J., K. Weir-Brown, R. M. Bannister, F. F. Chu, S. Reisz-Porszasz, Y. Fujii-Kuriyama, K. Sogawa, and O. Hankinson. 1992. Mechanism of action of a repressor of dioxin-dependent induction of Cyp1a1 gene transcription. *Mol. Cell. Biol.* 12:2115-23.

33. Gustafsson, J. A., D. J. Carlstedt, P. E. Stromstedt, A. C. Wikstrom, M. Denis, S. Okret, and Y. Dong. 1990. Structure, function and regulation of the glucocorticoid receptor. *Prog. Clin. Biol. Res.* 322:65-80.

34. Reyes, H., S. Reisz-Porszasz, and O. Hankinson. 1992. Identification of the Ah receptor nuclear translocator protein (Arnt) as a component of the DNA binding form of the Ah receptor. *Science* 256:1193-1195.

35. Chan, W. K., R. Chu, S. Jain, J. K. Reddy, and C. A. Bradfield. 1994. Baculovirus expression of the Ah receptor and Ah receptor nuclear translocator. Evidence for additional dioxin responsive element-binding species and factors required for signaling. *J. Biol. Chem.* 269:26464-26471.

36. McLane, K. E., and J. P. Whitlock, Jr. 1994. DNA sequence requirements for Ah receptor/Arnt recognition determined by in vitro transcription. *Receptor* 4:209-22.

37. Pollenz, R. S., C. A. Sattler, and A. Poland. 1994. The aryl hydrocarbon receptor and aryl hydrocarbon receptor nuclear translocator protein show distinct subcellular localizations in Hepa 1c1c7 cells by immunofluorescence microscopy. *Mol. Pharmacol.* 45:428-438.

38. Schmidt, J. V., and C. A. Bradfield. 1996. Ah receptor signaling pathways. *Annu. Rev. Cell Dev. Biol.* 12:55-89.

39. Gu, Y.-Z., J. Hogenesch, and C. Bradfield. 2000. The PAS superfamily: Sensors of environmental and developmental signals, p. 519-561, Annu. Rev. Pharmacol. Toxicol., vol. 40. Academic Press.

40. Jackson, F. R., T. A. Bargiello, S. H. Yun, and M. W. Young. 1986. Product of per locus of Drosophila shares homology with proteoglycans. *Nature* 320:185-8.

41. Nambu, J. R., J. O. Lewis, K. A. Wharton, Jr., and S. T. Crews. 1991. The Drosophila single-minded gene encodes a helix-loop-helix protein that acts as a master regulator of CNS midline development. *Cell* 67:1157-1167.

42. Fukunaga, B. N., M. R. Probst, S. Reisz-Porszasz, and O. Hankinson. 1995. Identification of functional domains of the aryl hydrocarbon receptor. *J. Biol. Chem.* 270:29270-8.

43. Lindebro, M. C., L. Poellinger, and M. L. Whitelaw. 1995. Protein-protein interaction via PAS domains: role of the PAS domain in positive and negative regulation of the bHLH/PAS dioxin receptor-Arnt transcription factor complex. *EMBO J.* 14:3528-39.

44. McGuire, J., P. Coumailleau, M. L. Whitelaw, J. A. Gustafsson, and L. Poellinger. 1995. The basic helix-loop-helix/PAS factor Sim is associated with hsp90. Implications for regulation by interaction with partner factors. *J. Biol. Chem.* 270:31353-7.

45. Hogenesch, J. B., W. K. Chan, V. H. Jackiw, R. C. Brown, Y. Z. Gu, M. Pray-Grant, G. H. Perdew, and C. A. Bradfield. 1997. Characterization of a subset of the basic-helix-loop-helix-PAS superfamily that interacts with components of the dioxin signaling pathway. *J. Biol. Chem.* 272:8581-93.

46. Zelzer, E., P. Wappner, and B. Z. Shilo. 1997. The PAS domain confers target gene specificity of Drosophila bHLH/PAS proteins. *Genes Dev.* 11:2079-89.

47. Pongratz, I., C. Antonsson, M. L. Whitelaw, and L. Poellinger. 1998. Role of the PAS domain in regulation of dimerization and DNA binding specificity of the dioxin receptor. *Mol. Cell. Biol.* 18:4079-88.

48. Sun, W., J. Zhang, and O. Hankinson. 1997. A mutation in the aryl hydrocarbon receptor (AHR) in a cultured mammalian cell line identifies a novel region of AHR that affects DNA binding. *J. Biol. Chem.* 272:31845-54.

49. Denis, M., S. Cuthill, A. C. Wikstrom, L. Poellinger, and J. A. Gustafsson. 1988. Association of the dioxin receptor with the Mr 90,000 heat shock protein: A structural kinship with the glucocorticoid receptor. *Biochem. Biophys. Res. Comm.* 155:801-807.

50. Perdew, G. H. 1988. Association of the Ah receptor with the 90-kDa heat shock protein. *J. Biol. Chem.* 263:13802-13805.

51. Poland, A., D. Palen, and E. Glover. 1994. Analysis of the four alleles of the murine aryl hydrocarbon receptor. *Mol. Pharmacol.* 46:915-21.

52. Coumailleau, P., L. Poellinger, J. A. Gustafsson, and M. L. Whitelaw. 1995. Definition of a minimal domain of the dioxin receptor that is associated with Hsp90 and maintains wild type ligand binding affinity and specificity. *J. Biol. Chem.* 270:25291-300.

53. Whitelaw, M. L., J. A. Gustafsson, and L. Poellinger. 1994. Identification of transactivation and repression functions of the dioxin receptor and its basic helix-loop-helix/PAS partner factor Arnt: Inducible versus constitutive modes of regulation. *Mol. Cell. Biol.* 14:8343-8355.

54. Perdew, G. H., and C. A. Bradfield. 1996. Mapping the 90 kDa heat shock protein binding region of the Ah receptor. *Biochem. Mol. Biol. Internat.* 39:589-93.

55. Jain, S., K. M. Dolwick, J. V. Schmidt, and C. A. Bradfield. 1994. Potent transactivation domains of the Ah receptor and the Ah receptor nuclear translocator map to their carboxyl termini. *J. Biol. Chem.* 269:31518-31524.

56. Whitelaw, M. L., J. McGuire, D. Picard, J. A. Gustafsson, and L. Poellinger. 1995. Heat shock protein hsp90 regulates dioxin receptor function *in vivo*. *Proc. Natl. Acad. Sci. USA* 92:4437-4441.

57. Ma, Q., L. Dong, and J. P. Whitlock, Jr. 1995. Transcriptional activation by the mouse Ah receptor. Interplay between multiple stimulatory and inhibitory functions. *J. Biol. Chem.* 270:12697-12703.

58. Whitelaw, M., I. Pongratz, A. Wilhelmsson, J. Gustafsson, and L. Poellinger. 1993. Ligand-dependent recruitment of the Arnt coregulator determines DNA recognition by the dioxin receptor. *Mol. Cell. Biol.* 13:2504-2514.

59. Blackwood, E. M., and R. N. Eisenman. 1991. Max: a helix-loop-helix zipper protein that forms a sequence-specific DNA-binding complex with Myc. *Science* 251:1211-1217.

60. Murre, C., P. S. McCaw, and D. Baltimore. 1989. A new DNA binding and dimerization motif in immunoglobulin enhancer binding, daughterless, MyoD, and myc proteins. *Cell* 56:777-83.

61. Reisz-Porszasz, S., M. R. Probst, B. N. Fukunaga, and O. Hankinson. 1994. Identification of functional domains of the aryl hydrocarbon receptor nuclear translocator protein (ARNT). *Mol. Cell. Biol.* 14:6075-6086.

62. Swanson, H. I., and J. H. Yang. 1996. Mapping the protein/DNA contact sites of the Ah receptor and the Ah receptor nuclear translocator. *J. Biol. Chem.* 271:31657-31665.

63.   Dong, L., Q. Ma, and J. P. Whitlock, Jr. 1996. DNA binding by the heterodimeric Ah receptor. Relationship to dioxin-induced CYP1A1 transcription in vivo. *J. Biol. Chem.* 271:7942-8.

64.   Fukunaga, B. N., and O. Hankinson. 1996. Identification of a novel domain in the aryl hydrocarbon receptor required for DNA binding. *J. Biol. Chem.* 271:3743-9.

65.   Swanson, H. I., W. K. Chan, and C. A. Bradfield. 1995. DNA binding specificities and pairing rules of the Ah receptor, ARNT, and SIM proteins. *J. Biol. Chem.* 270:26292-302.

66.   Bacsi, S. G., S. Reisz-Porszasz, and O. Hankinson. 1995. Orientation of the heterodimeric aryl hydrocarbon (dioxin) receptor complex on its asymmetric DNA recognition sequence. *Mol. Pharmacol.* 47:432-8.

67.   Davis, R. L., P.-F. Cheng, A. B. Lassar, and H. Weintraub. 1990. The MyoD DNA binding domain contains a recognition code for muscle-specific gene activation. *Cell* 60:733-746.

68.   Edmondson, D. G., and E. N. Olson. 1993. Helix-loop-helix proteins as regulators of muscle-specific transcription. *J. Biol. Chem.* 268:755-8.

69.   Ikuta, T., H. Eguchi, T. Tachibana, Y. Yoneda, and K. Kawajiri. 1998. Nuclear localization and export signals of the human aryl hydrocarbon receptor. *J. Biol. Chem.* 273:2895-904.

70.   Eguchi, H., T. Ikuta, T. Tachibana, Y. Yoneda, and K. Kawajiri. 1997. A nuclear localization signal of human aryl hydrocarbon receptor nuclear translocator/hypoxia-inducible factor 1beta is a novel bipartite type recognized by the two components of nuclear pore-targeting complex. *J. Biol. Chem.* 272:17640-7.

71.   Sogawa, K., K. Iwabuchi, H. Abe, and Y. Fujii-Kuriyama. 1995. Transcriptional activation domains of the Ah receptor and Ah receptor nuclear translocator. *Journal of Cancer Res. Clinical Oncol.* 121:612-20.

72.   Rowlands, J. C., I. J. McEwan, and J. A. Gustafsson. 1996. Trans-activation by the human aryl hydrocarbon receptor and aryl hydrocarbon receptor nuclear translocator proteins: direct interactions with basal transcription factors. *Mol. Pharmacol.* 50:538-48.

73.   Kumar, M. B., P. Ramadoss, R. K. Reen, J. P. Vanden Heuvel, and G. H. Perdew. 2001. The Q-rich subdomain of the human Ah receptor transactivation domain is required for dioxin-mediated transcriptional activity. *J. Biol. Chem.* 276:42302-10.

74.   Pongratz, I., G. G. Mason, and L. Poellinger. 1992. Dual roles of the 90-kDa heat shock protein hsp90 in modulating functional activities of the dioxin receptor. *J. Biol. Chem.* 267:13728-13734.

75.   Pratt, W. B. 1998. The hsp90-based chaperone system: involvement in signal transduction from a variety of hormone and growth factor receptors. *Proc. Soc. Exp. Biol. Med.* 217:420-34.

76.   Carver, L. A., V. Jackiw, and C. A. Bradfield. 1994. The 90-kDa heat shock protein is essential for Ah receptor signaling in a yeast expression system. *J. Biol. Chem.* 269:30109-30112.

77.   Wilhelmsson, A., S. Cuthill, M. Denis, A. C. Wikstrom, J. A. Gustafsson, and L. Poellinger. 1990. The specific DNA binding activity of the dioxin receptor is modulated by the 90 kd heat shock protein. *EMBO J.* 9:69-76.

78.   Kumar, M. B., and G. H. Perdew. 1999. Nuclear receptor coactivator SRC-1 interacts with the Q-rich subdomain of the AhR and modulates its transactivation potential. *Gene Expr.* 8:273-86.

79. Beischlag, T. V., S. Wang, D. W. Rose, J. Torchia, S. Reisz-Porszasz, K. Muhammad, W. E. Nelson, M. R. Probst, M. G. Rosenfeld, and O. Hankinson. 2002. Recruitment of the NCoA/SRC-1/p160 family of transcriptional coactivators by the aryl hydrocarbon receptor/aryl hydrocarbon receptor nuclear translocator complex. *Mol. Cell. Biol.* 22:4319-33.

80. Kobayashi, A., K. Numayama-Tsuruta, K. Sogawa, and Y. Fujii-Kuriyama. 1997. CBP/p300 functions as a possible transcriptional coactivator of Ah receptor nuclear translocator (Arnt). *J. Biochem. (Tokyo)* 122:703-10.

81. Lemon, B. D., and L. P. Freedman. 1999. Nuclear receptor cofactors as chromatin remodelers. *Curr. Opin. Genet. Dev.* 9:499-504.

82. Wang, S., and O. Hankinson. 2002. Functional involvement of the Brahma/SWI2-related gene 1 protein in cytochrome P4501A1 transcription mediated by the aryl hydrocarbon receptor complex. *J. Biol. Chem.* 277:11821-7.

83. Featherstone, M. 2002. Coactivators in transcription initiation: here are your orders. *Curr. Opin. Genet. Dev.* 12:149-55.

84. Swanson, H. I., and J. H. Yang. 1998. The aryl hydrocarbon receptor interacts with transcription factor IIB. *Mol. Pharmacol.* 54:671-7.

85. Klinge, C. M., S. C. Jernigan, K. E. Risinger, J. E. Lee, V. V. Tyulmenkov, K. C. Falkner, and R. A. Prough. 2001. Short heterodimer partner (SHP) orphan nuclear receptor inhibits the transcriptional activity of aryl hydrocarbon receptor (AHR)/AHR nuclear translocator (ARNT). *Arch. Biochem. Biophys.* 390:64-70.

86. Kobayashi, A., K. Sogawa, and Y. Fujii-Kuriyama. 1996. Cooperative interaction between AhR.Arnt and Sp1 for the drug-inducible expression of CYP1A1 gene. *J. Biol. Chem.* 271:12310-6.

87. Nguyen, T. A., D. Hoivik, J. E. Lee, and S. Safe. 1999. Interactions of nuclear receptor coactivator/corepressor proteins with the aryl hydrocarbon receptor complex. *Arch. Biochem. Biophys.* 367:250-7.

88. McGuire, J., K. Okamoto, M. L. Whitelaw, H. Tanaka, and L. Poellinger. 2001. Definition of a dioxin receptor mutant that is a constitutive activator of transcription: delineation of overlapping repression and ligand binding functions within the PAS domain. *J. Biol. Chem.* 276:41841-9.

89. Chen, H. S., S. S. Singh, and G. H. Perdew. 1997. The Ah receptor is a sensitive target of geldanamycin-induced protein turnover. *Arch. Biochem. Biophys.* 348:190-8.

90. Carver, L. A., and C. A. Bradfield. 1997. Ligand-dependent interaction of the aryl hydrocarbon receptor with a novel immunophilin homolog in vivo. *J. Biol. Chem.* 272:11452-6.

91. Ma, Q., and J. P. Whitlock, Jr. 1997. A novel cytoplasmic protein that interacts with the Ah receptor, contains tetratricopeptide repeat motifs, and augments the transcriptional response to 2,3,7,8-tetrachlorodibenzo-p-dioxin. *J. Biol. Chem.* 272:8878-84.

92. Meyer, B. K., M. G. Pray-Grant, J. P. Vanden Heuvel, and G. H. Perdew. 1998. Hepatitis B virus X-associated protein 2 is a subunit of the unliganded aryl hydrocarbon receptor core complex and exhibits transcriptional enhancer activity. *Mol. Cell. Biol.* 18:978-88.

93. Prokipcak, R. D., L. E. Faber, and A. B. Okey. 1989. Characterization of the Ah receptor for 2,3,7,8-tetrachlorodibenzo-p-dioxin: use of chemical crosslinking and a monoclonal antibody directed against a 59-kDa protein associated with steroid receptors. *Arch. Biochem. Biophys.* 274:648-58.

94.   Perdew, G. H. 1992. Chemical cross-linking of the cytosolic and nuclear forms of the Ah receptor in hepatoma cell line 1c1c7. *Biochem. Biophys. Res. Comm.* 182:55-62.

95.   Carver, L. A., J. J. LaPres, S. Jain, E. E. Dunham, and C. A. Bradfield. 1998. Characterization of the Ah Receptor-associated Protein, ARA9. *J. Biol. Chem.* 273:33580-6159.

96.   Meyer, B. K., and G. H. Perdew. 1999. Characterization of the AhR-hsp90-XAP2 core complex and the role of the immunophilin-related protein XAP2 in AhR stabilization. *Biochemistry* 38:8907-8917.

97.   Bell, D. R., and A. Poland. 2000. Binding of Aryl Hydrocarbon Receptor (AhR) to AhR-interacting Protein. THE ROLE OF hsp90. *J. Biol. Chem.* 275:36407-1.

98.   Chen, H. S., and G. H. Perdew. 1994. Subunit composition of the heteromeric cytosolic aryl hydrocarbon receptor complex. *J. Biol. Chem.* 269:27554-8.

99.   Petrulis, J. R., N. G. Hord, and G. H. Perdew. 2000. Subcellular Localization of the Aryl Hydrocarbon Receptor Is Modulated by the Immunophilin Homolog Hepatitis B Virus X-associated Protein 2. *J. Biol. Chem.* 275:37448-37453.

100.   Kazlauskas, A., L. Poellinger, and I. Pongratz. 2000. The immunophilin-like protein XAP2 regulates ubiquitination and subcellular localization of the dioxin receptor. *J. Biol. Chem.* 275:41317-24.

101.   LaPres, J. J., E. Glover, E. E. Dunham, M. K. Bunger, and C. A. Bradfield. 2000. ARA9 Modifies Agonist Signaling through an Increase in Cytosolic Aryl Hydrocarbon Receptor. *J. Biol. Chem.* 275:6153-6159.

102.   Meyer, B. K., J. R. Petrulis, and G. H. Perdew. 2000. Aryl hydrocarbon (Ah) receptor levels are selectively modulated by hsp90-associated immunophilin homolog XAP2. *Cell Stress Chaperones.* 5:243-54.

103.   Petrulis, J. R., A. Kusnadi, P. Ramadoss, B. Hollingshead, and G. H. Perdew. 2003. The hsp90 Co-chaperone XAP2 Alters Importin beta Recognition of the Bipartite Nuclear Localization Signal of the Ah Receptor and Represses Transcriptional Activity. *J. Biol. Chem.* 278:2677-85.

104.   Tai, P. K., M. W. Albers, H. Chang, L. E. Faber, and S. L. Schreiber. 1992. Association of a 59-kilodalton immunophilin with the glucocorticoid receptor complex. *Science* 256:1315-8.

105.   Kazlauskas, A., L. Poellinger, and I. Pongratz. 2002. Two distinct regions of the immunophilin-like protein XAP2 regulate dioxin receptor function and interaction with hsp90. *J. Biol. Chem.* 277:11795-801.

106.   Lamb, J. R., S. Tugendreich, and P. Hieter. 1995. Tetratrico peptide repeat interactions: to TPR or not to TPR? *Trends Biochem. Sci.* 20:257-9.

107.   Kazlauskas, A., L. Poellinger, and I. Pongratz. 1999. Evidence that the co-chaperone p23 regulates ligand responsiveness of the dioxin (Aryl hydrocarbon) receptor. *J. Biol. Chem.* 274:13519-24.

108.   Kazlauskas, A., S. Sundstrom, L. Poellinger, and I. Pongratz. 2001. The hsp90 chaperone complex regulates intracellular localization of the dioxin receptor. *Mol. Cell. Biol.* 21:2594-607.

109.   Cox, M. B., and C. A. Miller, 3rd. 2002. The p23 co-chaperone facilitates dioxin receptor signaling in a yeast model system. *Toxicol. Lett.* 129:13-21.

110.   Jones, P. B., D. R. Galeazzi, J. M. Fisher, and J. P. Whitlock, Jr. 1985. Control of cytochrome P1-450 gene expression by dioxin. *Science* 227:1499-502.

111.   Gonzalez, F. J. a. N., D.W. 1985. Autoregulation plus upstream positive and negative control regions associated with transcriptional activation of the mouse cytochrome P1-450 gene. *Nucl. Acids Res.* 13:7269-7288.

112. Sogawa, K., A. Fujisawa-Sehara, M. Ymane, and Y. Fujii-Kuriyama. 1986. Location of the regulatory elements responsible for drug induction in the rat cytochrome P-450c gene. *Proc. Natl. Acad. Sci. USA* 83:8044-8048.

113. Jones, P. B., Durrin, L.K., Galeazzi, D.R., and Whitlock, J.P., Jr. 1986. Control of cytochrome P1-450 gene expression: Analysis of a dioxin-responsive enhancer system. *Proc. Natl. Acad. Sci.* 83:2802-2806.

114. Hines, R. N., J. M. Mathis, and C. S. Jacob. 1988. Identification of multiple regulatory elements on the human cytochrome P450IA1 gene. *Carcinogenesis* 9:1599-1605.

115. Carrier, F., R. A. Owens, D. W. Nebert, and A. Puga. 1992. Dioxin-dependent activation of murine Cyp1a-1 gene transcription requires protein kinase C-dependent phosphorylation. *Mol. Cell. Biol.* 12:1856-63.

116. Durrin, L. K., P. B. C. Jones, J. M. Fisher, D. R. Galeazzi, and J. P. Whitlock. 1987. 2,3,7,8-Tetrachlorodibenzo-p-dioxin receptors regulate transcription of the cytochrome P1-450 gene. *J. Cell. Biochem.* 35:153-160.

117. Fujisawa-Sehara, A., K. Sogawa, M. Yamane, and Y. Fujii-Kuriyama. 1987. Characterization of xenobiotic responsive elements upstream from the drug-metabolizing cytochrome P-450c gene: a similarity to glucocorticoid regulatory elements. *Nucl. Acids Res.* 15:4179-4191.

118. Nebert, D. W., and J. E. Jones. 1989. Regulation of the mammalian cytochrome p1450 (CYP1A1) gene. *Int. J. Biochem.* 21.

119. Denison, M. S., J. M. Fisher, and J. J. Whitlock. 1989. Protein-DNA interactions at recognition sites for the dioxin-Ah receptor complex. *J. Biol. Chem.* 264:16478-82.

120. Favreau, L. V., and C. B. Pickett. 1991. Transcriptional regulation of the rat NAD(P)H:quinone reductase gene. Identification of regulatory elements controlling basal level expression and inducible expression by planar aromatic compounds and phenolic anitoxidants. *J. Biol. Chem.* 266:4556-4561.

121. Fujisawa-Sehara, A., M. Yamane, and Y. Fujii-Kuriyama. 1988. A DNA-binding factor specific for xenobiotic responsive elements of P-450c gene exists as a cryptic form in cytoplasm: Its possible translocation to the nucleus. *Proc. Natl. Acad. Sci. USA* 85:5859-5863.

122. Watson, A. J., and O. Hankinson. 1992. Dioxin- and Ah receptor-dependent protein binding to xenobiotic responsive elements and G-rich DNA studied by *in vivo* footprinting. *J. Biol. Chem.* 267:6874-6878.

123. Durrin, L. K., and Whitlock, J.P., Jr. 1987. *In situ* protein-DNA interactions at a dioxin-responsive enhancer associated with the cytochrome P1-450 gene. *Mol. Cell. Biol.* 7:3008-3011.

124. Neuhold, L. A., Y. Shirayoshi, K. Ozato, J. E. Jones, and D. W. Nebert. 1989. Regulation of mouse CYP1A1 gene expression by dioxin: Requirement of two cis-acting elements during induction. *Mol. Cell. Biol.* 9:2378-86.

125. Shen, E. S., and J. P. Whitlock, Jr. 1992. Protein-DNA interactions at a dioxin-responsive enhancer. Mutational analysis of the DNA-binding site for the liganded Ah receptor. *J. Biol. Chem.* 267:6815-9.

126. Yao, E. F., and M. S. Denison. 1992. DNA sequences determinants for binding of transformed Ah receptor to a dioxin-responsive enhancer. *Biochemistry* 31:5060-5067.

127. Fujii-Kuriyama, Y., A. Fujisawa-Sehara, and K. Sogawa. 1989. Regulatory mechanism of gene expression of methylcholanthrene-inducible cytochrome P-450. *Drug Metab. Rev.* 20:821-6.

128. Denison, M. S., J. M. Fisher, and J. P. Whitlock, Jr. 1988. The DNA recognition site for the dioxin-Ah receptor complex. Nucleotide sequence and functional analysis. *J. Biol. Chem.* 263:17221-4.

129. Nebert, D. W., A. Puga, and V. Vasiliou. 1993. Role of the Ah receptor and the dioxin-inducible [Ah] gene battery in toxicity, cancer, and signal transduction. *Ann. NY Acad. Sci.* 685:624-40.

130. Schrenk, D. 1998. Impact of dioxin-type induction of drug-metabolizing enzymes on the metabolism of endo- and xenobiotics. *Biochem. Pharmacol.* 55:1155-62.

131. Nebert, D. W., A. L. Roe, M. Z. Dieter, W. A. Solis, Y. Yang, and T. P. Dalton. 2000. Role of the aromatic hydrocarbon receptor and [Ah] gene battery in the oxidative stress response, cell cycle control, and apoptosis. *Biochem. Pharmacol.* 59:65-85.

132. Thomas, R. S., D. R. Rank, S. G. Penn, G. M. Zastrow, K. R. Hayes, K. Pande, E. Glover, T. Silander, M. W. Craven, J. K. Reddy, et al. 2001. Identification of toxicologically predictive gene sets using cDNA microarrays. *Mol. Pharmacol.* 60:1189-94.

133. Wang, G. L., B. H. Jiang, E. A. Rue, and G. L. Semenza. 1995. Hypoxia-inducible factor 1 is a basic-helix-loop-helix-PAS heterodimer regulated by cellular O_2 tension. *Proc. Natl. Acad. Sci. USA* 92:5510-5514.

134. Gradin, K., J. McGuire, R. H. Wenger, I. Kvietikova, M. L. Whitelaw, R. Toftgard, L. Tora, M. Gassmann, and L. Poellinger. 1996. Functional interference between hypoxia and dioxin signal transduction pathways: competition for recruitment of the Arnt transcription factor. *Mol. Cell. Biol.* 16:5221-31.

135. Chan, W. K., G. Yao, Y. Z. Gu, and C. A. Bradfield. 1999. Cross-talk between the aryl hydrocarbon receptor and hypoxia inducible factor signaling pathways. Demonstration of competition and compensation. *J. Biol. Chem.* 274:12115-23.

136. Nie, M., A. L. Blankenship, and J. P. Giesy. 2001. Interactions between aryl hydrocarbon receptor (AhR) and hypoxia signaling pathways. *Environ. Toxicol. Pharmacol.* 10:17-27.

137. Pollenz, R. S., N. A. Davarinos, and T. P. Shearer. 1999. Analysis of aryl hydrocarbon receptor-mediated signaling during physiological hypoxia reveals lack of competition for the aryl hydrocarbon nuclear translocator transcription factor. *Mol. Pharmacol.* 56:1127-37.

138. Safe, S., F. Wang, W. Porter, R. Duan, and A. McDougal. 1998. Ah receptor agonists as endocrine disruptors: antiestrogenic activity and mechanisms. *Toxicol. Lett.* 102-103:343-7.

139. Duan, R., W. Porter, I. Samudio, C. Vyhlidal, M. Kladde, and S. Safe. 1999. Transcriptional activation of c-fos protooncogene by 17beta-estradiol: mechanism of aryl hydrocarbon receptor-mediated inhibition. *Mol. Endocrinol.* 13:1511-21.

140. Klinge, C. M., J. L. Bowers, P. C. Kulakosky, K. K. Kamboj, and H. I. Swanson. 1999. The aryl hydrocarbon receptor (AHR)/AHR nuclear translocator (ARNT) heterodimer interacts with naturally occurring estrogen response elements. *Mol. Cell. Endocrin.* 157:105-19.

141. Klinge, C. M., K. Kaur, and H. I. Swanson. 2000. The aryl hydrocarbon receptor interacts with estrogen receptor alpha and orphan receptors COUP-TFI and ERRalpha1. *Arch. Biochem. Biophys.* 373:163-74.

142. Tian, Y., S. Ke, M. S. Denison, A. B. Rabson, and M. A. Gallo. 1999. Ah receptor and NF-kappaB interactions, a potential mechanism for dioxin toxicity. *J. Biol. Chem.* 274:510-5.

143. Ge, N.-L., and C. J. Elferink. 1998. A Direct Interaction between the Aryl Hydrocarbon Receptor and Retinoblastoma Protein. LINKING DIOXIN SIGNALING TO THE CELL CYCLE. *J. Biol. Chem.* 273:22708-22713.
144. Puga, A., S. J. Barnes, T. P. Dalton, C. Chang, E. S. Knudsen, and M. A. Maier. 2000. Aromatic hydrocarbon receptor interaction with the retinoblastoma protein potentiates repression of E2F-dependent transcription and cell cycle arrest. *J. Biol. Chem.* 275:2943-50.
145. Mimura, J., M. Ema, K. Sogawa, and Y. Fujii-Kuriyama. 1999. Identification of a novel mechanism of regulation of Ah (dioxin) receptor function. *Genes Dev.* 13:20-5.
146. Baba, T., J. Mimura, K. Gradin, A. Kuroiwa, T. Watanabe, Y. Matsuda, J. Inazawa, K. Sogawa, and Y. Fujii-Kuriyama. 2001. Structure and expression of the Ah receptor repressor gene. *J. Biol. Chem.* 276:33101-10.
147. Roberts, B. J., and M. L. Whitelaw. 1999. Degradation of the basic helix-loop-helix/Per-ARNT-Sim homology domain dioxin receptor via the ubiquitin/proteasome pathway. *J. Biol. Chem.* 274:36351-6.
148. Davarinos, N. A., and R. S. Pollenz. 1999. Aryl hydrocarbon receptor imported into the nucleus following ligand binding is rapidly degraded via the cytoplasmic proteasome following nuclear export. *J. Biol. Chem.* 274:28708-15.
149. Poland, A., and E. Glover. 1987. Variation in the molecular mass of the Ah receptor among vertebrate species and strains of rats. *Biochem. Biophys. Res. Comm.* 146:1439-1449.
150. Poland, A., and E. Glover. 1990. Characterization and strain distribution pattern of the murine Ah receptor specified by the Ah^d and Ah^{b-3} alleles. *Mol. Pharmacol.* 38:306-312.
151. Poland, A., and E. Glover. 1980. 2,3,7,8,-Tetrachlorodibenzo-p-dioxin: segregation of toxicity with the Ah locus. *Mol. Pharmacol.* 17:86-94.
152. Ema, M., N. Ohe, M. Suzuki, J. Mimura, K. Sogawa, I. Ikawa, and Y. Fujii-Kuriyama. 1994. Dioxin binding activities of polymorphic forms of mouse and human aryl hydrocarbon receptors. *J. Biol. Chem.* 269:27337-27343.
153. Chang, C., D. R. Smith, V. S. Prasad, C. L. Sidman, D. W. Nebert, and A. Puga. 1993. Ten nucleotide differences, five of which cause amino acid changes, are associated with the Ah receptor locus polymorphism of C57BL/6 and DBA/2 mice. *Pharmacogenetics* 3:312-21.
154. Thomas, R. S., S. G. Penn, K. Holden, C. A. Bradfield, and D. R. Rank. 2002. Sequence variation and phylogenetic history of the mouse Ahr gene. *Pharmacogenetics* 12:151-63.
155. Hahn, M. E., S. I. Karchner, M. A. Shapiro, and S. A. Perera. 1997. Molecular evolution of two vertebrate aryl hydrocarbon (dioxin) receptors (AHR1 and AHR2) and the PAS family. *Proc. Natl. Acad. Sci. USA* 94:13743-8.
156. Hahn, M. E. 1998. The aryl hydrocarbon receptor: a comparative perspective. *Comp. Biochem. Physiol. Part C Pharmacol., Toxicol., Endocrinol.* 121:23-53.
157. Hahn, M. E. 2002. Aryl hydrocarbon receptors: diversity and evolution. *Chem Biol Interact* 141:131-60.
158. Carver, L. A., J. B. Hogenesch, and C. A. Bradfield. 1994. Tissue specific expression of the rat Ah-receptor and ARNT mRNAs. *Nucl. Acids Res.* 22:3038-3044.
159. Powell-Coffman, J. A., C. A. Bradfield, and W. B. Wood. 1998. Caenorhabditis elegans orthologs of the aryl hydrocarbon receptor and its heterodimerization partner the aryl hydrocarbon receptor nuclear translocator. *Proc. Natl. Acad. Sci. USA* 95:2844-9.

160. Butler, R. A., M. L. Kelley, W. H. Powell, M. E. Hahn, and R. J. Van Beneden. 2001. An aryl hydrocarbon receptor (AHR) homologue from the soft-shell clam, Mya arenaria: evidence that invertebrate AHR homologues lack 2,3,7,8-tetrachlorodibenzo-p-dioxin and beta-naphthoflavone binding. *Gene* 278:223-34.

161. Duncan, D. M., E. A. Burgess, and I. Duncan. 1998. Control of distal antennal identity and tarsal development in Drosophila by spineless-aristapedia, a homolog of the mammalian dioxin receptor. *Genes Dev.* 12:1290-303.

162. Emmons, R. B., D. Duncan, P. A. Estes, P. Kiefel, J. T. Mosher, M. Sonnenfeld, M. P. Ward, I. Duncan, and S. T. Crews. 1999. The spineless-aristapedia and tango bHLH-PAS proteins interact to control antennal and tarsal development in Drosophila. *Development* 126:3937-45.

163. Wilson, C. L., and S. Safe. 1998. Mechanisms of ligand-induced aryl hydrocarbon receptor-mediated biochemical and toxic responses [see comments]. *Toxicol. Pathol.* 26:657-71.

164. Fernandez-Salguero, P., T. Pineau, D. M. Hilbert, T. McPhail, S. S. Lee, S. Kimura, D. W. Nebert, S. Rudikoff, J. M. Ward, and F. J. Gonzalez. 1995. Immune system impairment and hepatic fibrosis in mice lacking the dioxin-binding Ah receptor. *Science* 268:722-6.

165. Schmidt, J. V., G. H. Su, J. K. Reddy, M. C. Simon, and C. A. Bradfield. 1996. Characterization of a murine Ahr null allele: involvement of the Ah receptor in hepatic growth and development. *Proc. Natl. Acad. Sci. USA* 93:6731-6.

166. Mimura, J., K. Yamashita, K. Nakamura, M. Morita, T. N. Takagi, K. Nakao, M. Ema, K. Sogawa, M. Yasuda, M. Katsuki, et al. 1997. Loss of teratogenic response to 2,3,7,8-tetrachlorodibenzo-p-dioxin (TCDD) in mice lacking the Ah (dioxin) receptor. *Genes Cells* 2:645-54.

167. Andreola, F., P. M. Fernandez-Salguero, M. V. Chiantore, M. P. Petkovich, F. J. Gonzalez, and L. M. De Luca. 1997. Aryl hydrocarbon receptor knockout mice (AHR-/-) exhibit liver retinoid accumulation and reduced retinoic acid metabolism. *Cancer Res.* 57:2835-8.

168. Lahvis, G. P., and C. A. Bradfield. 1998. Ahr null alleles: distinctive or different? *Biochem. Pharmacol.* 56:781-7.

169. Peters, J. M., M. G. Narotsky, G. Elizondo, P. M. Fernandez-Salguero, F. J. Gonzalez, and B. D. Abbott. 1999. Amelioration of TCDD-induced teratogenesis in aryl hydrocarbon receptor (AhR)-null mice. *Toxicol. Sci.* 47:86-92.

170. Shimizu, Y., Y. Nakatsuru, M. Ichinose, Y. Takahashi, H. Kume, J. Mimura, Y. Fujii-Kuriyama, and T. Ishikawa. 2000. Benzo[a]pyrene carcinogenicity is lost in mice lacking the aryl hydrocarbon receptor. *Proc. Natl. Acad. Sci. USA* 97:779-82.

171. Sugihara, K., S. Kitamura, T. Yamada, S. Ohta, K. Yamashita, M. Yasuda, and Y. Fujii-Kuriyama. 2001. Aryl hydrocarbon receptor (AhR)-mediated induction of xanthine oxidase/xanthine dehydrogenase activity by 2,3,7,8-tetrachlorodibenzo-p-dioxin. *Biochem. Biophys. Res. Comm.* 281:1093-9.

172. Vorderstrasse, B. A., L. B. Steppan, A. E. Silverstone, and N. I. Kerkvliet. 2001. Aryl hydrocarbon receptor-deficient mice generate normal immune responses to model antigens and are resistant to TCDD-induced immune suppression. *Toxicol. Appl. Pharmacol.* 171:157-64.

173. Vorderstrasse, B. A., and N. I. Kerkvliet. 2001. 2,3,7,8-Tetrachlorodibenzo-p-dioxin affects the number and function of murine splenic dendritic cells and their expression of accessory molecules. *Toxicol. Appl. Pharmacol.* 171:117-25.

174. Abbott, B. D., J. E. Schmid, J. A. Pitt, A. R. Buckalew, C. R. Wood, G. A. Held, and J. J. Diliberto. 1999. Adverse reproductive outcomes in the transgenic Ah receptor-deficient mouse. *Toxicol. Appl. Pharm.* 155:62-70.

175. Lahvis, G., S. Lindell, R. Thomas, R. McCuskey, C. Murphy, E. Glover, M. Bentz, J. Southard, and C. Bradfield. 2000. Portosystemic shunts and persistent fetal vascular structures in Ah-receptor deficient mice. *Proc. Natl. Acad. Sci. USA* 97:10442-10447.

176. Robles, R., Y. Morita, K. K. Mann, G. I. Perez, S. Yang, T. Matikainen, D. H. Sherr, and J. L. Tilly. 2000. The aryl hydrocarbon receptor, a basic helix-loop-helix transcription factor of the PAS gene family, is required for normal ovarian germ cell dynamics in the mouse. *Endocrinology* 141:450-3.

177. Zaher, H., P. M. Fernandez-Salguero, J. Letterio, M. S. Sheikh, A. J. Fornace, A. B. Roberts, and F. J. Gonzalez. 1998. The involvement of aryl hydrocarbon receptor in the activation of transforming growth factor-beta and apoptosis. *Mol. Pharmacol.* 54:313-321.

178. Meyer, W. W., and J. Lind. 1966. The ductus venosus and the mechanism of its closure. *Arch. Dis. Child* 41:597-605.

179. Rudolph, A. M. 1983. Hepatic and ductus venosus blood flows during fetal life. *Hepatology* 3:254-8.

180. Kiserud, T. 1999. Hemodynamics of the ductus venosus. *Eur. J. Obstet. Gynecol. Reprod. Biol.* 84:139-47.

181. Denison, M. S., A. Pandini, S. R. Nagy, E. P. Baldwin, and L. Bonati. 2002. Ligand binding and activation of the Ah receptor. *Chem. Biol. Interact.* 141:3-24.

182. Sinal, C. J., and J. R. Bend. 1997. Aryl hydrocarbon receptor-dependent induction of cyp1a1 by bilirubin in mouse hepatoma hepa 1c1c7 cells. *Mol. Pharmacol.* 52:590-9.

183. Schaldach, C. M., J. Riby, and L. F. Bjeldanes. 1999. Lipoxin A4: a new class of ligand for the Ah receptor. *Biochemistry* 38:7594-600.

184. Phelan, D., G. M. Winter, W. J. Rogers, J. C. Lam, and M. S. Denison. 1998. Activation of the Ah receptor signal transduction pathway by bilirubin and biliverdin. *Arch. Biochem. Biophys.* 357:155-63.

185. Schrenk, D., D. Riebniger, M. Till, S. Vetter, and H. P. Fiedler. 1999. Tryptanthrins and other tryptophan-derived agonists of the dioxin receptor. *Adv. Exp. Med. Biol.* 467:403-8.

186. Heath-Pagliuso, S., W. J. Rogers, K. Tullis, S. D. Seidel, P. H. Cenijn, A. Brouwer, and M. S. Denison. 1998. Activation of the Ah receptor by tryptophan and tryptophan metabolites. *Biochemistry* 37:11508-15.

187. Wei, Y. D., L. Bergander, U. Rannug, and A. Rannug. 2000. Regulation of CYP1A1 transcription via the metabolism of the tryptophan- derived 6-formylindolo[3,2-b]carbazole. *Arch. Biochem. Biophys.* 383:99-107.

188. Seidel, S. D., G. M. Winters, W. J. Rogers, M. H. Ziccardi, V. Li, B. Keser, and M. S. Denison. 2001. Activation of the Ah receptor signaling pathway by prostaglandins. *J Biochem. Mol. Toxicol.* 15:187-96.

189. Savouret, J. F., M. Antenos, M. Quesne, J. Xu, E. Milgrom, and R. F. Casper. 2001. 7-ketocholesterol is an endogenous modulator for the arylhydrocarbon receptor. *J. Biol. Chem.* 276:3054-9.

190. Pongratz, I., P. E. Stromstedt, G. G. Mason, and L. Poellinger. 1991. Inhibition of the specific DNA binding activity of the dioxin receptor by phosphatase treatment. *J. Biol. Chem.* 266:16813-7.

191. Sadek, C. M., and B. L. Allen-Hoffmann. 1994. Suspension-mediated Induction of Hepa 1c1c7 Cypa1a-1 Expression Is Dependent on the Ah Receptor Signal Transduction Pathway. *J. Biol. Chem.* 269:31505-31509.

192. Park, S., E. C. Henry, and T. A. Gasiewicz. 2000. Regulation of DNA binding activity of the ligand-activated aryl hydrocarbon receptor by tyrosine phosphorylation. *Arch. Biochem. Biophys.* 381:302-12.

Chapter 8

THE HIF-1 FAMILY OF bHLH-PAS PROTEINS: MASTER REGULATORS OF OXYGEN HOMEOSTASIS

Gregg L. Semenza
The Johns Hopkins University School of Medicine, Baltimore, MD 21287

1.　　INTRODUCTION

　　Above all other elements, molecules, and compounds, O_2 is the critical environmental substrate that must be delivered to all cells of metazoan organisms on a virtually continuous basis for survival. The evolution of multicellular animals in which oxygen could no longer be acquired by all cells via simple diffusion necessitated the co-evolution of physiological systems specialized for O_2 delivery, which became progressively more complex as body size increased, most notably with the appearance of the vertebrates. The HIF-1 (hypoxia-inducible factor 1) family of bHLH-PAS transcription factors appears to have evolved as a specialized system for regulating oxygen homeostasis at the level of gene expression in metazoans. HIF-1 is present in simple invertebrates such as the roundworm *Caenorhabditis elegans,* which consists of $\sim 10^3$ cells and relies on simple diffusion for O_2 transport, more complex invertebrates such as the fruit fly *Drosophila melanogaster*, in which O_2 is distributed through the body via a set of specialized tracheal tubes, and in complex vertebrates such as *Homo sapiens*, which consist of $> 10^{13}$ cells that are supplied with O_2 via the combined functioning of highly complex and specialized circulatory and respiratory systems. In this chapter I will provide a brief summary of the members of the HIF-1 family, the novel molecular mechanisms by which they are regulated, and the roles they play in development and physiology as determined by the analysis of knockout mice. The important roles of HIF-1 family members in human disease pathophysiology have recently been reviewed elsewhere (1, 2).

2.   HIF-1 FAMILY MEMBERS

Analysis of the human erythropoietin gene identified a cis-acting hypoxia response element (HRE) in the 3'-flanking region that was required for increased transcription in response to reduced O_2 availability (3-5). HIF-1 was initially characterized as an HRE binding activity that was induced in cells subjected to hypoxia (6). Protein purification by ion-exchange and DNA-affinity chromatography revealed that HIF-1 was composed of two subunits, designated HIF-1α and HIF-1β (7). Protein microsequencing lead to the isolation of cDNA sequences encoding the two subunits (8). HIF-1β was found to be identical to a bHLH-PAS protein previously designated as the aryl hydrocarbon receptor nuclear translocator (ARNT). HIF-1α was shown to be a novel member of the bHLH-PAS family of proteins (Figure 1). As described in detail below, the expression and activity of the HIF-1α subunit is exquisitely regulated by the cellular O_2 concentration, whereas the HIF-1β subunit is constitutively expressed.

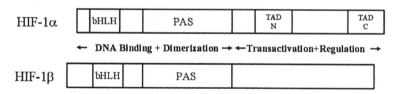

Figure 1. Structure of HIF-1. HIF-1 is a heterodimer composed of the bHLH-PAS proteins HIF-1α and HIF-1β (ARNT). The amino-terminal halves of these proteins are required for dimerization and DNA binding. The carboxyl-terminal half of HIF-1α contains two transactivation domains (TAD-N and TAD-C) and sequences that mediate O2-dependent regulation of protein stability and transcriptional activity.

As described throughout this book, the majority of bHLH-PAS proteins function as heterodimeric DNA-binding transcription factors. These heterodimers share in common the utilization of members of the ARNT family (ARNT, ARNT2, and ARNT3) as one member of the heterodimer (9-11). Whereas ARNT (HIF-1β) is ubiquitously expressed, ARNT2 and ARNT3 show a more restricted pattern of expression, most notably in the central nervous system. Just as ARNT2 and ARNT3 were identified on the basis of their sequence similarity to ARNT, database searches identified two proteins with sequence similarity to HIF-1α, that were designated HIF-2α (also known as endothelial PAS domain protein (EPAS1), HIF-1α-like factor (HLF), HIF-1α-related factor (HRF), and member of PAS domain family 2 (MOP2)), and HIF-3α (12-16). Whereas HIF-1α is ubiquitously expressed, HIF-2α and HIF-3α show more restricted patterns of expression. All heterodimeric combinations of ARNT and HIF-1α family members have been demonstrated to form *in vitro*, whereas only heterodimers composed of HIF-1α and HIF-1β have been purified from cells.

HIF-1α and HIF-2α are bipartite polypeptides. The amino-terminal half is composed of the bHLH and PAS domains whereas the carboxyl-terminal half is composed of the transactivation and regulatory domains (4, 8, 17-20). HIF-1 and HIF-2, i.e. heterodimers containing HIF-1α or HIF-2α, respectively, activate transcription of target genes that have an HRE containing one or more copies of the core recognition sequence 5'-RCGTG-3'. A splice variant of HIF-3α designated inhibitory PAS domain protein (IPAS) appears to function as a negative regulator of HIF-1α (21, 22).

3. HIF BIOCHEMISTRY

3.1 O_2-Dependent Regulation

The unique characteristic of the HIF-1 family of proteins is that their expression is precisely and instantaneously regulated as a function of the cellular O_2 concentration (23, 24). Investigation of the molecular basis for this regulation revealed a novel O_2-dependent mechanism of post-translational regulation (Figure 2). HIF-1α and HIF-2α are subject to modification by a novel class of dioxygenases that utilize O_2 and 2-oxoglutarate (α-ketoglutarate) as substrates to hydroxylate one or more proline residues embedded within the consensus peptide sequence LXXLAP (A, alanine; L, leucine; P, proline; X, any amino acid) (Epstein et al. 2001). These enzymes contain the sequence HXD (D, aspartate; H, histidine) in their active site which binds a molecule of Fe (II) that catalyzes the prolyl hydroxylation of HIF-1α. In this reaction, 2-oxoglutarate is converted to succinate and CO_2 is generated. Three members of this family have been designated PHD (proline hydroxylase-domain protein) 1, 2, and 3 or, alternatively, as HPH (HIF-1α proline hydroxylase) 3, 2, and 1, respectively (25, 26). In the case of HIF-1α, proline residues 402 and 564 are subject to hydroxylation (27-29). Remarkably, O_2 appears to be a rate-limiting substrate for these enzymes *in vivo* such that reduced O_2 availability leads to reduced hydroxylation of HIF-1α family members.

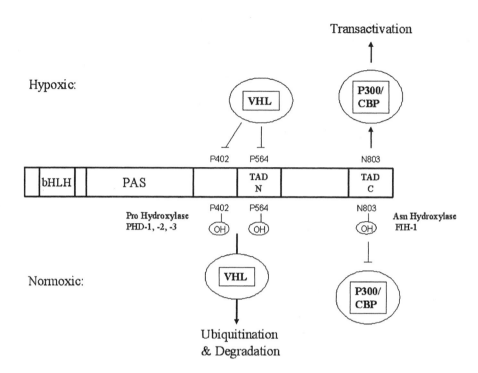

Figure 2. O$_2$-dependent regulation of HIF-1α protein stability and transcriptional activity. Under normoxic conditions, is hydroxylated on proline residues 402 and 564 by the prolyl hydroxylases PHD-1, -2, and -3. Hydroxylation of P402 and P564 is required for binding to VHL, the recognition component of an E3 ubiquitin-protein ligase that targets HIF-1α for ubiquitination and proteasomal degradation. Asparagine residue 803 is also hydroxylated under normoxic conditions by the asparagines hydroxylase FIH-1. Hydroxylation of N803 blocks the interaction of TAD-C with the co-activators p300 and CBP which is required for HIF-1 transcriptional activity.

Proline hydroxylation of HIF-1α or HIF-2α is required for their interaction with the von Hippel-Lindau tumor suppressor (VHL) protein (27, 28). VHL is the recognition component of an E3 ubiquitin-protein ligase complex. Ubiquitination of HIF-1α and HIF-2α by this complex targets these proteins for degradation by the proteasome (30, 31). Thus, the expression of HIF-1α family members is regulated by O$_2$-dependent hydroxylation, ubiquitination, and degradation reactions that are inhibited under hypoxic conditions, leading to rapid accumulation of the proteins (Figure 2). This system is remarkably conserved among metazoan species (26) from *C. elegans* to *H. sapiens* whose body sizes vary by more than 10 orders of magnitude.

The activity of the transactivation domains of HIF-1α and HIF-2α are also regulated by the cellular O$_2$ concentration independent of the protein expression (4, 19, 20). The amino-terminal transactivation domain encompasses P564 and is thus bound by VHL in an O_2-dependent manner. VHL, in addition to its role as the recognition component of an E3 ubiquitin-protein ligase, also recruits histone deacetylases that function as negative regulators of transcription (32). The carboxyl-terminal transactivation domain is regulated independently. Remarkably, the molecular basis of this regulation is also O_2-dependent hydroxylation (Figure 2), but in this case an asparagine residue (N803 in HIF-1α) is modified by a dioxygenase distinct from the proline hydroxylases (33). Factor Inhibiting HIF-1 (FIH-1) was identified in a yeast two-hybrid assay as a protein that interacts with amino acid residues 757-826 of HIF-1α and inhibits its ability to transactivate target genes (32). FIH-1 also stably interacts with VHL. Most remarkably, FIH-1 is the asparagine hydroxylase that modifies N803 of HIF-1α thus preventing interaction of the carboxyl-terminal transactivation domain with the coactivators p300 and CBP (33, 34). This enzymatic activity also requires O_2, Fe(II), and 2-oxoglutarate, similar to the proline hydroxylases that also modify HIF-1α. Thus, hydroxylation is a post-translational modification that is similar to other post-translational modifications such as phosphorylation in providing a mechanism for regulating (positively or negatively) protein-protein interactions. However, the hydroxylation of HIF-1α family members is unique in its dependence upon the cellular O_2 concentration.

3.2 O_2-Independent Regulation of HIF-1α

The regulation of HIF-1α expression and activity described above is a direct response to changes in the cellular O_2 concentration because of the putative high K_m of the proline and asparagine hydroxylases for O_2, although thus far only non-quantitative studies of the proline hydroxylases and no studies of O_2-activity relationship for the asparagine hydroxylase FIH-1 have been reported. Iron chelators, such as desferrioxamine, and divalent metal cations, such as Co (II) also induce HIF-1α expression (35) and activity (4, 19) by inhibiting hydroxylase activity. HIF-1 activity is not induced significantly by inhibiting oxidative phosphorylation, again indicating that hypoxia rather than a sequela, such as ATP depletion, is the stimulus for HIF-1 activation.

The first indication that a second pathway for HIF-1 activation existed was the discovery that HIF-1α expression and HIF-1 transcriptional activation of target genes (those encoding VEGF and enolase 1) were induced in cells transformed with the *v-src* oncogene (36). Subsequently, HIF-1α expression was demonstrated in several human prostate cancer cell lines under non-hypoxic conditions (37) which could be eliminated by treatment of the cells with inhibitors of the signal transduction pathway leading from the epidermal growth factor receptor (EGFR) tyrosine kinase to phosphatidylinositol-3-kinase (PI-3-kinase) and the downstream serine/threonine kinase AKT (also known as protein kinase B) and FRAP (FKBP-rapamycin associated protein; also known as mammalian target of rapamycin).

An analysis of human breast cancer cells in which the EGFR family member HER2[neu] was activated revealed that induction of *VEGF* expression in these cells was due to increased translation of HIF-1α mRNA into protein that was also dependent upon activity of the PI-3-kinase – AKT – FRAP pathway (38). In human colon cancer cells stimulated with insulin-like growth factor 1 (IGF-1) or IGF-2, IGF-1 receptor activation leads to increased HIF-1α protein and VEGF mRNA expression as a result of signaling via both the PI-3-kinase and p42/p44 (ERK) MAP kinase pathways (39).

Both the PI-3-kinase and MAP kinase signal transduction pathways are known to phosphorylate key regulators of translational initiation and, by doing so, stimulate the synthesis of a subset of proteins within growth factor-treated cells. Thus, in contrast to hypoxia, which induces HIF-1α expression by inhibiting degradation of the protein, growth factor – receptor-tyrosine kinase signaling induces HIF-1α expression by stimulating synthesis of the protein (Figure 3). In addition, the ERK signal-transduction pathway has been shown to induce HIF-1α transactivation function (40, 41). Thus, growth factors can also induce both HIF-1α expression and activity. However, hypoxia activates HIF-1 in all cell types whereas the effects of growth factor stimulation appear to be cell type specific. These findings indicate that HIF-1 expression cannot be considered as a surrogate marker of hypoxia since other stimuli can also induce its expression. In particular, these signaling pathways are constitutively activated in cancer cells and lead to constitutive activation of HIF-1 that plays important roles in tumor biology (2).

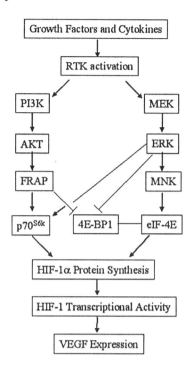

Figure 3. Activation of HIF-1 expression by kinase cascades. Activation of the phosphatidylinositol-3-kinase (PI3K) or MAP kinase (ERK) signal transduction pathway by receptor tyrosine kinase (RTK) signaling leads to the phosphorylation of the translational regulatory proteins p70 S6 kinase, 4E-BP1, and eIF-4E. As a result, translation of a subset of mRNAs within the cell, including HIF-1α mRNA, is selectively upregulated. This pathway provides by a mechanism for increased HIF-1α protein synthesis that is cell-type- and ligand-specific.

3.3   Target Genes

Currently, over 60 genes are known to be regulated by HIF-1 is and the number is continuing to grow rapidly (Table 1). These target genes can be grouped according to the physiological role played by their protein products. One group includes proteins that increase O_2 delivery to hypoxic tissue. For example, erythropoietin increases the number of red cells and thus increases the blood O_2-carrying capacity. In addition, the delivery of iron to the bone marrow for incorporation into hemoglobin within erythrocytes is also controlled by HIF-1 via transactivation of the genes encoding transferrin and transferrin receptor. Finally, VEGF stimulates the formation of new blood vessels to deliver O_2-carrying erythrocytes to hypoxic tissue. A second group of HIF-1 target genes encodes proteins such as glucose transporters and glycolytic enzymes that allow cells to adapt their energy metabolism to reduced O_2 availability by increasing their rate of anaerobic ATP generation via glycolysis. A third group of target genes encode proteins that function as survival factors that inhibit hypoxia-induced cell death, including erythropoietin, insulin-like growth factor 2, nitric oxide synthase 2, and VEGF.

Table 1. HIF-1 Target Genes.

1. Targets for Activation

Aminopeptidase A (42)	Adenylate kinase 3 (43)
α_{1B} adrenergic receptor (44)	Adrenomedullin (45)
Aldolase A (46, 47)	Aldolase C (46)
Carbonic anhydrase 9 (42)	Carbonic anhydrase 12 (42)
Cathepsin D (48)	Ceruloplasmin (49)
Collagen type V, a1 (42)	Cyclin G1 (48)
DEC1 (42)	Endocrine gland-derived VEG (50)
Endothelin-1 (51)	Enolase 1 (46)
Erythropoietin (18)	ETS-1 (52)
Fibronectin 1 (48)	Glucose transporter 1 (43, 46, 47)
Glucose transporter 3 (46)	Glyceraldehyde-3-P dehydrogenase (46, 47)
Heme oxygenase-1 (53)	Hexokinase 1 (46)
Hexokinase 2 (46)	Insulin-like growth factor (IGF) 2 (54)
IGF binding protein 1	IGF factor binding protein 2
IGF factor binding protein 3	Intestinal trefoil factor
Keratin 14 (48)	Keratin 18 (48)
Keratin 19 (48)	Lactate dehydrogenase A (46, 47)
LDL receptor-related protein 1 (42)	Matrix metalloproteinase 2 (48)
MIC2 (42)	Multi-drug resistance 1 (55)
Nitric oxide synthase 2 (56, 57)	NIP3 (58, 59)
NIX (58)	P21 (60)
P35srj (61)	Phosphofructokinase L (46)
Phosphoglycerate kinase 1 (46, 47, 60)	PFKBF3 (62)
Plasminogen activator inhibitor 1 (63)	Prolyl-4-hydroxylase α (I) (64)
Pyruvate kinase M (PKM) (46)	RTP801 (65)
Transferrin (66)	Transferrin receptor (67, 68)
Tranforming Growth Factor α (48)	Transforming Growth Factor $\beta3$ (69)
Transglutaminase 2 (42)	Triosephosphate isomerase (46)
Urokinase plasminogen activator rec. (48)	Vascular endothelial growth fac. (46, 47, 60)
VEGF receptor-1 (FLT-1) (70)	VEGF receptor-2 (FLK-1) (71)
Vimentin (48)	

2. Targets for Repression

α-fetoprotein (72)	PPAR-α (73)

Whereas the glucose transporters, glycolytic enzymes, and VEGF are induced by hypoxia in most cell types, other genes are only induced by hypoxia in a limited number of cell types. A reasonable prediction (to the closest order of magnitude) is that approximately 1% of all human genes may demonstrate HIF-1-dependent hypoxia-inducible expression in one or more cell types. A knowledge of the complete battery of genes whose expression is regulated by HIF-1 in a given cell type is necessary (but not sufficient) for a complete understanding of the molecular basis of physiological and/or pathophysiological responses to hypoxia or ischemia. Indeed, the great challenge of defining the role of a transcription factor that controls the expression of such a large battery of genes is to determine among these which gene products contribute protective effects, which gene products contribute pathogenic effects, and which gene products are irrelevant to the particular biological process under investigation.

4. HIF GENETICS

4.1 HIF-1α Expression is Essential for Embryonic Development

To investigate the role of HIF-1 in embryogenesis, a null mutation was engineered in the mouse *Hif1a* gene by homologous recombination in embryonic stem cells by replacement of exon 2, encoding the bHLH domain of HIF-1α, with a gene encoding neomycin resistance (46). The knockout allele was transmitted through the germline and mice heterozygous for the mutation developed normally. However, mating of heterozygotes revealed that embryos homozygous for the mutant allele arrested in their development by E9.0 and died by E10.5 with failure of neural tube closure and major cardiac malformations. The embryonic lethality of mice lacking HIF-1α expression was subsequently confirmed by analysis of mice homozygous for two independently derived knockout alleles that also targeted exon 2 of the *Hif1a* gene (47, 60).

Vascular development in stage-matched wild-type and *Hif1a*[-/-] embryos was analyzed by whole mount immunohistochemistry using an anti-PECAM-1 antibody (46). These studies revealed that vasculogenesis initiated properly in *Hif1a*[-/-] embryos. However, at the stage of remodeling of the primary vascular plexus in wild-type embryos (E8.75-E9.25), striking vascular regression occurred in the cephalic and branchial regions of the embryo, leading to the formation of enormous dilated endothelial-lined structures similar to those observed in *Vegf*[+/-] embryos. In addition to the striking vascular regression, dramatic mesenchymal cell death occurred in the cephalic and branchial regions (46). The first defect noted in *Hif1a*[-/-]

embryos by scanning electron microscopy was the loss of cells at the neurosomatic junction at E8.5-E8.75, indicating death of cephalic neural crest cells, the progenitors of cephalic mesenchymal cells, which in turn represents the population from which vascular pericytes are derived (74). The establishment of tight-junction interactions with pericytes is critical for endothelial cell survival at this stage of embryogenesis. Thus, the endothelial cell death observed in HIF-1α-null embryos may occur as a cell-autonomous effect of HIF-1α deficiency in endothelial cells or may occur secondary to the loss of supporting cells, although these pathogenic mechanisms are not mutually exclusive.

The findings described above suggest that in the absence of HIF-1α, certain cell populations within the embryo cannot survive the normally low O_2 concentrations that exist in the midgestation embryo. Immunoblot analysis of wild-type embryos revealed a marked increase in HIF-1α expression between E8.5 and E9.5, *i.e.*, the period during which the development of HIF-1α-null embryos fails (46). At this stage, the size of the embryo increases to the point at which O_2 can no longer be delivered to all cells by diffusion and the establishment of a functioning circulatory system is essential for further embryonic development and survival. The analysis of HIF-1α-null embryos indicates that HIF-1α is required for cardiovascular development and the survival of endothelial and mesenchymal cells.

4.2   Cell-lineage-specific Effects of HIF-1α Deficiency

The death of HIF-1α-null embryos at midgestation precluded an analysis of the effect of complete HIF-1α deficiency on later developmental processes. One approach to circumvent the lethal effect of HIF-1α deficiency on embryogenesis is to generate conditional knockout alleles in which the targeted recombination event is restricted to a single cell type by the use of Cre-Lox technology. The analysis of mice in which Cre-mediated *Hif1a* gene inactivation occurred only in chondrocytes revealed an important role for HIF-1 in bone formation (75). These studies revealed that the avascular cartilaginous growth plate of long bones is a highly hypoxic environment. In this milieu, the survival of chondrocytes is dramatically reduced in the absence of HIF-1α. These studies again point to a critical role for HIF-1 in promoting cell survival during embryogenesis. Cre-Lox technology will allow the role of HIF-1 in a variety of cell types to be studied and it will be interesting to determine how many other cell types are dependent upon HIF-1α for survival and/or normal differentiation.

Another approach that can be utilized to circumvent the lethality of complete HIF-1α deficiency is to generate chimeric mice in which only certain cell types are derived from *Hif1a*$^{-/-}$ cells. *Rag2*$^{-/-}$ mice lack B and T lymphocytes due to loss of an enzyme that is required for rearrangement of the *Ig* and *Tcr* genes encoding immunoglobulins and T cell receptors, respectively. Injection of wild-type ES cells into a *Rag2$^{-/-}$* mouse blastocyst results in rescue of lymphocyte development. This system can be used to determine whether a gene product is required for lymphocyte development by injecting ES cells that are homozygous for a loss-of-function allele at the relevant gene locus. The injection of *Hif1a$^{-/-}$* ES cells into *Rag2$^{-/-}$* blastocysts resulted in the development of chimeric mice with normal T cell numbers, similar to the injection of wild-type ES cells. However, major defects in B lymphocyte development occurred in *Hif1a$^{-/-}$* → *Rag2$^{-/-}$* chimeric mice (76). The maturation and proliferation of conventional B-2 lymphocytes in the bone marrow was impaired. In contrast, excess peritoneal B-1 lymphocytes with abnormal expression of cell surface markers developed. In addition, *Hif1a$^{-/-}$* → *Rag2$^{-/-}$* mice manifested an autoimmune syndrome characterized by the accumulation of anti-dsDNA antibodies and rheumatoid factor in serum, deposits of IgG and IgM in the kidney, and proteinuria. These studies revealed dramatic, lineage-specific effects of HIF-1α deficiency on B lymphocyte development.

4.3 Effects of Partial HIF-1α Deficiency on Cardiovascular and Respiratory Physiology

As described above, *Hif1a$^{+/-}$* mice develop normally and are viable and fertile. However, these mice have impaired physiological responses when subjected to chronic hypoxia (77). Chronic hypoxia can occur either as an environmental challenge (*e.g.,* ascent to high altitude) or as a medical complication (*e.g.,* in patients with chronic lung disease or congenital heart disease). Chronic hypoxia induces production of erythropoietin, which stimulates red blood cell production resulting in polycythemia. The increases in the blood hemoglobin and hematocrit levels observed in wild-type mice after 3 weeks at 10% O_2 were significantly reduced in *Hif1a$^{+/-}$* littermates. Weight loss is also observed in animals subjected to chronic hypoxia. Compared to wild-type littermates, *Hif1a$^{+/-}$* mice manifested significantly increased weight loss. Thus, partial HIF-1α deficiency is associated with impaired ability to increase blood O_2-carrying capacity and to maintain body weight under conditions of chronic hypoxia.

In patients with chronic lung disease an uncommon but often fatal complication is the development of hypoxic pulmonary hypertension. This disease can be modeled in wild-type mice exposed to 10% O_2 for 3 weeks, which develop increased tone and medial wall thickness within pulmonary arterioles. The changes increase pulmonary arterial resistance, resulting in right ventricular hypertrophy. All of these cardiovascular responses to chronic hypoxia were impaired in $Hif1a^{+/-}$ mice (77).

Pulmonary artery smooth muscle (PASM) cell hypertrophy can be analyzed by the electrophysiological parameter capacitance, which is a measure of cell volume. Studies of PASM cells that were isolated from wild-type mice after exposure to 10% O_2 for 3 weeks revealed increased capacitance, whereas PASM cells isolated from hypoxic $Hif1a^{+/-}$ mice showed no increase in capacitance (78). Isolated PASM cells from hypoxic wild-type mice also manifested depolarization of the resting membrane potential due to decreased activity of voltage-gated potassium K_v channels. These hypoxia-induced alterations in membrane potential and K_v currents were markedly impaired in PASM cells from $Hif1a^{+/-}$ mice. Thus, partial deficiency of HIF-1α has dramatic effects on the two major responses of PASM cells to chronic hypoxia, hypertrophy and depolarization, which result in increased pulmonary arterial tone and medial wall thickness, respectively, that are hallmarks of pulmonary hypertension. Since these changes occurred in vivo it remains to be determined whether they represent cell autonomous responses of PASM cells to hypoxia or whether they occur as a result of hypoxia-induced production of vasoactive molecules such as endothelin-1 by PA endothelial cells.

The studies described thus far have demonstrated that HIF-1 plays important roles in cardiac and vascular development and in the postnatal control of red blood cell production and pulmonary vascular remodeling in response to hypoxia. Thus, all three major components of the circulatory system (heart, vessels, and blood) are regulated by HIF-1. The other major physiological system involved in O_2 delivery in mammals is the respiratory system. Acute changes in blood PO_2 are sensed by the carotid body, a specialized chemoreceptor organ located at the bifurcation of the carotid artery. The type II or glomus cells in the carotid body depolarize in response to hypoxia and afferents projecting to the brainstem initiate the reflex pathway resulting in ventilatory responses that increase O_2 uptake.

When wild-type mice are exposed to 10% O_2 for 3 days and returned to room air, they manifest an augmented ventilatory response to a subsequent acute hypoxic challenge, indicating a physiological adaptation to hypoxia. $Hif1a^{+/-}$ mice manifest normal ventilatory responses to acute hypoxia but manifest no ventilatory adaptation in response to chronic hypoxia (79). Ventilatory adaptation to chronic hypoxia is mediated by the carotid body in wild-type mice, suggesting abnormal carotid body function in $Hif1a^{+/-}$ mice. When carotid bodies isolated from wild-type mice were exposed to either cyanide or hypoxia, a marked increase in carotid sinus nerve activity was recorded. Remarkably, carotid bodies from $Hif1a^{+/-}$ mice depolarized in response to cyanide but showed no neural activity in response to hypoxia.

This profound but selective physiological defect was not associated with any histological abnormality of carotid body size or glomus cell number.

In *Hif1a*$^{+/-}$ mice, the ventilatory responses to acute hypoxia were mediated by other chemoreceptors, such as the aortic bodies, that utilize vagal afferents to signal to the central nervous system. Whereas vagotomy had little effect on ventilatory responses to acute hypoxia in wild-type littermates, it significantly impaired these responses in *Hif1a*$^{+/-}$ mice. This reliance on chemoreceptors other than the carotid body is similar to the adaptation that occurs in normal animals following carotid denervation. Thus partial HIF-1α deficiency has a dramatic effect on carotid body neural activity and ventilatory adaptation to chronic hypoxia. Furthermore, these studies indicate differences in the regulation of carotid and non-carotid body chemoreceptor function as reflected in the variable effects of partial HIF-1α deficiency.

The remarkable effects of partial HIF-1α deficiency on physiological responses to hypoxia are not matched in any system other than the dramatic vascular failure and embryonic lethality associated with partial VEGF deficiency (80, 81). The profound effects of partial HIF-1α deficiency may reflect the fact that HIF-1 coordinates physiological responses by regulating the expression of multiple genes encoding proteins that are required for these responses, resulting in a synergistic effect of partial loss-of-function for multiple system components. An alternative and not mutually exclusive possibility is based upon the demonstration that HIF-1α expression increases exponentially as O_2 concentrations decline, providing a mechanism for a graded transcriptional response to hypoxia. The decreased levels of HIF-1α expressed at any given level of hypoxia in *Hif1a*$^{+/-}$ mice would thus result in a suboptimal response to the stimulus. The complete failure of carotid body glomus cells to depolarize in response to hypoxia indicates that a threshold level of HIF-1α may be necessary for some physiological responses to occur, although it is also possible that an underlying defect in carotid body development that was not apparent on histological examination may be responsible.

4.4 Effects of Complete HIF-2α Deficiency on Mouse Development

HIF-2α is expressed in a restricted number of tissue types including developing lung and kidney and in vascular endothelial cells. Homologous recombination in ES cells was performed to inactivate the mouse *Epas1* gene encoding HIF-2α by replacement of exon 2 encoding the bHLH domain. Using the same strategy two groups reported strikingly different results. The first group reported that embryonic lethality of *Epas1*$^{-/-}$ mice occurred between E12.5 and E16.5 and was associated with severe bradycardia secondary to reduced catecholamine levels that could be rescued by maternal administration of the catecholamine precursor dihydroxyphenylserine (82). In wild-type mice, in situ hybridization revealed expression of HIF-2α mRNA in the organ of Zuckerkandl which is the source of catecholamines in the mouse embryo. These results revealed an unexpected role for HIF-2α in the regulation of embryonic catecholamine production.

In contrast to these results, a second group reported that HIF-2α-null embryos died between E9.5 and E13.5 with severe vascular defects (83). Vasculogenesis occurred normally in these embryos but there was a failure to remodel the initial vascular plexus into vessels of varying diameter (*i.e.,* large-diameter proximal vessels giving rise to progressively smaller distal radicles) based upon blood flow. The knockout mice were generated on different mouse strains in the two labs and the striking difference in the phenotype associated with complete loss of function for HIF-2α is most likely a reflection of genetic variation at one or more other loci. One obvious candidate modifier locus is *Hif1a* but studies designed to test this hypothesis have not been reported.

4.5   Effects of Complete ARNT Deficiency on Mouse Development

Embryos completely lacking ARNT expression die by E10.5 with vascular defects (84). Despite the use of ARNT as a common subunit for multiple bHLH-PAS proteins, the defects identified in ARNT-null embryos appeared to reflect a deficiency of dimerization partner for HIF-1α and HIF-2α. Several phenotypic differences between HIF-1α- and ARNT-deficient embryos were instructive. ARNT-null embryos were remarkable for the degree of failure of yolk sac angiogenesis and abnormal placental formation, suggesting that HIF-1α and HIF-2α both play a role for in these processes. HIF-1α-null embryos manifested a more dramatic failure of vascular development and cell death in the cephalic region. This may reflect the expression in the developing central nervous system of ARNT2, which may provide an alternative dimerization partner for HIF-1α and HIF-2α in the cephalic region of ARNT-null embryos. Although it is difficult to interpret the effects of ARNT deficiency because of its multiple dimerization partners, a failure to form active HIF-1 and HIF-2 heterodimers is a likely consequence due to the restricted expression patterns of ARNT2 and ARNT3.

Analysis of ARNT-null embryos revealed defective placental development characterized by reduced labyrinthine and spongiotrophoblast layers and an increased number of trophoblast giant cells (85). When trophoblast stem cells were cultured ex vivo, exposure of the cells to hypoxic conditions favored their differentiation into spongiotrophoblasts as opposed to giant cells (85). Several recent studies have demonstrated that critical aspects of placental development are O_2-regulated (69, 86) and the analyses of ARNT-deficient embryos described above suggest that these processes are mediated by HIF-1 and HIF-2. This hypothesis is supported by immunohistochemical analyses of human placentae (87, 88).

Another experimental approach to understanding the role of specific gene products in developmental processes that has not been previously discussed is the differentiation of cultured ES cells into embryoid bodies (EBs) which contain a variety of cell types found in the developing embryo. Differentiation into specific cell lineages can be promoted by culturing the EBs in specific cytokines or growth factors. Using this approach, differentiation of wild-type ES cells under hypoxic culture conditions promoted the appearance of hematopoietic progenitor cells, an effect that was not observed when ARNT-null ES cells were used (89). Furthermore, yolk sac cultures from ARNT-null embryos yielded decreased numbers of hematopoietic colonies. Expression of VEGF in ARNT-null EBs was decreased and deficient hematopoietic colony formation could be corrected by addition of VEGF to the culture medium. Taken together these data suggest that heterodimers of ARNT with HIF-1α or HIF-2α play an important role in yolk sac hematopoiesis.

5. SUMMARY

Analysis of mice deficient for HIF-1α and HIF-2α has revealed multiple essential roles for these proteins in the development of specialized physiological systems that have evolved in mammals for the purpose of maintaining O_2 homeostasis. Use of mice that are deficient for either HIF-1α or HIF-2α almost certainly underestimates the effects of these factors for two reasons. First, embryonic lethality precludes analysis of effects at later developmental stages. Second, HIF-1α and HIF-2α are co-expressed in a number of tissues and in some cases may function in a redundant manner, which may account for some of the phenotypic effects of ARNT deficiency as described above. Additional important homeostatic effects of HIF-1 family members that have not been discussed in detail in this review occur at the cellular level, *e.g.,* in the regulation of energy metabolism according to O_2 availability. In addition, HIF-1 family members play critical roles in the pathophysiology of ischemic and neoplastic disorders, which represent the most common causes of mortality in the U.S. population. Additional studies using conditional knockouts and other experimental strategies are necessary to further define the molecular mechanisms by which these factors function as master regulators of oxygen homeostasis and exert their considerable influence over physiology and medicine.

REFERENCES

1.   Semenza, G. L. 2001. Hypoxia-inducible factor 1: oxygen homeostasis and disease pathophysiology. Trends Mol. Med. 7:345-50.
2.   Semenza, G. L. 2002. HIF-1 and tumor progression: pathophysiology and therapeutics. Trends Mol. Med. 8:S62-7.
3.   Beck, I., S. Ramirez, R. Weinmann, and J. Caro. 1991. Enhancer element at the 3'-flanking region controls transcriptional response to hypoxia in the human erythropoietin gene. J. Biol. Chem. 266:15563-6.
4.   Pugh, C. W., C. C. Tan, R. W. Jones, and P. J. Ratcliffe. 1991. Functional analysis of an oxygen-regulated transcriptional enhancer lying 3' to the mouse erythropoietin gene. Proc. Natl. Acad. Sci. USA 88:10553-7.
5.   Semenza, G. L., M. K. Nejfelt, S. M. Chi, and S. E. Antonarakis. 1991. Hypoxia-inducible nuclear factors bind to an enhancer element located 3' to the human erythropoietin gene. Proc. Natl. Acad. Sci. USA 88:5680-4.
6.   Semenza, G. L., and G. L. Wang. 1992. A nuclear factor induced by hypoxia via de novo protein synthesis binds to the human erythropoietin gene enhancer at a site required for transcriptional activation. Mol. Cell. Biol. 12:5447-54.
7.   Wang, G. L., and G. L. Semenza. 1995. Purification and characterization of hypoxia-inducible factor 1. J. Biol. Chem. 270:1230-7.
8.   Wang, G. L., B. H. Jiang, E. A. Rue, and G. L. Semenza. 1995. Hypoxia-inducible factor 1 is a basic-helix-loop-helix-PAS heterodimer regulated by cellular O2 tension. Proc. Natl. Acad. Sci. USA 92:5510-4.
9.   Hirose, K., M. Morita, M. Ema, J. Mimura, H. Hamada, H. Fujii, Y. Saijo, O. Gotoh, K. Sogawa, and Y. Fujii-Kuriyama. 1996. cDNA cloning and tissue-specific expression of a novel basic helix-loop-helix/PAS factor (Arnt2) with close sequence similarity to the aryl hydrocarbon receptor nuclear translocator (Arnt). Mol. Cell. Biol. 16:1706-13.
10.   Hoffman, E. C., H. Reyes, F. F. Chu, F. Sander, L. H. Conley, B. A. Brooks, and O. Hankinson. 1991. Cloning of a factor required for activity of the Ah (dioxin) receptor. Science 252:954-8.
11.   Takahata, S., K. Sogawa, A. Kobayashi, M. Ema, J. Mimura, N. Ozaki, and Y. Fujii-Kuriyama. 1998. Transcriptionally active heterodimer formation of an Arnt-like PAS protein, Arnt3, with HIF-1a, HLF, and clock. Biochem. Biophys. Res. Commun. 248:789-94.
12.   Ema, M., S. Taya, N. Yokotani, K. Sogawa, Y. Matsuda, and Y. Fujii-Kuriyama. 1997. A novel bHLH-PAS factor with close sequence similarity to hypoxia-inducible factor 1alpha regulates the VEGF expression and is potentially involved in lung and vascular development. Proc. Natl. Acad. Sci. USA 94:4273-8.
13.   Flamme, I., T. Frohlich, M. von Reutern, A. Kappel, A. Damert, and W. Risau. 1997. HRF, a putative basic helix-loop-helix-PAS-domain transcription factor is closely

related to hypoxia-inducible factor-1 alpha and developmentally expressed in blood vessels. Mech. Dev. 63:51-60.

14. Gu, Y. Z., S. M. Moran, J. B. Hogenesch, L. Wartman, and C. A. Bradfield. 1998. Molecular characterization and chromosomal localization of a third alpha-class hypoxia inducible factor subunit, HIF3alpha. Gene Expr. 7:205-13.

15. Hogenesch, J. B., W. K. Chan, V. H. Jackiw, R. C. Brown, Y. Z. Gu, M. Pray-Grant, G. H. Perdew, and C. A. Bradfield. 1997. Characterization of a subset of the basic-helix-loop-helix-PAS superfamily that interacts with components of the dioxin signaling pathway. J. Biol. Chem. 272:8581-93.

16. Tian, H., S. L. McKnight, and D. W. Russell. 1997. Endothelial PAS domain protein 1 (EPAS1), a transcription factor selectively expressed in endothelial cells. Genes Dev. 11:72-82.

17. Huang, L. E., J. Gu, M. Schau, and H. F. Bunn. 1998. Regulation of hypoxia-inducible factor 1alpha is mediated by an O2-dependent degradation domain via the ubiquitin-proteasome pathway. Proc. Natl. Acad. Sci. USA 95:7987-92.

18. Jiang, B. H., E. Rue, G. L. Wang, R. Roe, and G. L. Semenza. 1996. Dimerization, DNA binding, and transactivation properties of hypoxia-inducible factor 1. J. Biol. Chem. 271:17771-8.

19. Jiang, B. H., J. Z. Zheng, S. W. Leung, R. Roe, and G. L. Semenza. 1997. Transactivation and inhibitory domains of hypoxia-inducible factor 1alpha. Modulation of transcriptional activity by oxygen tension. J. Biol. Chem. 272:19253-60.

20. O'Rourke, J. F., Y. M. Tian, P. J. Ratcliffe, and C. W. Pugh. 1999. Oxygen-regulated and transactivating domains in endothelial PAS protein 1: comparison with hypoxia-inducible factor-1alpha. J. Biol. Chem. 274:2060-71.

21. Makino, Y., R. Cao, K. Svensson, G. Bertilsson, M. Asman, H. Tanaka, Y. Cao, A. Berkenstam, and L. Poellinger. 2001. Inhibitory PAS domain protein is a negative regulator of hypoxia-inducible gene expression. Nature 414:550-4.

22. Makino, Y., A. Kanopka, W. J. Wilson, H. Tanaka, and L. Poellinger. 2002. Inhibitory PAS domain protein (IPAS) is a hypoxia-inducible splicing variant of the hypoxia-inducible factor-3alpha locus. J. Biol. Chem. 277:32405-8.

23. Jewell, U. R., I. Kvietikova, A. Scheid, C. Bauer, R. H. Wenger, and M. Gassmann. 2001. Induction of HIF-1alpha in response to hypoxia is instantaneous. FASEB J. 15:1312-4.

24. Jiang, B. H., G. L. Semenza, C. Bauer, and H. H. Marti. 1996. Hypoxia-inducible factor 1 levels vary exponentially over a physiologically relevant range of O2 tension. Am. J. Physiol. 271:C1172-80.

25. Bruick, R. K., and S. L. McKnight. 2001. A conserved family of prolyl-4-hydroxylases that modify HIF. Science 294:1337-40.

26. Epstein, A. C., J. M. Gleadle, L. A. McNeill, K. S. Hewitson, J. O'Rourke, D. R. Mole, M. Mukherji, E. Metzen, M. I. Wilson, A. Dhanda, et al. 2001. C. elegans EGL-9 and mammalian homologs define a family of dioxygenases that regulate HIF by prolyl hydroxylation. Cell 107:43-54.

27. Ivan, M., K. Kondo, H. Yang, W. Kim, J. Valiando, M. Ohh, A. Salic, J. M. Asara, W. S. Lane, and W. G. Kaelin, Jr. 2001. HIFalpha targeted for VHL-mediated destruction by proline hydroxylation: implications for O2 sensing. Science 292:464-8.

28. Jaakkola, P., D. R. Mole, Y. M. Tian, M. I. Wilson, J. Gielbert, S. J. Gaskell, A. Kriegsheim, H. F. Hebestreit, M. Mukherji, C. J. Schofield, et al. 2001. Targeting of

HIF-alpha to the von Hippel-Lindau ubiquitylation complex by O2-regulated prolyl hydroxylation. Science 292:468-72.

29. Masson, N., C. Willam, P. H. Maxwell, C. W. Pugh, and P. J. Ratcliffe. 2001. Independent function of two destruction domains in hypoxia-inducible factor-alpha chains activated by prolyl hydroxylation. EMBO J. 20:5197-206.

30. Maxwell, P. H., M. S. Wiesener, G. W. Chang, S. C. Clifford, E. C. Vaux, M. E. Cockman, C. C. Wykoff, C. W. Pugh, E. R. Maher, and P. J. Ratcliffe. 1999. The tumour suppressor protein VHL targets hypoxia-inducible factors for oxygen-dependent proteolysis. Nature 399:271-5.

31. Tanimoto, K., Y. Makino, T. Pereira, and L. Poellinger. 2000. Mechanism of regulation of the hypoxia-inducible factor-1 alpha by the von Hippel-Lindau tumor suppressor protein. EMBO J. 19:4298-309.

32. Mahon, P. C., K. Hirota, and G. L. Semenza. 2001. FIH-1: a novel protein that interacts with HIF-1alpha and VHL to mediate repression of HIF-1 transcriptional activity. Genes Dev. 15:2675-86.

33. Lando, D., D. J. Peet, J. J. Gorman, D. A. Whelan, M. L. Whitelaw, and R. K. Bruick. 2002. FIH-1 is an asparaginyl hydroxylase enzyme that regulates the transcriptional activity of hypoxia-inducible factor. Genes Dev. 16:1466-71.

34. Hewitson, K. S., L. A. McNeill, M. V. Riordan, Y. M. Tian, A. N. Bullock, R. W. Welford, J. M. Elkins, N. J. Oldham, S. Bhattacharya, J. M. Gleadle, et al. 2002. Hypoxia-inducible factor (HIF) asparagine hydroxylase is identical to factor inhibiting HIF (FIH) and is related to the cupin structural family. J. Biol. Chem. 277:26351-5.

35. Wang, G. L., and G. L. Semenza. 1993. Desferrioxamine induces erythropoietin gene expression and hypoxia-inducible factor 1 DNA-binding activity: implications for models of hypoxia signal transduction. Blood 82:3610-5.

36. Jiang, B. H., F. Agani, A. Passaniti, and G. L. Semenza. 1997. V-SRC induces expression of hypoxia-inducible factor 1 (HIF-1) and transcription of genes encoding vascular endothelial growth factor and enolase 1: involvement of HIF-1 in tumor progression. Cancer Res. 57:5328-35.

37. Zhong, H., K. Chiles, D. Feldser, E. Laughner, C. Hanrahan, M. M. Georgescu, J. W. Simons, and G. L. Semenza. 2000. Modulation of hypoxia-inducible factor 1alpha expression by the epidermal growth factor/phosphatidylinositol 3-kinase/PTEN/AKT/FRAP pathway in human prostate cancer cells: implications for tumor angiogenesis and therapeutics. Cancer Res. 60:1541-5.

38. Laughner, E., P. Taghavi, K. Chiles, P. C. Mahon, and G. L. Semenza. 2001. HER2 (neu) signaling increases the rate of hypoxia-inducible factor 1alpha (HIF-1alpha) synthesis: novel mechanism for HIF-1-mediated vascular endothelial growth factor expression. Mol. Cell. Biol. 21:3995-4004.

39. Fukuda, R., K. Hirota, F. Fan, Y. D. Jung, L. M. Ellis, and G. L. Semenza. 2002. Insulin-like growth factor 1 induces hypoxia-inducible factor 1-mediated vascular endothelial growth factor expression, which is dependent on MAP kinase and phosphatidylinositol 3-kinase signaling in colon cancer cells. J. Biol. Chem. 277:38205-11.

40. Sodhi, A., S. Montaner, V. Patel, M. Zohar, C. Bais, E. A. Mesri, and J. S. Gutkind. 2000. The Kaposi's sarcoma-associated herpes virus G protein-coupled receptor up-regulates vascular endothelial growth factor expression and secretion through mitogen-activated protein kinase and p38 pathways acting on hypoxia-inducible factor 1alpha. Cancer Res. 60:4873-80.

41. Richard, D. E., E. Berra, E. Gothie, D. Roux, and J. Pouyssegur. 1999. p42/p44 mitogen-activated protein kinases phosphorylate hypoxia-inducible factor 1alpha (HIF-1alpha) and enhance the transcriptional activity of HIF-1. J. Biol. Chem. 274:32631-7.

42. Wykoff, C. C., C. W. Pugh, P. H. Maxwell, A. L. Harris, and P. J. Ratcliffe. 2000. Identification of novel hypoxia dependent and independent target genes of the von Hippel-Lindau (VHL) tumour suppressor by mRNA differential expression profiling. Oncogene 19:6297-305.

43. Wood, S. M., M. S. Wiesener, K. M. Yeates, N. Okada, C. W. Pugh, P. H. Maxwell, and P. J. Ratcliffe. 1998. Selection and analysis of a mutant cell line defective in the hypoxia-inducible factor-1 alpha-subunit (HIF-1alpha). Characterization of hif-1alpha-dependent and -independent hypoxia-inducible gene expression. *J. Biol. Chem.* 273:8360-8.

44. Eckhart, A. D., N. Yang, X. Xin, and J. E. Faber. 1997. Characterization of the alpha1B-adrenergic receptor gene promoter region and hypoxia regulatory elements in vascular smooth muscle. Proc. Natl. Acad. Sci. USA 94:9487-92.

45. Cormier-Regard, S., S. V. Nguyen, and W. C. Claycomb. 1998. Adrenomedullin gene expression is developmentally regulated and induced by hypoxia in rat ventricular cardiac myocytes. J. Biol. Chem. 273:17787-92.

46. Iyer, N. V., L. E. Kotch, F. Agani, S. W. Leung, E. Laughner, R. H. Wenger, M. Gassmann, J. D. Gearhart, A. M. Lawler, A. Y. Yu, et al. 1998. Cellular and developmental control of O2 homeostasis by hypoxia-inducible factor 1 alpha. Genes Dev. 12:149-62.

47. Ryan, H. E., J. Lo, and R. S. Johnson. 1998. HIF-1 alpha is required for solid tumor formation and embryonic vascularization. EMBO J. 17:3005-15.

48. Krishnamachary, B., S. Berg-Dixon, B. Kelly, F. Agani, D. Feldser, G. Ferreira, N. Iyer, J. LaRusch, B. Pak, P. Taghavi, et al. 2003. Regulation of colon carcinoma cell invasion by hypoxia-inducible factor 1. Cancer Res 63:1138-43.

49. Mukhopadhyay, C. K., B. Mazumder, and P. L. Fox. 2000. Role of hypoxia-inducible factor-1 in transcriptional activation of ceruloplasmin by iron deficiency. J. Biol. Chem. 275:21048-54.

50. LeCouter, J., J. Kowalski, J. Foster, P. Hass, Z. Zhang, L. Dillard-Telm, G. Frantz, L. Rangell, L. DeGuzman, G. A. Keller, et al. 2001. Identification of an angiogenic mitogen selective for endocrine gland endothelium. Nature 412:877-84.

51. Hu, J., D. J. Discher, N. H. Bishopric, and K. A. Webster. 1998. Hypoxia regulates expression of the endothelin-1 gene through a proximal hypoxia-inducible factor-1 binding site on the antisense strand. Biochem. Biophys. Res. Commun. 245:894-9.

52. Oikawa, M., M. Abe, H. Kurosawa, W. Hida, K. Shirato, and Y. Sato. 2001. Hypoxia induces transcription factor ETS-1 via the activity of hypoxia-inducible factor-1. Biochem. Biophys. Res. Commun. 289:39-43.

53. Lee, P. J., B. H. Jiang, B. Y. Chin, N. V. Iyer, J. Alam, G. L. Semenza, and A. M. Choi. 1997. Hypoxia-inducible factor-1 mediates transcriptional activation of the heme oxygenase-1 gene in response to hypoxia. J. Biol. Chem. 272:5375-81.

54. Feldser, D., F. Agani, N. V. Iyer, B. Pak, G. Ferreira, and G. L. Semenza. 1999. Reciprocal positive regulation of hypoxia-inducible factor 1alpha and insulin-like growth factor 2. Cancer Res. 59:3915-8.

55. Comerford, K. M., T. J. Wallace, J. Karhausen, N. A. Louis, M. C. Montalto, and S. P. Colgan. 2002. Hypoxia-inducible factor-1-dependent regulation of the multidrug resistance (MDR1) gene. Cancer Res. 62:3387-94.

56.   Melillo, G., T. Musso, A. Sica, L. S. Taylor, G. W. Cox, and L. Varesio. 1995. A hypoxia-responsive element mediates a novel pathway of activation of the inducible nitric oxide synthase promoter. J. Exp. Med. 182:1683-93.

57.   Palmer, L. A., G. L. Semenza, M. H. Stoler, and R. A. Johns. 1998. Hypoxia induces type II NOS gene expression in pulmonary artery endothelial cells via HIF-1. Am. J. Physiol. 274:L212-9.

58.   Sowter, H. M., P. J. Ratcliffe, P. Watson, A. H. Greenberg, and A. L. Harris. 2001. HIF-1-dependent regulation of hypoxic induction of the cell death factors BNIP3 and NIX in human tumors. Cancer Res. 61:6669-73.

59.   Bruick, R. K. 2000. Expression of the gene encoding the proapoptotic Nip3 protein is induced by hypoxia. Proc. Natl. Acad. Sci. USA 97:9082-7.

60.   Carmeliet, P., Y. Dor, J. M. Herbert, D. Fukumura, K. Brusselmans, M. Dewerchin, M. Neeman, F. Bono, R. Abramovitch, P. Maxwell, et al. 1998. Role of HIF-1alpha in hypoxia-mediated apoptosis, cell proliferation and tumour angiogenesis. Nature 394:485-90.

61.   Bhattacharya, S., C. L. Michels, M. K. Leung, Z. P. Arany, A. L. Kung, and D. M. Livingston. 1999. Functional role of p35srj, a novel p300/CBP binding protein, during transactivation by HIF-1. Genes Dev 13:64-75.

62.   Minchenko, A., I. Leshchinsky, I. Opentanova, N. Sang, V. Srinivas, V. Armstead, and J. Caro. 2002. Hypoxia-inducible factor-1-mediated expression of the 6-phosphofructo-2-kinase/fructose-2,6-bisphosphatase-3 (PFKFB3) gene. Its possible role in the Warburg effect. J. Biol. Chem. 277:6183-7.

63.   Kietzmann, T., U. Roth, and K. Jungermann. 1999. Induction of the plasminogen activator inhibitor-1 gene expression by mild hypoxia via a hypoxia response element binding the hypoxia-inducible factor-1 in rat hepatocytes. Blood 94:4177-85.

64.   Takahashi, Y., S. Takahashi, Y. Shiga, T. Yoshimi, and T. Miura. 2000. Hypoxic induction of prolyl 4-hydroxylase alpha (I) in cultured cells. J. Biol. Chem. 275:14139-46.

65.   Shoshani, T., A. Faerman, I. Mett, E. Zelin, T. Tenne, S. Gorodin, Y. Moshel, S. Elbaz, A. Budanov, A. Chajut, et al. 2002. Identification of a novel hypoxia-inducible factor 1-responsive gene, RTP801, involved in apoptosis. Mol. Cell. Biol. 22:2283-93.

66.   Rolfs, A., I. Kvietikova, M. Gassmann, and R. H. Wenger. 1997. Oxygen-regulated transferrin expression is mediated by hypoxia-inducible factor-1. J. Biol. Chem. 272:20055-62.

67.   Lok, C. N., and P. Ponka. 1999. Identification of a hypoxia response element in the transferrin receptor gene. J. Biol. Chem. 274:24147-52.

68.   Tacchini, L., L. Bianchi, A. Bernelli-Zazzera, and G. Cairo. 1999. Transferrin receptor induction by hypoxia. HIF-1-mediated transcriptional activation and cell-specific post-transcriptional regulation. J. Biol. Chem. 274:24142-6.

69.   Caniggia, I., H. Mostachfi, J. Winter, M. Gassmann, S. J. Lye, M. Kuliszewski, and M. Post. 2000. Hypoxia-inducible factor-1 mediates the biological effects of oxygen on human trophoblast differentiation through TGFbeta(3). J. Clin. Invest. 105:577-87.

70.   Gerber, H. P., F. Condorelli, J. Park, and N. Ferrara. 1997. Differential transcriptional regulation of the two vascular endothelial growth factor receptor genes. Flt-1, but not Flk-1/KDR, is up-regulated by hypoxia. J. Biol. Chem. 272:23659-67.

71.   Brusselmans, K., F. Bono, P. Maxwell, Y. Dor, M. Dewerchin, D. Collen, J. M. Herbert, and P. Carmeliet. 2001. Hypoxia-inducible factor-2alpha (HIF-2alpha) is involved in the apoptotic response to hypoglycemia but not to hypoxia. J. Biol. Chem. 276:39192-6.

72. Mazure, N. M., C. Chauvet, B. Bois-Joyeux, M. A. Bernard, H. Nacer-Cherif, and J. L. Danan. 2002. Repression of alpha-fetoprotein gene expression under hypoxic conditions in human hepatoma cells: characterization of a negative hypoxia response element that mediates opposite effects of hypoxia inducible factor-1 and c-Myc. Cancer Res 62:1158-65.

73. Narravula, S., and S. P. Colgan. 2001. Hypoxia-inducible factor 1-mediated inhibition of peroxisome proliferator-activated receptor alpha expression during hypoxia. J Immunol 166:7543-8.

74. Kotch, L. E., N. V. Iyer, E. Laughner, and G. L. Semenza. 1999. Defective vascularization of HIF-1alpha-null embryos is not associated with VEGF deficiency but with mesenchymal cell death. Dev. Biol. 209:254-67.

75. Schipani, E., H. E. Ryan, S. Didrickson, T. Kobayashi, M. Knight, and R. S. Johnson. 2001. Hypoxia in cartilage: HIF-1alpha is essential for chondrocyte growth arrest and survival. Genes Dev. 15:2865-76.

76. Kojima, H., H. Gu, S. Nomura, C. C. Caldwell, T. Kobata, P. Carmeliet, G. L. Semenza, and M. V. Sitkovsky. 2002. Abnormal B lymphocyte development and autoimmunity in hypoxia-inducible factor 1alpha -deficient chimeric mice. Proc. Natl. Acad. Sci. USA 99:2170-4.

77. Yu, A. Y., L. A. Shimoda, N. V. Iyer, D. L. Huso, X. Sun, R. McWilliams, T. Beaty, J. S. Sham, C. M. Wiener, J. T. Sylvester, et al. 1999. Impaired physiological responses to chronic hypoxia in mice partially deficient for hypoxia-inducible factor 1alpha. J. Clin. Invest. 103:691-6.

78. Shimoda, L. A., D. J. Manalo, J. S. Sham, G. L. Semenza, and J. T. Sylvester. 2001. Partial HIF-1alpha deficiency impairs pulmonary arterial myocyte electrophysiological responses to hypoxia. Am. J. Physiol. Lung Cell. Mol. Physiol. 281:L202-8.

79. Kline, D. D., Y. J. Peng, D. J. Manalo, G. L. Semenza, and N. R. Prabhakar. 2002. Defective carotid body function and impaired ventilatory responses to chronic hypoxia in mice partially deficient for hypoxia-inducible factor 1 alpha. *Proc. Natl. Acad. Sci.* USA 99:821-6.

80. Carmeliet, P., V. Ferreira, G. Breier, S. Pollefeyt, L. Kieckens, M. Gertsenstein, M. Fahrig, A. Vandenhoeck, K. Harpal, C. Eberhardt, et al. 1996. Abnormal blood vessel development and lethality in embryos lacking a single VEGF allele. Nature 380:435-9.

81. Ferrara, N., K. Carver-Moore, H. Chen, M. Dowd, L. Lu, K. S. O'Shea, L. Powell-Braxton, K. J. Hillan, and M. W. Moore. 1996. Heterozygous embryonic lethality induced by targeted inactivation of the VEGF gene. Nature 380:439-42.

82. Tian, H., R. E. Hammer, A. M. Matsumoto, D. W. Russell, and S. L. McKnight. 1998. The hypoxia-responsive transcription factor EPAS1 is essential for catecholamine homeostasis and protection against heart failure during embryonic development. Genes Dev. 12:3320-4.

83. Peng, J., L. Zhang, L. Drysdale, and G. H. Fong. 2000. The transcription factor EPAS-1/hypoxia-inducible factor 2alpha plays an important role in vascular remodeling. Proc. Natl. Acad. Sci. USA 97:8386-91.

84. Maltepe, E., J. V. Schmidt, D. Baunoch, C. A. Bradfield, and M. C. Simon. 1997. Abnormal angiogenesis and responses to glucose and oxygen deprivation in mice lacking the protein ARNT. Nature 386:403-7.

85. Adelman, D. M., M. Gertsenstein, A. Nagy, M. C. Simon, and E. Maltepe. 2000. Placental cell fates are regulated in vivo by HIF-mediated hypoxia responses. Genes Dev. 14:3191-203.

86. Genbacev, O., Y. Zhou, J. W. Ludlow, and S. J. Fisher. 1997. Regulation of human placental development by oxygen tension. Science 277:1669-72.

87. Rajakumar, A., and K. P. Conrad. 2000. Expression, ontogeny, and regulation of hypoxia-inducible transcription factors in the human placenta. Biol. Reprod. 63:559-69.

88. Rajakumar, A., K. A. Whitelock, L. A. Weissfeld, A. R. Daftary, N. Markovic, and K. P. Conrad. 2001. Selective overexpression of the hypoxia-inducible transcription factor, HIF-2alpha, in placentas from women with preeclampsia. Biol. Reprod. 64:499-506.

89. Adelman, D. M., E. Maltepe, and M. C. Simon. 1999. Multilineage embryonic hematopoiesis requires hypoxic ARNT activity. Genes Dev. 13:2478-83.

Chapter 9

HORMONES, OBESITY, LEARNING, AND BREATHING - THE MANY FUNCTIONS OF MAMMALIAN *SINGLE-MINDED* GENES

Chen-Ming Fan
Carnegie Institution of Washington, Baltimore, MD 21211

1. INTRODUCTION

There are two mammalian *Single-minded* (*Sim*) genes. Both of them have been shown to play important roles during normal mammalian embryonic development. Each of them has also been directly implicated in contributing to the genetic causes of specific human conditions such as obesity and learning disability. While much is inferred about the genes' functions by means of mutation and over-expression studies in transgenic mouse models, little is known about what causes their dysfunction to lead to developmental abnormalities.

Below, I'll review the various functional assays of the two mammalian *Sim* genes and discuss their relevance and implication in mammalian physiology.

2. CLONING OF THE MAMMALIAN *SINGLE-MINDED* GENES

Evolutionary comparisons of sequence and expression of gene homologs between different species have been a main force in unveiling mammalian gene function. However, the first mammalian *Sim* gene was not cloned using a low stringency hybridization fishing for the *Drosophila sim* homolog. Rather, it was identified through a serendipitous effort of searching for coding exon sequences on human chromosome 21. Triplication of this chromosome causes the pathological features of Down Syndrome (DS) in human patients, the most common autosomal aneuploidy among live births.   DS patients have characteristic craniofacial malformation, mental retardation, myotonia, heart cushion defects, and increased rate of leukemia.   Rare DS patients carrying chromosomal rearrangements resulting in only partial chromosome triplication allowed determination of abnormal functional segments of chromosome21.   It is believed, somewhat controversially, that a minimal region of chromosome 21 (21q22.2), HC21, contains genes that contribute to many of the DS clinical pathologies (1).   When HC21 was subjected to an exon-trapping experiment to identify novel genes, two groups independently isolated several exons that encode sequences highly homologous to the *Drosophila sim* gene (2, 3).

Due to the importance of *sim* function in the fly, parallel independent efforts were made to isolate the mouse *Sim* genes based on sequence homology - please refer the fundamental importance of the fly *sim* gene in Chapter 4 (and reviewed in (4, 5).   Several groups have independently isolated a gene highly homologous to the fly *sim* gene (6-9). This gene was initially called the mouse *Sim* gene and is localized to chromosome 16 in a region syntenic to human chromosome 21 (more precisely, at 21q22.2, between *Tiam1* and *Ets2*). Worth noting, the triplication of portions of mouse chromosome 16 (e.g. Ts65Dn, (10) has long been used to make an animal model of DS.

It soon became clear that there are two *Sim* homologs in the mouse genome. One of them, named *Sim1* (6, 7), was found to have the most homology with the fly *sim* gene, more so than the initially isolated mouse *Sim* gene. *Sim1* is localized to mouse chromosome 10, corresponding to a syntenic region on human chromosome 6 (between *Fyn* and *Ros1*) (6, 7). Later, using the mouse sequence information, the second human *Sim* homolog, orthologous to *Sim1*, was cloned and named *SIM1* and the first cloned human *Sim* gene was renamed *SIM2* (11). Needless to say, each pair of the human and mouse *Sim* orthologs is more homologous than the paralogs – *Sim1*s are more similar to each other than to *Sim2*s. The human and mouse genome sequence drafts have been released and it does not appear that a third *sim* homolog exists.

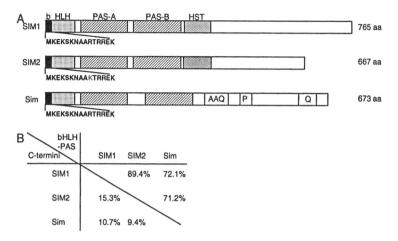

Figure 1. A. Diagrams depict the structures of SIM1, SIM2, and Sim protein. The size of each protein is indicated at right. The basic (b, black box), helix-loop-helix (HLH, stippled box), PAS-A and PAS-B (striped boxes), and the HST domain) grey boxes) are as indicated. AAQ, P, and Q motifs (clear boxes) of Sim are also indicated. The basic domain sequences (black bold letters) are listed below each gene and the R->K aa. change in SIM2 is in grey. B. Percentage comparison between the regions compared between the three proteins: the bHLH-PAS domains (top right corner) and the C- termini (lower left corner).

The predicted SIM1 protein (765 aa) is larger than both SIM2 (667 aa) and Sim (673 aa) (Figure 1A). The assignment of the number 1 and 2 of the mouse *Sim* genes is based on the fact that within the basic domain (the DNA binding domain), SIM1 is identical to Sim, while SIM2 has a conserved Lysine change in place of the Arginine at position 10. Aside from that, both SIM1 and SIM2 are highly homologous to the fly Sim protein throughout their basic-helix-loop-helix and PAS domains, as well as their histidine-threonine-serine (HST) rich domain immediately following the PAS domain: 71.2% between Sim and SIM1 and 71.2% between Sim and SIM2 (Figure 1B). Within this first half of the coding region (approximately 350 aa.), SIM1 and SIM2 are more homologous to each other (89.4%) than to Sim, consistent with their later duplication in the genome.

Two points in Sim family sequence alignment are worth noting: (i) Unlike any other bHLH-PAS proteins, the Sim family members comprise the only class with their basic domains at the extreme N-termini; (ii) There is practically no homology among SIM1, SIM2, and Sim in their C-termini (10.7% between Sim and SIM1, 9.4% between Sim and SIM2, and 15.3% between SIM1 and SIM2). However, the last 10 aa of SIM1 and SIM2 do have 8 identities, although the significance of this terminal conservation is not clear. There is also a human *SIM2* cDNA derived from an alternatively spliced transcript that predicts a truncated form of SIM2 (or SIM2s) of 570 aa. This open reading frame transcript terminates early and does not contain the 8 conserved amino acids at its C-terminal. Consequently, Northern analysis yields multiple *Sim2* (and *SIM2*) transcripts (6.0, 4.4, 3.0, and 2.5 kb), while *Sim1* (and *SIM1*) is consistently represented by one single 9 kb transcript (7, 12). Even more perplexing is that the C-termini have no obvious homology to any known genes in the database that could decipher their transcriptional activity. This presents a quandary, as the fly *sim* is a well-known transcriptional activator and this function is rendered, in part, by the Q-rich motifs located at its C-terminus (4). This has thus prompted the investigation of the transcriptional properties of the mammalian *Sim* genes.

3.   ARE SIM1 AND SIM2 TRANSCRIPTIONAL ACTIVATORS OR REPRESSORS?

Deletion analysis of SIM1 and SIM2 supported the notion that the general rules for bHLH-PAS proteins apply: The SIM1 and SIM2 basic domains are their DNA binding domains, their HLH-PAS domains are the bipartite dimerization domains with the general dimerization partners, and their C-termini contain the transcriptional regulatory domains (6, 8, 13, 14). Both SIM1 and SIM2 can associate with ARNT, ARNT2 (15) and BMAL1 (16), the three general dimerization partners, either using the yeast-two-hybrid assay or the co-immunoprecipitation assay (6, 8, 13, 14, 17). Together with ARNT, the partner of HIF-1α and AHR (see Chapters 6 and 8), SIM1 and SIM2 can bind to the canonical fly Sim/Tango (Tango is the fly equivalent of ARNT and ARNT2, (18)) binding site CME (CNS-midline element, the core sequence: TACGTG). However, they have very poor affinities (of μM range) to these sites (14). Even using PCR-mediated random oligo selection, the optimized binding sites for SIM1/ARNT and SIM2/ARNT are only subtle variations of the fly Sim binding site CME and of low affinity (14). Despite the low affinity observed in vitro, CME drives Sim/Tango-mediated CNS-midline specific expression in the fly at a reasonable expression level (4, 18). Curiously, the CME does not appear to mediate in vivo expression at *Sim1* or *Sim2* expression sites in the mouse (unpublished). When the C-termini of SIM1 and SIM2 were fused to the Gal4-DNA-binding domain and tested with reporter genes driven by the Gal4-binding sites both in yeast and mammalian cell culture, both of them acted as transcriptional repressors (6, 13); Figure 2. In those studies, the SIM2 C-terminus displayed more repression activity than the SIM1 C-terminus. Thus, contrary to the fly Sim, they behave likely transcriptional repressors. It was not clear, however, what is the molecular mechanism(s) mediating their repressive activity.

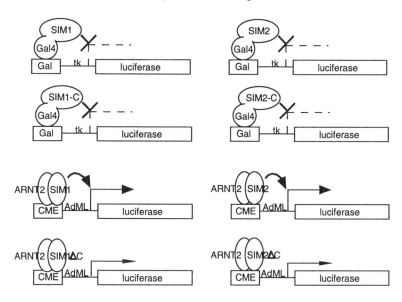

Figure 2. Transcriptional activities of SIM1 and SIM2. Left diagrams are luciferase reporter assays for SIM1activites in various assays: full-length protein fused to Gal-4 on a Gal4-binding site, SIM1-C-terminus fusion to Gal4 on a Gal4-bnding site, full-length SIM1 together with ARNT2 (or ARNT) on a CME site, and SIM1-C-terminal truncation (ΔC) with ARNT2 (or ARNT). Right diagrams are the same assays for SIM2. Note that the promoter used for the top two panels is the minimal tk promoter while for the bottom two panels is the AdML minimal promoter. Transcription repression is indicated by T signs, activation is indicated by curved arrow, and the lack of "additional contributory" activities from SIM1/SIM2ΔC proteins are not further indicated other than the basal activity of the minimal promoter.

The plot thickened when it was found that in different cell types tested and using the CME sites linked to a different reporter gene, SIM1/ARNT, SIM1/ARNT2, SIM2/ARNT and SIM2/ARNT2 functioned as transcription activators (19); Figure 2). In this study, ARNT and ARNT2 appear to be equivalent in their ability to cooperate with SIM1 and SIM2 or activation. A simple explanation for this difference in SIM1/SIM2 activity from the Gal4-fusion experiments is that CME binding of SIM1/SIM2 physically alters their transcription activity. Or, the topology of the Gal4-fused SIM1/SIM2 C-terminus blocks access to the basal transcription machinery. In these studies, the basal promoters used and the cell types tested are also different, making the conclusion less straightforward. A side by side testing of SIM1/ARNT2 using a CME-AdML-Luc and a CME-SV40-Luc reporters in transiently transfected 293 and Neuro2a cells in my laboratory revealed that SIM1/ARNT2 activates the CME-AdML-Luc reporter, but represses the CME-SV40-Luc reporter (Liu and Fan, unpublished). This implies that promoter compatibility also contributes to Sim1's transcriptional function.

Thus, a systematic survey of promoters and cell types may be required to sort out these controversial results obtained in different laboratories and systems. Because the human *SIM* genes are practically identical to the mouse genes, their transcriptional activity is also not absolutely certain at this point. One prediction is that the SIM1-C-terminus interacts with different basal promoter factors on different gene promoters in different cell types that dictate the assessed outcome.

Lastly, although SIM1 and SIM2 are able to associate with HSP90 in an in vitro co-immunoprecipitation assay (13), the importance of such an association is presently unclear. First, transiently expressed SIM1 and SIM2 proteins are mostly, if not exclusively, nuclear localized. Second, so far the antibodies to SIM1 protein only detect the epitope in the nucleus in several mouse tissues surveyed. Third, there is as yet no ligand for SIM1 and SIM2 to displace HSP90 and alter their transcriptional activity, a scenario found for the Ahr/HSP90. Furthermore, both mammalian ARNT and ARNT2 are also nuclear-localized in cell lines as well as in tissues. Together, these lines of evidence suggest that there is no additional regulation at the nuclear entry level for SIM1 and SIM2.

4.   EXPRESSION PATTERN OF THE *SINGLE-MINDED* GENES

The expression patterns of *Sim1* and *Sim2* have been examined by *in situ* hybridization of embryos and adult animals. The primary expression sites of *Sim1* and *Sim2* expression are in the CNS at early embryogenesis. In day 9.5 mouse embryos, both *Sim1* and *Sim2* are expressed at the basal diencephalic region (6, 7) (Figure 3B, C). Applying their expression domains to the forebrain prosomeric model proposed by Puelles (7), shows that their initial expression domains correspond to prosomere 2-4 (p2-p4). This embryonic domain eventually gives rise to the mammillary body located in the posterior hypothalamus. Consistently, *Sim1* and *Sim2* are detected in the histologically mature mammillary body (6, 7). The human *SIM2* transcript was also documented to be expressed in the mammillary body (3). Later at E10.5, E12.5 and E16.5, *Sim1* and *Sim2* (combined expression) are also found to be expressed in many other regions, e.g. the cortex, amygdala, anterior hypothalamus, lateral hypothalamus, midbrain, spinal cord, as well as non-CNS structures such as kidney, muscle, cartilage, dermis and hair follicles (Figure 3D).

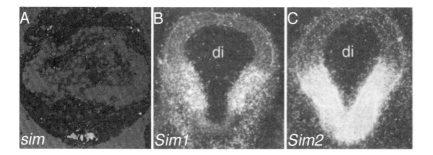

D. Expression patterns of *Sim1* and *Sim2*.

	Sim1	Sim2
E8.5-E10.5	diencephalon:p1-2 basal	diencephalon:p1-2 basal &midline
	zona limitans	zona limitans
	mesonephric: tubules	mesonephric: ducts
	dermomyotome: lateral	
E11.5-E12.5	lateral mammallary	lateral mammallary
	anterior hypothalamus	anterior hypothalamus
	lateral hypothalamus	lateral hypothalamus
	zona limitans	zona limitans
	amygdala	cortex
	dermis	cartilages:
	metanephrose: tubules	rib, vertebra, limbs, brachial arches
		metanephrose: ducts
E16.5-Newborn	lateral mammallary	lateral mammallary
	zona limitans	zona limitans
	anterior hypothalamus: preoptic, aPV, PVN, SON	anterior hypothalamus: preoptic, aPV, PVN
	lateral hypothalamus	lateral hypothalamus
	midbrain-basal plate	midbrain-basal plate
	amygdala-NLOT	cortex
	spinal cord - basal region	cartilages:
	head mesenchyme	rib, vertebra, muscle, limb, digit, palate, mandible, hyoid, trachea, pleural wall
	dermis	
	genital eminence	oral epithelium
	kidney distal tubules	kidney collecting ducts

Figure 3. A. Expression of fly *sim* in the midline cells of the ventral nerve cord (ventral whole-mount view). B. Expression of mouse Sim1 gene in the ventral diencephalon at E9.5 (coronal section) is next to the midline cells. C. Expression of mouse Sim2 gene in the ventral diencephalon at E9.5 (coronal section) includes the midline cells. D. Summary of Sim1 and Sim2 expression during development.

Despite the ambiguity of the tissue equivalency inference between the mouse and the fly tissues, the embryonic expression patterns of the mouse *Sims* are reminiscent of the fly *sim*. For example, Figure 3A-C shows that the *Drosophila sim* transcripts and protein are detected in the CNS midline cells located in the ventral wall. Similarly, the mouse *Sim2* expression in the ventral diencephalons including the midline. In contrast, *Sim1* is expressed in cells next to the midline of both the diencephalons and the spinal cord. *Sim1* and *Sim2* are also detected in the kidney and specific groups of muscles, possibly related to *sim* expression in the malpighian tubules and myoblasts (1, 7, 20). Of course, *Sim1* and *Sim2* are also expressed in tissues that do not have equivalent counterparts in the fly, such as the dermis, hair follicles, and cartilages. Although seemingly trivial, the expression data often provides a foundation for the analysis of homologous function in mammals. Because the mammalian genes are often recruited into acting in various different tissues, the multiplicity of *Sim1* and *Sim2* expression sites presents tremendous difficulties to their functional categorization. Below, I'll discuss the overexpression studies and the mutant phenotype of *Sim1* and *Sim2*, which illustrate this point quite well.

5.   IN VIVO FUNCTION REVEALED BY OVEREXPRESSION

Since *Sim2* is located at a chromosomal region associated with DS pathology, this aspect of its possible malfunction was tested in the mouse. Two studies have been conducted. One involved a transgenic animal model that used a general overexpression system utilizing an actin promoter to express *Sim2* (21). The model in the other study used a bacterial artificial chromosome (BAC) containing the entire *Sim2* gene with some, if not all, of its regulatory elements (12). The first group described a mild learning disability due to general overexpression, while the second group documented delayed fear response in the BAC-directed transgenic mice. As *Sim2* is expressed in the cortex from late embryogenesis to adulthood, it is conceivable that the affected behaviors in these animal models are directly due to *Sim2*'s abnormal function in cortical development. However, neither study found any morphological abnormalities of the cortex documented in human DS patient and the Ts65Dn mouse model. It is to be noted that in both cases the possibility remained that unknown ectopic sites of overexpressed *Sim2* in the CNS (due to transgenesis) are the cause of the phenotypes. Alternatively, overexpression of *Sim2* in other regions of the CNS may indirectly contribute to the learning dysfunction. In both scenarios, *Sim2* overexpression did not cause any obvious craniofacial defects or myotonia in the mouse - even though *Sim2* is highly expressed in these tissues and they are two other hallmarks of DS pathology. Whether the behavioral defects are due to *Sim2's* repressor activity and what the downstream mediators are to cause these behavioral defects by overexpression are not known. It would be of interest to test whether *Sim1* or even *sim*, when overexpressed under the same conditions, may also cause the same phenotype.

The earliest expression of *Sim2* in the caudal diencephalic ventral midline (Figure 2) and the zona limitans coincides with that of the *Sonic hedgehog (Shh)* gene (22). *Shh* encodes a secreted factor, and is homologous to the fly *hedgehog (hh)* gene. In human and mouse, *Shh* mutations cause holoprosencephaly (fusion of the two hemispheres of the forebrain due to under development and growth), a mild form of which also causes mental retardation (23, 24). When *Sim2* is ectopically expressed in the dorsal neural tube under the control of a *Wnt1*-enhancer in transgenic mouse embryos, *Shh* is ectopically activated at the dorsal midbrain region and causes exencephaly (opening of the brain), presumably due to overgrowth of the dorsal neuroepithelium (22). This data suggested that *Sim2* is involved in regulating *Shh's* diencephalic expression *in vivo*.

The in vivo relevance of this data becomes puzzling with the following observations. First, *Sim1* and *sim* also cause ectopic expression of *Shh* to the similar extent as *Sim2* when using the same *Wnt1*-enhancer expression cassette. This is the only evidence thus far that all *Sim* homologs may have the same transcriptional activity – despite the fact that they have distinct activities in vitro and possess no similarity in their C-termini. Since the fly Sim has consistently been shown to be an activator, it is reasonable to assume that Sim1 and Sim2 can act as activators in this context (the dorsal midbrain). Second, *Sim2* mutants as well as *Sim1/Sim2* double mutants display normal *Shh* expression at the diencephalic midline and zona limitans, indicating that the *Sim* genes play a small role in regulating *Shh* normally. Third, a dominant-negative form of ARNT lacking its basic domain inhibits endogenous *Shh* transcription in the ventral diencephalic midline region (22). This result supports the contention that a bHLH-PAS protein neither SIM1 nor SIM2) can dimerize with ARNT and activate *Shh* expression in the diencephalon. Since SIMs can activate and dominant negative ARNT can inhibit *Shh* transcription, this presumptive *Shh* regulatory bHLH-PAS protein must have similar DNA binding property to that of the SIMs. The hunt for this bHLH-PAS factor is ongoing.

The above overexpression experiments, although they are of high interest and provided intriguing information regarding the "possible functions" of *Sim1* and *Sim2*, did not provide conclusive answers to the normal function of *Sim1* and *Sim2*. To decipher their normal function in the context of the organism, *Sim1* and *Sim2* mutant mice have been generated by homologous recombination-ES cell knock-out technology. Both null mutants are perinatal lethal with no visible outward structural defects. Below, I'll review these studies.

6.   THE *SIM1* MUTANT MOUSE

Most *Sim1* mutant mice die within a few days of birth with some 5-10% living up to 28 days. Those that die within days of birth suckle milk normally and do not exhibit obvious physiological abnormalities (17). The cause of lethality has not been ascertained. Those that live beyond 20 days start to display severe intentional tremor during movement. They also display peculiar hindquarter, involuntary tail spinning seizures after bursts of movement, and eventually die during one of these eclectic cycles of behavior (unpublished).

Expression studies showed that *Sim1* is expressed in the anterior neuroendocrine hypothalamus, including the paraventricular nucleus (PVN), anterior periventricular nucleus (aPV) and supraoptic nucleus (SON) (Figure 4). Comparative histological examinations revealed that the *Sim1* mutant brain has no structural PVN, aPV and SON. PVN and SON are two neuroendocrine centers within the hypothalamus (25). The PVN contains two generally defined types of neurons: Cells that have a large cell body are called the magnocellular neurons and cells that have a small cell body, the parvocellular neurons. SON contains exclusively the magnocellular neurons. In fact, the magnocellular neurons in the PVN and SON arise from the same progenitors. Within each group, there are distinct sets of cells expressing unique hormones that maintain homeostasis by regulating pituitary function. The magnocellular neurons project to the posterior pituitary where they secrete hormones directly into the bloodstream, while the parvocellular neurons secrete their primary hormones into the median eminence, which carries them into the anterior pituitary to stimulate secondary hormone secretion (26-29).

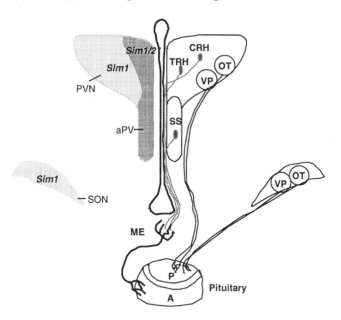

Figure 4. Diagram of the hypothalamic-pituitary axis. The organization of PVN, SON, and aPV is as indicated. The left side depicts the expression patterns of Sim1 and Sim2, while the right side depicts the neuroendocrine cell types (VP,OT, TRH,CRH, and SS cells, as labeled) distributed within these nuclei. Note that Sim1 (lightly shaded area) is expressed in the entire aPV, PVN and SON, while Sim2 (dark shaded area) is expressed in the aPV and a sub-region of the PVN. The TRH, SS and CRH cells project their axons to the median eminence (ME), which carries the hormones to the anterior (A) pituitary. The OT and VP neurons project directly to the posterior (P) pituitary. In the Sim1 mutant, the anatomical structures of these three nuclei are missing and none of these cell types ever terminally differentiate to express the hormones.

Time course analyses indicate that the aPV/PVN/SON precursor cells never terminally differentiate into hormone producing neuroendocrine cells, not even at the earliest time point (E12.5). At least five major cell types have been examined: three types of parvocellular cells, CRH (corticotropin releasing hormone)-, TRH (thyrotropin releasing hormone)-, SS (somatostatin)-producing neurons, and two types of magnocellular cells, VP (vasopressin)- and OT (oxytocin)-producing neurons. The genesis and functions of these neurons have been extensively investigated (27, 29-41). However, the precise regional signals and factors that control their production remain unclear and their normal physiological functions are vast and complex. It is tempting to conclude that the absence of these five cells may be the cause of mutant lethality: VP maintains cell volume and water content /electro-potentials of the blood and a mutation in *Vp* causes diabetes insipidous; OT mainly functions in parturation and lactation and a mutation in *Ot* causes the loss of maternal instinct for rearing progenies; the CRH pathway leads to elevated glucocorticoids release from the adrenal glands which then mediate anxiety and fear response, and often negatively affect feeding behavior; TRH stimulates the thyroid hormone pathway, which is important for energy metabolism and brain synaptic development; SS, on the other hand, inhibits growth hormone secretion in the anterior pituitary, and thus acts negatively in body growth. Intriguingly, mutations in each hormone gene individually do not cause perinatal lethality. It is still possible that simultaneous loss of all five cells types, and possibly other hormone-producing cells (since the entire PVN is missing by histology), could indeed lead to lethality.

PVN and SON cells are presumed to derive from progenitors within the alar plate of prosomere 5 around E10.5-12.5 (30, 31, 33, 34, 38). *Sim1* expression in the pia, but not the ventricular zone of the corresponding region starting at E10.5, is consistent with its involvement in progenitor specification/differentiation, but not in proliferation (Figure 4). Its persistent PVN/SON expression into adulthood also suggests its later role in regulating their function. Regional-specific expression of *Sim1* also makes it an excellent marker for following the location of these progenitor cells throughout development. In the *Sim1* mutant, *Sim1* transcripts are still detectable by a 5'UTR probe, allowing the mutant cells to be followed. Between E10.5 and E12.5, *Sim1* expression is normally activated in the mutant, suggesting that mutant PVN/SON progenitors are generated and specified. After that, *Sim1*-expressing mutant cells are no longer seen at the normal PVN/SON position. The loss of expression is not due to programmed cell death, suggesting that the mutant cells either change fate, or they mis-migrate to an abnormal position. Thus, *Sim1* appears to contribute only to the terminal differentiation step and possibly the migration step of these cells, but not their proliferation or early specification.

Mice mutant for *Brn2*, a POU-homeobox transcription factor, also have a PVN/SON defect -(42, 43). As opposed to the *Sim1* mutants, *Brn2* mutants appear to be only deficient of the CRH, VP and OT neurons in the PVN and VP and OT neurons in the SON, while retaining the TRH and SS neurons. In addition, *Brn2* expression in the hypothalamus, although partially overlapping with that of *Sim1*, is more general and is not restricted to the PVN/SON region (Figure3). Importantly, its expression is specifically not present in a subdomain within the *Sim1* domain, even at an early time point (E12.5) (17). This result has the following implications: (i) *Sim1* acts through *Brn2* to control (at least) the CRH, VP, and OT cell types in their terminal differentiation. In the case of the CRH cells, *Brn2* has been shown to bind to its promoter and activate its transcription (44); (ii) The *Brn2*-negative domain in the *Sim1* mutant defines the prospective PVN/SON, the posterior-dorsal quadrant of the *Sim1* domain (17). One can envision two scenarios: One is that the entire aPV/PVN/SON comes from this *Brn2*-negative domain, implying that *Brn2*, although expressed in this domain, does not participate in TRH nor SS cell differentiation, the other is that only the CRH, VT and OT neurons come from this region, while the other cell types of the PVN arise from the remaining portion of the *Sim1* domain. The latter scenario proposes that TRH and SS cells do not share the same progenitor population with the CRH/OT/VP cells. The two possibilities are indeed difficult to resolve, mainly because of the extensive morphogenetic movement and non-tangential neuronal migration in the forebrain during this period of development. This is further complicated by the fact that each of these neuronal cell types appears at a different time (sometimes in several waves for one cell type) spanning from E12.5 (e.g. the first wave of the TRH-positive cells) to E17.5 (e.g. the OT-positive cells). However, if the first scenario is correct, there must exist a specific set of transcription factors expressed at different times specifically to direct distinct cell types (downstream of *Brn2*). If the latter is true, there must exist a specific set of transcription factors for each of cell types expressed in different sub-domains within the larger *Sim1* domain (parallel with *Brn2*). Lastly, although the genetic hierarchy is clear, *Sim1* has not been shown to activate *Brn2* transcription directly (19).

Sim1 and *Brn2* mutants also have an underdeveloped posterior pituitary (42, 43) where the pituicytes are located. Pituicytes are support cells in the posterior pituitary. It was suggested that the VP, OT, or another unidentified neuronal cell type that projects into the posterior pituitary also deliver a trophic factor for the pituicytes. Whether OT and VP themselves can function as trophic factors for the pituicytes has not been directly examined.

The *Brn2* homolog in the fly, named *drifter*, is also expressed in a subset of *sim*-positive CNS midline cells (45). Drifter is thought to be important for the proper migration of a subset of these midline cells. This leads to the speculation that both *Sim1* and *Brn2* may also have cell migration defects, provided this POU-domain protein acts conservatively throughout evolution. Only by comparing a permanently marked mutant progenitor population, e.g. by Cre-mediated tissue-specific cell marking, may one be able to answer this question. Another intriguing observation is that in the insect, CRH-reactive cells have been found in the CNS midline cells (5), a place where *sim* gene function is required. The fact that *Brn2* binds to the CRH promoter provides a foundation for speculating that *drifter* may also act in an evolutionarily-conserved manner to control CRH-related peptide gene expression in insects. Whether *sim* also controls other neuroendocrine-like lineages in the fly has yet to be determined.

The finding that *Sim1* and *Brn2* are in the same pathway that control PVN/SON development has important implications in their possible roles in maintaining metabolic rate, feeding behavior, urination, lactation, anxiety, and fear response. Unfortunately, both mutants are lethal, precluding the assessment of their physiological roles in the adult. Unless the temporal- and tissue-specific KO mice of them in the adult PVN/SON are viable, one will not be able to determine their roles in the adult.

7.   *ARNT2* FUNCTION IN APV/PVN/SON

Among the known general dimerization partners for bHLH-PAS proteins, ARNT2 is the best candidate dimer-partner for Sim1. This is because *Arnt2* is particularly upregulated at the PVN/SON region throughout embryogenesis to adult, corresponding to *Sim1* expression. *Arnt2* mutants were analyzed by three independent groups. One group used the c^{112k} of the *albino* (*c*) deletion series generated in the 1960's (46). *Arnt2* is deleted in the c^{112k} allele. The other two groups generated *Arnt2*-specific KO mice (47, 48). In all cases, the results support the same conclusion: lack of *Arnt2* function leads to histologically abnormal PVN and SON, a phenotype identical to the *Sim1* mutant. Coincidentally, at E12.5, the early time point of PVN/SON precursor genesis, *Brn2* expression is also lost at the anterior-dorsal quadrant of the *Sim1* domain (the prospective PVN/SON domain) in the *Arnt2* mutant. Importantly, *Sim1* expression at this time point is not affected in the *Arnt2* mutant. In conjunction with the fact that SIM1 can cooperate with ARNT2 to activate reporter gene expression and quantitatively co-immunoprecipitate ARNT2, these results together strongly suggest that SIM1 and ARNT2 act as a pair of transcription factors essential for PVN/SON development.

A few aspects of this paradigm have not been fully explored:

(i) *Arnt2* is generally expressed in the brain, but specifically up-regulated and maintained in the prospective PVN/SON region. The upstream regulator responsible for this regional up-regulation of *Arnt2* and the regional-specific expression of *Sim1* is unknown.

(ii) It has not yet been determined whether *Arnt2* expression in the PVN/SON precursor is normal in the *Sim1* mutant. One possible scenario is that *Sim1* cooperates with *Arnt2* to autoregulate, and thus maintain, their own expression. This is a possible explanation of the eventual loss of *Sim1* expression in the *Sim1* mutant.

(iii) It is not certain whether Sim1/ARNT2 acts as an activator or a repressor. Interestingly, *Brn2* expression is missing in both mutants, yet this observation does not necessarily favor the activator over the repressor model. It is in fact not at all clear whether Sim1/ARNT2 directly binds to

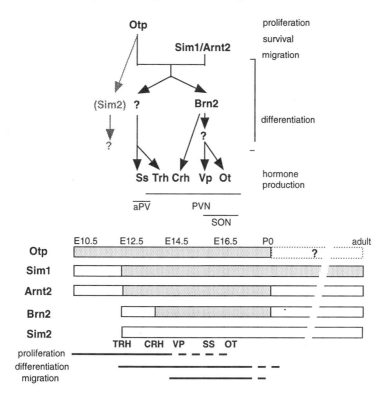

Figure -5. Top panel represents the working model of the genetic pathway of Sim1, Arnt2, Otp, Brn2, and Sim2 in neuroendocrine hypothalamus development. Arrows indicate the hierarchical relationships between the genes and cell types. The steps of each gene's action are labeled at right. Bottom panel diagrams the timing (labeled at top) of each gene's (labeled at left) expression (clear box) and action (shaded box). The time points of the appearance of each neuroendocrine cell types are label at bottom.

the promoters and activates the transcription of *Trh*, *Ss*, *Vp* and *Ot* genes, or even *Brn2*.

(iv) *Arnt2* may also participate in the transcriptional regulation of these five genes and regulate the homeostasis state of the adult, whose mutations are generated through conditional inactivation that bypasses perinatal lethality.

(v) Perhaps most intriguingly, *Arnt* does not compensate for the loss of *Arnt2*, even though all in vitro assays so far demonstrate their equivalency in conjunction with Sim1. Keith et al. (48) showed clearly that in the CNS, ARNT protein levels are the same or higher than those of ARNT2, based on their binding activities to the hypoxia response element (Chapter 8). Is this because there is an additional component to the dimerization choices of Sim1 to ARNT2 vs. ARNT in the PVN/SON? Is it because the defect is purely dosage-dependent and *Arnt* mutants, if they survived to birth, would also lack a PVN/SON? Or, is it simply that within the PVN/SON, ARNT2 and ARNT protein levels are post-translationally regulated, resulting in the exclusive presence of ARNT2?

8.   THE HOMEODOMAIN GENE *ORTHOPEDIA* (*OTP*)

Subsequent to these reports, a mouse homolog of the fly *orthopedia* gene, *Otp*, was inactivated by two independent groups (49, 50). Two main defects were found. One is that the *Otp* mutant has no visible structures of PVN and SON. Another is that the *Otp* mutant has no SS in the arcuate nucleus (located just ventral-caudal to the PVN), while other markers of the arcuate nucleus remain intact. The knock-in, in which LacZ is transcribed and maintained in place of *Otp* in the mutant cells, made it possible to follow these *Otp* mutant cells. Not only are there fewer precursor cells being generated, there is also increased programmed cell death and incorrect cell migration (49). However, since some of the precursors are generated, as in the case of *Sim1* and *Arnt2* mutants, it is not clear which of the three documented defects is the main cause of the absence of PVN/SON structure. None of these defects have been found in the *Sim1* and *Arnt2* mutants. However, these three genes do act similarly in one step: terminal differentiation of SS, TRH, VP, OT, and CRH cells. Furthermore, *Sim1* expression in the prospective PVN in the *Otp* mutant is unaltered and vice versa, indicating that *Sim1* and *Otp* transcription are independent of each other's function and that they function independently in a parallel capacity (49).

Otp and *Sim1* expression in prosomere 5 at E12.5 are highly similar, further suggesting that this is indeed the region of PVN/SON precursors. Since they act in parallel, it is surprising that the *Otp* mutant also loses *Brn2* expression at the same time as the *Sim1* mutant in the prospective PVN/SON. They must therefore converge into the same pathway for generating CRH, VP, and OT neurons. Is *Brn2* the final nodal point for the differentiation program of these cells? As to whether they also converge into a single downstream regulator that controls TRH and SS cell differentiation is not yet determined. The evolutionary conservation of the relationship between *sim* and *otp* is unknown. *Drosophila otp* is not expressed in the embryonic CNS midline cells, but *sim* and *otp* may overlap in other cell types, such as the anal pad (51). Thus far, no specific protein-protein interactions between OTP and SIM1 or between OTP and ARNT2 have been reported. Of course, it is possible that their interaction(s) will only occur through a novel bridging factor on the appropriate promoters/enhancers and cannot be easily detected by conventional methods.

9.   *SIM1* HETEROZYGOTES ARE OBESE

An 8-year old patient with severe obesity was reported to carry a balanced translocation disrupting the *SIM1* gene by separating its bHLH domain from its PAS domain (52). This child has early-onset (by the age of 2) obesity, increased linear growth, and hyperphagia with no decrease of energy expenditure. There are also other reports of morbid obesity in children with chromosomal deletions in the 6q16 region where *SIM1*is located (53-55). These findings suggested that haplo-insufficiency of *SIM1* is responsible for the obesity in these children. They also prompted the investigation of the role of *SIM1* in regulating energy metabolism using the animal model. *Sim1* is important for PVN/SON development (17). It is perhaps not surprising that the *Sim1* mutant has imbalanced homeostasis in some physiological aspects – however, its lethality precludes analysis in the adult. Fortunately, *Sim1* heterozygous animals survive to adulthood and develop obesity (56).

Below, I summarize the findings derived from the mouse data (56) and compare them to the human patients:

(i) Similar to the human patients with *SIM1* heterozygosity (by translocation or deletion), *Sim1* heterozygous mice develop obesity, hyperphagia and increased growth without decreased energy metabolism.

(ii) In contrast to the human patients, *Sim1* heterozygous mice do not develop prepubescent obesity. Rather, they display accelerated weight gain past 4-weeks of age with linear growth, hyperinsulinemia, and hyperleptinemia.

(iii) The brown fat and white fat are increased in the *Sim1* mouse model by different mechanisms: the brown fat increases tissue mass by increasing cell size, while the white fat increases its volume by increasing its cell number.

(iv) As their energy expenditure is not decreased, the *Sim1* mice are distinct from the other obesity models such as the Leptin and the MC4R mutant animals.

Body weight regulation is achieved by a simple principle: the balance between food intake and energy expenditure. Food intake is influenced by two opposing forces, satiety and appetite. This is well reviewed elsewhere (57-59). Suffice to say, several hypothalamic nuclei control appetite, satiety, and metabolic rate. The primary focus of the obesity field has been the arcuate nucleus, where the POMC-producing cells are located and regulated by Leptin signaling. POMC-derived peptide hormones negatively regulate appetite. They in fact project to the PVN and modulate PVN activity (60). Other lines of evidence also support the role of the PVN in appetite control: first, selective lesions in the PVN cause hyperphagia and obesity without affecting energy expenditures (61-64); second, infusion of orexigenic signals or anorexic signals into the PVN lead to food intake increase and decrease, respectively (54, 65); third, PVN neurons express NPYR and MSH receptors and integrate both the NPY(feeding) and MSH (stop feeding) signals (66, 67). The *Sim1* heterozygous PVN is approximately 20% smaller than that of the wild type siblings by histology. It was proposed that the obesity phenotype is a reflection of a haploid-insufficient activity of *Sim1* resulting in an under-developed PVN. It was noted, however, that several orexigenic genes (e.g. 5HT-2C (68), Tubby (58), and galanin (69)) expressed in the PVN are present in the *Sim1* heterozygous mice, suggesting that the obesity phenotype may be due to a general reduction of the PVN orexigenic neurons than a specific target gene that is misregulated in the heterozygotes (56). Until all genes are surveyed, one cannot exclude the possibility that there exists an unknown obesity-related pathway dysregulated in the *Sim1* heterozygous. Of note, the size of the human patient's PVN has not been examined by MRI.

The proposal that obesity can result from under-development of the PVN has a profound implication for the genetic causes of obesity (56): alteration of gene activity needed for PVN development can potentially cause obesity. This would predict that *Arnt2*, *Otp*, and possibly *Brn2* are all candidate loci for being associated with obesity. For example, reduced CRH-neurons and CRH secretion are possible mechanisms for hyperphagia. One drawback of this hypothesis involves the TRH neurons in the *Sim1* heterozygous PVN. TRH is one of the main regulators of energy metabolism via thyroid hormone regulatory pathway. If the hypo-development of the PVN also includes a deficit of the TRH neuronal population, the reduced secretion of TRH should lead to reduced metabolic rate/energy expenditure in the *Sim1* heterozygotes, which is not observed, unless, of course, TRH neurons are spared from the haploid insufficiency of *Sim1*.

10.   *SIM2* MUTANT AND SIM2 DIMERIZATION PARTNER(S)

Sim2 mutants are born alive at a Mendelian ratio and cannot be distinguished from their siblings initially. Shortly afterwards, they display severe respiratory distress manifested by dyspnea, aerophagia, cyanosis, and reduced lung inflation, and are presumed to die from the severe breathing disorder (70). In an initial report of the *Sim2* mutant in the 129sv/BL6 mixed background (71), the mutant was described to have an un-developed nasal cavity and cleft palate, explaining the respiratory difficulty. The sensory-motor cortex was also reported to have reduced cell number. However, none of these defects were found when the *Sim2* mutant allele was backcrossed into the BL-6 background, suggesting that *Sim2* activity is sensitive to genetic background. Curiously, these mutant animals still die from respiratory difficulties.

Sim2 transcripts are not found in the lung proper, nor in the classical CNS and PNS nuclei that regulate breathing patterns (70). Histological, molecular and electronic microscopic studies support the claim that the mutant lungs are normal (70, 71). Instead, *Sim2* expression is found in many tissues of the airway and surrounding the plural cavity, e.g. the nasal cavity, trachea, ribs, vertebrae, sternum, diaphragm, and pleural mesenchyme. Furthermore, the *Sim2* mutant has a thinner than normal diaphragm, broken pleural wall mesenchymes, multiple exostosis on the ribs, and scoliosis. The rib-specific exostosis and the scoliosis observed in the *Sim2* mutant are similar to rare human congenital conditions (20, 72, 73). These defects imply that *Sim2* normally regulate the growth of cartilaginous elements. *In vivo* BrdU incorporation data are consistent with the hypothesis that the irregular cartilage structures in the mutant result from local growth dysregulation. The poor efficacy of lung inflation of the mutant is, at present, thought to be a consequence of compromised mechanical force needed for lung inflation caused by defects in these essential mesenchymal components.

Although the craniofacial and cortical defects found in the 129sv/BL background are consistent with its possible involvement in contributing to aspects of the DS pathology (71), the overexpression studies do not support its involvement in malformation of the craniofacial structures (12, 21). Since the mutant animals die before reaching adulthood, the normal role of *Sim2* in cognitive function will require future cortex-specific mutants. As in the mutant analysis in the BL6 background, the lack of craniofacial defects in the overexpression models may also be due to the genetic backgrounds of those animals. Since the severity of the diaphragm defect, exostosis, and scoliosis found in the BL6 background is variable and dosage-dependent, it is possible that mutants in the 129sv/BL6 background and the overexpression transgenic background do have milder defects in these regions. If *Sim2* genetically interacts with another gene(s), it is possible that this modifier locus (or loci) is also located on mouse chromosome 16 (or chromosome 21 in human). As ARNT is the main general dimerization partner expressed in the mesodermal tissues, Sim2 may utilize ARNT in the mesoderm and ARNT2 in the CNS to operate. Both of them are also the potential genes of *Sim2* modifiers.

11.   IS THERE A GENETIC INTERACTION BETWEEN SIM1 AND SIM2?

Both *Sim1* and *Sim2* are expressed in the PVN, yet the initial histological analysis did not reveal any PVN defects in the *Sim2* mutant (70). This is perhaps not surprising. First, *Sim1* is expressed at a much higher level than *Sim2* in the PVN. Second, *Sim1* is expressed in the entire PVN as well as the SON, while *Sim2* is expressed only in a small subset of cells in the PVN and not the SON (50, 70). The simplest explanation for why *Sim2* mutants have no assessable PVN defect could be compensation by *Sim1*. This hypothesis would argue for their similarity in action despite the differences in their C-terminal sequences. Alternatively, *Sim2*-positive PVN cells are affected in the mutant, but the crude histological analysis was not able to reveal the deficiency of only a small group of cells. If *Sim2* acts in some aspects of PVN development, it likely does not operate in the VP and the OT lineages, as the same precursor population gives rise to the VP and OT neurons in the SON and *Sim2* is not expressed in the SON. The challenge to finding a PVN-related phenotype for the *Sim2* mutant is then to identify a marker(s) for the *Sim2*-positive cell population for molecular assessment. Recent comparative expression data obtained in my laboratory indicates that the *Sim2* expression pattern is highly similar to and overlaps with those of the TRH and SS cell populations, but not to those of the CRH, VP and OT cell populations (unpublished). Currently, we are investigating whether there is indeed a defect related to the first two types of neurons in the *Sim2* mutant.

Since both *Sim1* and *Sim2* mutants are lethal, *Sim1/Sim2* double heterozygous animals were generated and evaluated for physiological defects. The double heterozygotes were measured to be no more obese than the *Sim1* heterozygotes in the first 7 months of their lives (70), suggesting that *Sim2* is not likely to contribute to the obesity pathway in which *Sim1* acts upon. However, unless other combinations of compound mutants were generated by Cre-mediated temporal- and tissue-specific gene inactivation bypassing their embryonic requirement, one cannot be absolutely certain that *Sim2* has no contribution at all. Keep in mind, however, both genes may still have essential roles postnatal and their inactivation may provoke death regardless of the time of gene inactivation. Moreover, no other physiological and behavioral tests have been conducted on the doubly heterozygous animals (or even *Sim2* heterozygotes) to assess other possible defects. Even if *Sim2* plays no role at all in the PVN, it may still function in other regions of the CNS. The other regions of *Sim1* and *Sim2* co-expression in the brain are the basal plate of pre-tegmentum, the mammillary body of the hypothalamus and the zona limitans. Analysis of these regions in the double mutant is necessary to determine the defects, if any.

12. FUTURE STUDIES

Co-expression, co-precipitation and identical mutant phenotype data together indicate that Sim1 and ARNT2 are a bona fide bHLH-PAS protein pair that orchestrates the developmental program of the PVN and SON, joining the other two genetically-defined in vivo pairs of bHLH-PAS dimers: ARNT/HIF-1α regulating the hypoxia response, and CLOCK/BMAL1 regulating circadian rhythms. Unlike HIF-1α/ARNT and AHR/ARNT, little is known how the *Sim* genes function to elicit their downstream target genes that directly mediate their particular functional output. Thus, identification of the interaction partners that dictate their transcriptional activities as well as their downstream target genes will be the key to the future understanding of their actions in mammals. In light of the fact they all have the PAS domain, a domain that has been shown to interact with exogenous chemical ligands in the case of AHR, they are suitable and accessible molecular targets for small molecule stimulation or inhibition. These future presumptive ligands may be the key to modulate the activities of the bHLH-PAS proteins and rebalance bodily homeostasis clinically, be it weight, appetite, learning, or memory.

ACKNOWLEDGEMENTS

The author would like to thank Mr. John Lovejoy for his critical reading and extensive correction of this chapter.

REFERENCES

1. Delabar, J. M., D. Theophile, Z. Rahmani, Z. Chettouh, J. L. Blouin, M. Prieur, B. Noel, and P. M. Sinet. 1993. Molecular mapping of twenty-four features of Down syndrome on chromosome 21. *Eur. J. Hum. Genet.* 1:114-24.
2. Chen, H., R. Chrast, C. Rossier, A. Gos, S. E. Antonarakis, J. Kudoh, A. Yamaki, N. Shindoh, H. Maeda, S. Minoshima, et al. 1995. Single-minded and Down syndrome. *Nature Genetics* 10:9-10.
3. Dahmane, N., G. Charron, C. Lopes, M. L. Yaspo, C. Maunoury, L. Decorte, P. M. Sinet, B. Bloch, and J. M. Delabar. 1995. Down syndrome-critical region contains a

gene homologous to Drosophila sim expressed during rat and human central nervous system development. *Proc. Natl. Acad. Sci. USA* 92:9191-9195.

4. Crews, S. T. 1998. Control of cell lineage-specific development and transcription by bHLH-PAS proteins. *Genes Dev.* 12:607-620.

5. Crews, S. T., and C.-M. Fan. 1999. Remembrance of things PAS: regulation of development by bHLH-PAS proteins. *Curr. Opin. Genet. Dev.* 9:580-587.

6. Ema, M., M. Morita, S. Ikawa, M. Tanaka, Y. Matsuda, O. Gotoh, Y. Saijoh, H. Fujii, H. Hamada, and Y. Fujii-Kuriyama. 1996. Two new members of murine *Sim* gene family are transcriptional repressors and show different expression patterns during mouse embryogenesis. *Mol. Cell. Biol.* 16:5865-5875.

7. Fan, C.-M., E. Kuwana, A. Bulfone, C. F. Fletcher, N. G. Copeland, N. A. Jenkins, S. Crews, S. Martinez, L. Puelles, J. L. R. Rubenstein, et al. 1996. Expression patterns of two murine homologs of *Drosophila single-minded* suggest possible roles in embryonic patterning and in the pathogenesis of Down Syndrome. *Mol. Cell. Neuro.* 7:1-16.

8. Moffett, P., M. Dayo, M. Reece, M. K. McCormick, and J. Pelletier. 1996. Characterization of msim, a murine homologue of the Drosophila sim transcription factor. *Genomics* 35:144-55.

9. Yamaki, A., S. Noda, J. Kudoh, N. Shindoh, H. Maeda, S. Minoshima, K. Kawasaki, Y. Shimizu, and N. Shimizu. 1996. The mammalian single-minded (SIM) gene: mouse cDNA structure and diencephalic expression indicate a candidate gene for Down syndrome. *Genomics* 35:136-43.

10. Reeves, R. H., N. G. Irving, T. H. Moran, A. Wohn, C. Kitt, S. S. Sisodia, C. Schmidt, R. T. Bronson, and M. T. Davisson. 1995. A mouse model for Down syndrome exhibits learning and behaviour deficits. *Nature Genetics* 11:177-184.

11. Chrast, R., H. S. Scott, H. Chen, J. Kudoh, C. Roisser, S. Minoshima, Y. Wang, N. Shimizu, and S. E. Antonarakis. 1997. Cloning of two human homologs of the *Drosophila single-minded* gene SIM1 on Chromosome 6q and SIM2 on 21q within the Down Syndrome chromosomal region. *Genome Res.* 7:615-624.

12. Chrast, R., H. S. Scott, R. Madani, L. Huber, D. P. Wolfer, M. Prinz, A. Aguzzi, H. P. Lipp, and S. E. Antonarakis. 2000. Mice trisomic for a bacterial artificial chromosome with the single-minded 2 gene (Sim2) show phenotypes similar to some of those present in the partial trisomy 16 mouse models of Down syndrome. *Hum. Mol. Genet.* 9:1853-64.

13. Probst, M. R., C.-M. Fan, M. Tessier-Lavigne, and O. Hankinson. 1997. Two murine homologs of the *Drosophila* Single-minded protein that interact with the mouse aryl hydrocarbon receptor nuclear translocator protein. *J. Biol. Chem.* 272:4451-4457.

14. Swanson, H. I., W. K. Chan, and C. A. Bradfield. 1995. DNA binding specificities and pairing rules of the Ah receptor, ARNT, and SIM proteins. *J. Biol. Chem.* 280:26292-26302.

15. Jain, S., E. Maltepe, M. M. Lu, C. Simon, and C. A. Bradfield. 1998. Expression of ARNT, ARNT2, HIF1 alpha, HIF2 alpha and Ah receptor mRNAs in the developing mouse. *Mech. Dev.* 73:117-23.

16. Ikeda, M., and M. Nomura. 1997. cDNA cloning and tissue-specific expression of a novel basic helix-loop-helix/PAS protein (BMAL1) and identification of alternatively spliced variants with alternative translation initiation site usage. *Biochem. Biophys. Res. Commun.* 233:258-64.

17. Michaud, J. L., T. Rosenquist, N. R. May, and C.-M. Fan. 1998. Development of neuroendocrine lineages requires the bHLH-PAS transcription factor SIM1. *Genes Dev.* 12:3264-3275.

18. Sonnenfeld, M., M. Ward, G. Nystrom, J. Mosher, S. Stahl, and S. Crews. 1997. The *Drosophila tango* gene encodes a bHLH-PAS protein that is orthologous to mammalian Arnt and controls CNS midline and tracheal development. *Development* 124:4583-4594.

19. Moffett, P., and J. Pelletier. 2000. Different transcriptional properties of mSim-1 and mSim-2. *FEBS Lett.* 466:80-86.

20. de Turckheim, M. C., J. M. Clavert, and M. Paira. 1991. Costal exostoses, complicated in the neonatal period, by brachial plexus paralysis. A distinct entity of exostoses? *Ann. Pediatr. (Paris)* 38:23-5.

21. Ema, M., S. Ikegami, T. Hosoya, J. Mimura, H. Ohtani, K. Nakao, K. Inokuchi, M. Katsuki, and Y. Fujii-Kuriyama. 1999. Mild impairment of learning and memory in mice overexpressing the mSim2 gene located on chromosome 16: an animal model of Down's syndrome. *Hum. Mol. Genet.* 8:1409-1415.

22. Epstein, D. J., L. Martinu, J. L. Michaud, K. M. Losos, C. Fan, and A. L. Joyner. 2000. Members of the bHLH-PAS family regulate shh transcription in forebrain regions of the mouse CNS. *Development* 127:4701-9.

23. Belloni, E., M. Muenke, E. Roessler, G. Traverso, J. Siegel-Bartelt, A. Frumkin, H. F. Mitchell, H. Donis-Keller, C. Helms, A. V. Hing, et al. 1996. Identification of Sonic hedgehog as a candidate gene responsible for holoprosencephaly. *Nat. Genet.* 14:353-6.

24. Roessler, E., E. Belloni, K. Gaudenz, P. Jay, P. Berta, S. W. Scherer, L. C. Tsui, and M. Muenke. 1996. Mutations in the human Sonic Hedgehog gene cause holoprosencephaly. *Nat. Genet.* 14:357-60.

25. Kandel, E. R., J. H. Scwartz, and T. M. JEssell. 2000. Principles of Neural Science, 4th ed, vol. Ch. 49. McGraw-Hill, New York, New York.

26. Du Vigneaud, V. 1995. Hormones of posterior pituitary glands: Oxytocin and vasopressin. *Harvey Lecture Ser.* L:1-26.

27. Sawchenko, P. E., T. Imaki, and W. Vale. 1992. Co-localization of neuroactive substances in the endocrine hypothalamus. *Ciba Found. Symp.* 168:16-30; discussion 30-42.

28. Swanson, L. W. 1986. Organization of mammalian neuroendocrine system, p. 317-363. In V. B. Mountcastle, F. E. Bloom, and S. R. Geiger (ed.), Handbook of Physiology, Section 1: Ther nervous system, Vol. IV, Intrinsic regulatory systems of brain. American Physiology Society, Bethesda, MD.

29. Swanson, L. W., and P. E. Sawchenko. 1983. Hypothalamic integration: organization of the paraventricular and supraoptic nuclei. *Annu. Rev. Neurosci.* 6:269-324.

30. Altman, J., and S. A. Bayer. 1978. Development of the diencephalon in the rat. II. Correlation of the embryonic development of the hypothalamus with the time of origin of its neurons. *J. Comp. Neurol.* 182:973-93.

31. Burgunder, J. M., and T. Taylor. 1989. Ontogeny of thyrotropin-releasing hormone gene expression in the rat diencephalon. *Neuroendocrinology* 49:631-40.

32. Hyodo, S., C. Yamada, T. Takezawa, and A. Urano. 1992. Expression of provasopressin gene during ontogeny in the hypothalamus of developing mice. *Neuroscience* 46:241-50.

33. Jing, X., A. K. Ratty, and D. Murphy. 1998. Ontogeny of the vasopressin and oxytocin RNAs in the mouse hypothalamus. *Neurosci. Res.* 30:343-9.

34.   Karim, M. A., and J. C. Sloper. 1980. Histogenesis of the supraoptic and paraventricular neurosecretory cells of the mouse hypothalamus. *J. Anat.* 130:341-7.

35.   Keegan, C. E., J. P. Herman, I. J. Karolyi, K. S. O'Shea, S. A. Camper, and A. F. Seasholtz. 1994. Differential expression of corticotropin-releasing hormone in developing mouse embryos and adult brain. *Endocrinology* 134:2547-55.

36.   Muglia, L., L. Jacobson, P. Dikkes, and J. A. Majzoub. 1995. Corticotropin-releasing hormone deficiency reveals major fetal but not adult glucocorticoid need. *Nature* 373:427-32.

37.   Nishimori, K., L. J. Young, Q. Guo, Z. Wang, T. R. Insel, and M. M. Matzuk. 1996. Oxytocin is required for nursing but is not essential for parturition or reproductive behavior. *Proc. Natl. Acad. Sci. USA* 93:11699-704.

38.   Okamura, H., K. Fukui, E. Koyama, H. L. Tsutou, T. Tsutou, H. Terubayashi, H. Fujisawa, and Y. Ibata. 1983. Time of vasopressin neuron origin in the mouse hypothalamus: examination by combined technique of immunocytochemistry and [3H]thymidine autoradiography. *Brain Res.* 285:223-6.

39.   Schmale, H., and D. Richter. 1984. Single base deletion in the vasopressin gene is the cause of diabetes insipidus in Brattleboro rats. *Nature* 308:705-9.

40.   Seasholtz, A. F., S. A. Bourbonaiser, C. E. Harnden, and S. A. Camper. 1991. Nucleotide sequence and expression of mouse corticotropin releasing hormone gene. *Mol. Cell. Neurosci.* 2:266-273.

41.   Shiosaka, S., K. Takatsuki, M. Sakanaka, S. Inagaki, H. Takagi, E. Senba, Y. Kawai, H. Iida, H. Minagawa, Y. Hara, et al. 1982. Ontogeny of somatostatin-containing neuron system of the rat: immunohistochemical analysis. II. Forebrain and diencephalon. *J. Comp. Neurol.* 204:211-24.

42.   Nakai, S., H. Kawano, T. Yudate, M. Nishi, J. Kuno, A. Nagata, K. Jishage, H. Hamada, H. Fujii, K. Kawamura, et al. 1995. The POU domain transcription factor Brn-2 is required for the determination of specific neuronal lineages in the hypothalamus of the mouse. *Genes Dev.* 9:3109-21.

43.   Schonemann, M. D., A. K. Ryan, R. J. McEvilly, S. M. O'Connell, C. A. Arias, K. A. Kalla, P. Li, P. E. Sawchenko, and M. G. Rosenfeld. 1995. Development and survival of the endocrine hypothalamus and posterior pituitary gland requires the neuronal POU domain factor Brn-2. *Genes Dev.* 9:3122-35.

44.   Li, P., X. He, M. R. Gerrero, M. Mok, A. Aggarwal, and M. G. Rosenfeld. 1993. Spacing and orientation of bipartite DNA-binding motifs as potential functional determinants for POU domain factors. *Genes Dev.* 7:2483-96.

45.   Anderson, M. G., G. L. Perkins, P. Chittick, R. J. Shrigley, and W. A. Johnson. 1995. *drifter*, a *Drosophila* POU-domain transcription factor, is required for correct differentiation and migration of tracheal cells and midline glia. *Genes & Dev.* 9:123-137.

46.   Michaud, J. L., C. DeRossi, N. R. May, B. C. Holdener, and C. M. Fan. 2000. ARNT2 acts as the dimerization partner of SIM1 for the development of the hypothalamus. *Mech Dev.* 90:253-261.

47.   Hosoya, T., Y. Oda, S. Takahashi, M. Morita, S. Kawauchi, M. Ema, M. Yamamoto, and Y. Fujii-Kuriyama. 2001. Defective development of secretory neurones in the hypothalamus of Arnt2-knockout mice. *Genes Cells* 6:361-74.

48.   Keith, B., D. M. Adelman, and M. C. Simon. 2001. Targeted mutation of the murine arylhydrocarbon receptor nuclear translocator 2 (Arnt2) gene reveals partial redundancy with Arnt. *Proc. Natl. Acad. Sci. USA* 98:6692-7.

49. Acampora, D., M. P. Postiglione, V. Avantaggiato, M. Di Bonito, F. M. Vaccarino, J. Michaud, and A. Simeone. 1999. Progressive impairment of developing neuroendocrine cell lineages in the hypothalamus of mice lacking the Orthopedia gene. *Genes Dev.* 13:2787-800.

50. Wang, W., and T. Lufkin. 2000. The murine Otp homeobox gene plays an essential role in the specification of neuronal cell lineages in the developing hypothalamus. *Dev. Biol.* 227:432-49.

51. Simeone, A., M. R. D'Apice, V. Nigro, J. Casanova, F. Graziani, D. Acampora, and V. Avantaggiato. 1994. Orthopedia, a novel homeobox-containing gene expressed in the developing CNS of both mouse and Drosophila. *Neuron* 13:83-101.

52. Holder, J. L., N. F. Butte, and A. R. Zinn. 2000. Profound obesity associated with a balanced translocation that disrupts the SIM1 gene. *Hum. Mol. Genet.* 9:101-8.

53. Gilhuis, H. J., C. M. van Ravenswaaij, B. J. Hamel, and F. J. Gabreels. 2000. Interstitial 6q deletion with a Prader-Willi-like phenotype: a new case and review of the literature. *Eur. J. Paediatr. Neurol.* 4:39-43.

54. Turleau, C., G. Demay, M. O. Cabanis, G. Lenoir, and J. de Grouchy. 1988. 6q1 monosomy: a distinctive syndrome. *Clin. Genet.* 34:38-42.

55. Villa, A., M. Urioste, J. M. Bofarull, and M. L. Martinez-Frias. 1995. De novo interstitial deletion q16.2q21 on chromosome 6. *Am. J. Med. Genet.* 55:379-83.

56. Michaud, J. L., F. Boucher, A. Melnyk, F. Gauthier, E. Goshu, E. Levy, G. A. Mitchell, J. Himms-Hagen, and C. M. Fan. 2001. Sim1 haploinsufficiency causes hyperphagia, obesity and reduction of the paraventricular nucleus of the hypothalamus. *Hum. Mol. Genet.* 10:1465-73.

57. Elmquist, J. K., C. F. Elias, and C. B. Saper. 1999. From lesions to leptin: hypothalamic control of food intake and body weight. *Neuron* 22:221-32.

58. Rosenbaum, M., R. L. Leibel, and J. Hirsch. 1997. Obesity. *N. Engl. J. Med.* 337:396-407.

59. Schwartz, M. W., S. C. Woods, D. Porte, Jr., R. J. Seeley, and D. G. Baskin. 2000. Central nervous system control of food intake. *Nature* 404:661-71.

60. Sawchenko, P. E. 1998. Toward a new neurobiology of energy balance, appetite, and obesity: the anatomists weigh in. *J. Comp. Neurol.* 402:435-41.

61. Leibowitz, S. F., N. J. Hammer, and K. Chang. 1981. Hypothalamic paraventricular nucleus lesions produce overeating and obesity in the rat. *Physiol. Behav.* 27:1031-40.

62. Sims, J. S., and J. F. Lorden. 1986. Effect of paraventricular nucleus lesions on body weight, food intake and insulin levels. Behav. *Brain Res.* 22:265-81.

63. Tokunaga, K., M. Fukushima, J. W. Kemnitz, and G. A. Bray. 1986. Comparison of ventromedial and paraventricular lesions in rats that become obese. *Am. J. Physiol.* 251:R1221-7.

64. Weingarten, H. P., P. K. Chang, and T. J. McDonald. 1985. Comparison of the metabolic and behavioral disturbances following paraventricular- and ventromedial-hypothalamic lesions. *Brain Res. Bull.* 14:551-9.

65. Kotz, C. M., M. K. Grace, J. Briggs, A. S. Levine, and C. J. Billington. 1995. Effects of opioid antagonists naloxone and naltrexone on neuropeptide Y-induced feeding and brown fat thermogenesis in the rat. Neural site of action. *J. Clin. Invest.* 96:163-70.

66. Cowley, M. A., N. Pronchuk, W. Fan, D. M. Dinulescu, W. F. Colmers, and R. D. Cone. 1999. Integration of NPY, AGRP, and melanocortin signals in the hypothalamic paraventricular nucleus: evidence of a cellular basis for the adipostat. *Neuron* 24:155-63.

67.   Kotz, C. M., C. F. Wang, J. E. Briggs, A. S. Levine, and C. J. Billington. 2000. Effect of NPY in the hypothalamic paraventricular nucleus on uncoupling proteins 1, 2, and 3 in the rat. *Am. J. Physiol. Regul. Integr. Comp. Physiol.* 278:R494-8.

68.   Wright, D. E., K. B. Seroogy, K. H. Lundgren, B. M. Davis, and L. Jennes. 1995. Comparative localization of serotonin1A, 1C, and 2 receptor subtype mRNAs in rat brain. *J. Comp Neurol.* 351:357-73.

69.   Mercer, J. G., C. B. Lawrence, and T. Atkinson. 1996. Regulation of galanin gene expression in the hypothalamic paraventricular nucleus of the obese Zucker rat by manipulation of dietary macronutrients. *Brain Res. Mol. Brain Res.* 43:202-8.

70.   Goshu, E., H. Jin, R. Fasnacht, M. Sepenski, J. L. Michaud, and C. M. Fan. 2002. Sim2 mutants have developmental defects not overlapping with those of Sim1 mutants. *Mol. Cell. Biol.* 22:4147-57.

71.   Shamblott, M. J., E. M. Bugg, A. M. Lawler, and J. D. Gearhart. 2002. Craniofacial abnormalities resulting from targeted disruption of the murine Sim2 gene. *Dev. Dyn.* 224:373-80.

72.   Giampietro, P. F., C. L. Raggio, and R. D. Blank. 1999. Synteny-defined candidate genes for congenital and idiopathic scoliosis. Am. J. Med. Genet. 83:164-77.

73.   Porter, D. E., and A. H. Simpson. 1999. The neoplastic pathogenesis of solitary and multiple osteochondromas. *J. Pathol.* 188:119-25.

Chapter 10

PAS PROTEINS IN THE MAMMALIAN CIRCADIAN CLOCK

John B. Hogenesch and Steve A. Kay
The Genomics Institute of the Novartis Research Foundation, San Diego, CA, 92121; The Scripps Research Institute, San Diego, CA,92037

1. INTRODUCTION

Many aspects of physiology are regulated in a temporal manner by an internal clock that anticipates the time of day and directs physiological processes accordingly (1). In mammals, cellular processes such as cholesterol and heme biosynthesis, temperature and hormonal rhythms, and the sleep wake cycle are all "timed" by the circadian clock to occur at appropriate time periods in the absence of environmental cues (2). Remarkably, in mammals these processes are controlled by a small nucleus in the hypothalamus, the suprachiasmatic nucleus (SCN), which harbors the "master" oscillator, or core clock. This master oscillator, in turn, synchronizes peripheral clocks that regulate physiology in a tissue- and time-dependent manner. This timekeeping mechanism is composed of three essential elements: (i) a light input pathway that allows for resetting with a changing environment, (ii) a core oscillator that keeps track of time, and (iii) the output of the clock that dictates circadian controlled physiology (Figure 1). The past thirty years of study has elucidated many components of the core circadian oscillator, including a basic model for clock mechanism in flies and mammals (3). In fact, the first clues to mammalian circadian mechanism began with "P" in PAS and the study of the fruit fly.

In pioneering work in the late 1960s, Seymour Benzer and colleagues were seeking to test an important new hypothesis, that genes could encode behavior. They did so in the model system, *Drosophila melanogaster*, and focused their research on the study of behavioral timing and the relationship to the light:dark cycle. Eclosion, the emergence of the larvae from their pupal cases, was known to occur preferentially in the morning (4). Flies were also known to be more active during the hours surrounding dusk and dawn, times when food and water sources are more plentiful. After several attempts, a simple test was devised that could measure locomotor activity over time, by placing flies in a tube with a mechanical sensor to measure their movement inside the tube. A genetic screen taking advantage of this highly quantitative measure of activity, culminated in the identification of several fly lines that had long, short, and even arrhythmic locomotor behaviors (5). They named the locus, *period* (*per*), after the length of activity onset, near 24 hours for wild type flies, but significantly shorter for *per^s* (short), or longer for *per^l* (long), or arrhythmic as in case of the *per^0* flies (5).

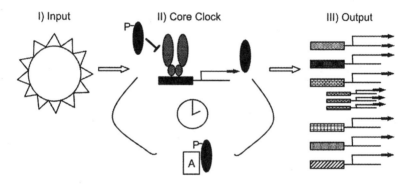

Figure 1. A schematic of the circadian system. I) A light input system signals to the clock resetting the oscillator to a precise 24-hour clock each day. II) The core oscillator, a transcriptional/translational feedback loop composed of transcriptional activators (in grey), repressors (black), and accessory factors (A). The transcriptional activators drive transcription of their target genes, which include repressor proteins. Repressor proteins are synthesized, and are usually post-translationally modified (P) to regulate stability. Accessory factors function to modify, translocate, and otherwise refine the oscillator. III) Transcriptional outputs of the clock include many target genes. Some of these are themselves transcription factors, with their own specific set of target genes.

Unfortunately, it would take nearly a decade and the development of molecular cloning techniques to identify the molecular nature of the *per* deficits. Two efforts, one led by Jeff Hall and Michael Rosbash at Brandeis University, and the other led by Michael Young at the Rockefeller would use the phenotypic assays of Konopka and Benzer as well as genetic and molecular techniques to physically map the *per* locus to the X chromosome. Subsequently, they were able to positionally clone the *per* locus, which revealed a gene that encoded a very long polypeptide of 1250 amino acids

(6, 7). With very little in the way of sequence analysis tools at the time, the knowledge of the molecular sequence of *per* shed little light on the nature of the circadian clock. However, the very existence of the deduced Per amino acid sequence as well as its functional role would prove critical for decoding the molecular clockworks of both flies and mammals.

2. FROM HUMBLE ORIGINS

In this era of fully sequenced genomes, it's difficult to appreciate the historical impact of molecular biology, sequence analysis tools, the internet, and public sequence repositories. These elements, however, were critical in defining protein domains—regions of amino acid similarity that oftentimes share analogous function—that contributed greatly to our understanding of circadian rhythms (and many other disciplines). Subsequent to the cloning of *per*, the *Drosophila* protein Sim, derived from the *single-minded* locus, and the mammalian factor ARNT were identified and all three proteins were found to share a region of approximately 275 amino acids termed the PAS domain (for Per, Arnt, and Sim) (8). Further sequence analysis revealed that two direct repeats of approximately forty amino acids, the PAS-A repeat and PAS-B repeat, were subdomains of the larger PAS domain, and the latter was followed by a region of enhanced conservation termed the PAC domain referring to the C-terminal end of the PAS domain. Furthermore, careful sequence analysis of the "A" and "B" repeats showed that they occurred in proteins from other phyla. For example, bacterial two-component signal transduction pathways utilize sensor kinase proteins. Certain sensor kinase proteins, such as Fixl, which binds and utilizes heme to detect molecular oxygen levels in nitrogen-fixing bacteria, harbor these short repeats. In fact, this shorter domain also occurs in other circadian signal transduction pathways; the *Neurospora crassa* protein White Collar 1, for example, harbors PAS repeats as well as the PAC signatures (9).

Additional clues to the functioning of the circadian oscillator were present in the structures of the other PAS members, Sim and ARNT. These proteins harbored a second domain known as the basic-helix-loop-helix (bHLH) domain N-terminal to their PAS domain (10). The HLH domain had earlier been identified as a hallmark of a family of transcription factors that regulated proliferation and differentiation. These bHLH proteins formed homodimers or heterodimers with their HLH domains, and bound DNA in a sequence dependent fashion with their basic amino acid regions. In fact, certain HLH family members such as the repressor protein Id lack the basic region and function as repressors by interfering with the formation of functional DNA binding and trans-activating complexes. The identification of a second bHLH-PAS protein, the aryl hydrocarbon receptor (Ahr), and subsequent studies showed that it formed a heterodimeric complex with Arnt to elicit responses to levels of xenobiotic metabolizing enzymes in response to polyaromatic hydrocarbons (11, 12). The observation that Sim, Ahr, and Arnt, therefore, were DNA binding transcription factors, while Per had apparently no such capacity, suggested that Per may function as a transcriptional repressor.

Genetics and experimental biology were also contributing to the understanding of circadian mechanisms in the fly. Later, after the initial description of *per*, it was noted that the mRNA of *per* was expressed in a rhythmic manner, with a period length corresponding to the genotype of the *per* locus (13). For example, *per^s* flies with a short locomotor activity period also had a short period in the rhythmic expression of their mRNA. *per^l* flies had long period mRNA expression, and *per^0* flies did not have an appreciable rhythm for *per* message. Subsequently, it was observed that Per protein levels also had a corresponding rhythm (14). It was also noticed that additional copies of the *per* locus led to higher Per protein levels and shorter period lengths (15). These observations suggested that *per* was acting to regulate its own expression. Later, the Weitz and Young laboratories independently identified a second component, *timeless (tim)*, of the fly clock by further forward genetics and protein-protein interaction studies (16, 17). The Per and Tim proteins formed dimers that translocated from the cytoplasm in the day to the nucleus at night to negatively regulate their own transcriptional levels (16). Their protein decay was mediated in part by the phosphorylation status of Per and Tim, as both were progressively phosphorylated and degraded in the late subjective night (18, 19). Specifically, Tim accumulation and heterodimerization blocked phosphorylation and degradation of Per, thereby allowing the Per/Tim complex proteins to repress their own transcription. While this negative feedback loop was being investigated, clues to the identity of the positive elements of the circadian axis in flies came from analysis of the *per* and *tim* promoters. A short piece of DNA harboring an E-box element was shown to confer cycling to a reporter gene (20). It would require the description of the mammalian constituents of the bHLH-PAS family to further the model for fly rhythms.

3. THE CORE CLOCK

3.1 The Central Feedback Loop

The completion of the mouse and human genome projects and development of sensitive mining techniques has completed the cast of mammalian bHLH-PAS proteins. This has revealed a family of transcription factors numbering twenty-two with several distinct roles (Figure 2). These twenty-two include response regulator-like proteins such as Ahr (and the related Ahrr), which sense polyaromatic hydrocarbons, HIF-1α (and the related HIF-2α and HIF-3α), which regulate transcription in response to low oxygen levels (reviewed in (10) and chapters by Yao et al. and Semenza in this volume). As Arnt functions as a critical component of both the dioxin receptor complex as well as the hypoxia inducible complex, it has been termed a "general" factor. Several coactivators have likewise been identified, including steroid receptor coactivator 1 (and the related Tif2 and Rac3), which was identified based on its ability to coactivate nuclear hormone receptor complexes (21). In addition, three orthologs of the *Drosophila* Per protein were identified in mammals, Per1, Per2, and Per3, each of which is without the N-terminal bHLH domains found in other family members (22-26). Thus, bHLH-PAS proteins could play distinct functional roles, and participate in varied signal transduction pathways that regulate responses to environmental change.

Forward genetics in the fly revealed Per as the first protein associated with regulation of circadian behavior. Unfortunately, amino acid homology between the *Drosophila* Per protein and its mammalian orthologs was low - hampering their eventual identification more than a decade after *Drosophila* Per was characterized. In the interim, Joseph Takahashi's laboratory decided to apply the powerful tool of forward genetics to the study of mouse circadian biology. In collaboration with William Dove, ethyl nitrosourea (ENU) mutagenized mice were screened for period length deficits on running wheels. Serendipitously, the twenty-fifth mouse screened displayed a long period phenotype and was designated *Clock* (27). Despite screening nearly five hundred additional mice, they did not find another mutant. The *Clock* mouse was identified by virtue of its long period of locomotor activity, 25 hours, in constant darkness. When the *Clock* mice were bred to homozygosity, their period length of locomotor activity got even longer, 27 hours, before the animals no longer consolidated their locomotor activity patterns and went arrhythmic. At the time, positional cloning in the mouse was a daunting task - the genome sequence had not yet begun and few well-spaced markers were available.

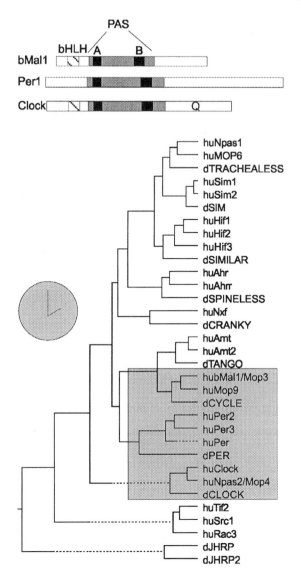

Figure 2. (Top) A schematic of the domain structure of Bmal1, Per1, and Clock. Protein sequences were analyzed by PFAM searches. The bHLH domain, responsible for interaction and DNA binding, is depicted by a hatched box. The PAS domain is shaded in grey, while the characteristic "A" and "B" repeats are in black and indicated. The "Q" in Clock indicates the transcriptionally active glutamine-rich domain, which is deleted in the *Clock/Clock* mutant mouse by mis-splicing. (Bottom) Phylogeny of the PAS family of transcription factors. All protein sequences were obtained from Genbank, the PAS domains were aligned using the Clustal algorithm, and a phylogeny was constructed using the Megalign program (DNAstar, Madison, WI). PAS family members implicated in *Drosophila* or mammalian clock function are shaded and indicated by a clock-face.

Nevertheless, Takahashi's group persevered and in a heroic effort identified the *Clock* locus and its protein product - a bHLH-PAS protein homologous to another bHLH-PAS protein, Npas2/Mop4 (28, 29). The mutation resulted in mis-splicing of the protein product and deletion of a C-terminal exon encoding poly-glutamine (Q). These Q-rich regions had been previously shown to support transactivation properties in other transcription factor families, and coupled with *Clock's* intact bHLH and PAS domains and heterozygous phenotype, suggested that *Clock* may function as an antimorph (30).

Because bHLH-PAS proteins function as obligate heterodimers (Arnt binds DNA only as a reluctant homodimer), the search was on for Clock's partner. Two groups led by Chris Bradfield and Chuck Weitz applied protein-protein interaction screening (the so-called yeast two hybrid) and independently identified Brain and muscle Arnt-like protein1 (Bmal1) or Member of PAS domain family 3 (Mop3) as the partner of Clock (31, 32). To no one's surprise, Bmal1 was also a bHLH-PAS protein, first identified just two months prior to the identification of Clock (33, 34). Later, *Bmal1's* critical role in the circadian clock was cemented by the observation that mice deficient in *Bmal1* levels were completely arrhythmic in constant darkness (35). Bradfield's group further characterized the Clock/Bmal1 complex in vitro, using a selection technique to trap the E-box enhancer bound by the complex. This observation was important because Paul Hardin had earlier demonstrated that a short sequence of the *per* promoter in flies harbored an E-box sequence critical for rhythmicity in flies (33). Weitz's group went on to show that Clock/Bmal1 complexes regulated E-box sequences present in the promoter regions of the *mPer1* locus (31). Both groups, then arrived at a similar conclusion, that Clock and Bmal1 activated expression of the *mPer1* locus, which somehow fed back and inhibited the activity of the Clock/Bmal1 complex with a period length of nearly 24 hours.

Startlingly, the positive arm of the circadian clock was conserved over nearly six hundred million years of evolution, between flies and mice. Shortly after the discovery of the mouse Clock protein, the fly version was identified by biochemical and genetic means. Simultaneously, Jeff Hall and Michael Rosbash's groups had been characterizing additional loci identified by chemical mutagenesis and forward genetics in the fly. Two mutants, *Jrk* and *Cycle*, had been identified, and when positional cloning was complete, it was apparent that *Jrk* was the *Drosophila Clock* and *Cycle* was the *Drosophila Bmal1/Mop3* (36-38). Importantly, both *Jrk* and *Cycle* phenotypes reflected that of their mammalian orthologs—both *Jrk/+* or *Cycle/+* flies were long period, while *Jrk/Jrk* or *Cycle/Cycle* flies were arrhythmic (36, 37). Importantly, *per* levels were reduced in both *Jrk* (or *dClock*) and *Cycle* flies, consistent with the notion that these genes directly transactivated the genes encoding their repressors, Per and Tim. The Kay group used cell-based assays in *Drosophila* S2 cells to show that dClock activated the E-box present in the *per* and *tim* structural genes, while the *Jrk* mutant version of dClock activated much more weakly, consistent with its putative role as a hypomorph (38). Furthermore, they demonstrated that dClock and dCycle heterodimerized like their mammalian versions, and activated transcription on E-box driven reporter genes. These efforts, therefore, closed the (or a) circadian loop in flies.

While the circadian loop was closed in flies, the loop was still very much open in mammals. The role of the Per orthologs in repression of the circadian clock in mammals was proving more complicated than initially hypothesized. The mammalian *Per* genes, *mPer1* and *mPer2*, were definitively implicated in circadian clock function when their compound deletion was shown to cause arrhythmic locomotor activity in the mouse (39). However, biochemically, they could only partially repress Clock/Bmal1 complex activity. A partial solution to this problem came from an unexpected angle of incidence. Cryptochrome (Cry) proteins had been hypothesized to be components of the light input pathway in mammals (40). They had been previously shown in plants to function as blue light photoreceptors and participate in light resetting of the plant circadian system (41, 42). Furthermore, a mutation in the fly *cryptochrome* gene called *cry^{baby}* resulted in deficits in light resetting of the oscillator in flies (43). In addition, the fly *cryptochrome* was also shown to play a role in light resetting of the circadian oscillator (44). Van der Horst and colleagues sought to test whether the two mammalian cryptochrome genes, *Cry1* and *Cry2*, likewise played a role in light resetting of the circadian oscillator in mammals and generated mice deficient in one or both genes (45). Surprisingly, when either gene was missing, mice displayed either a short period (*mCry1*) or long period (*mCry2*) locomotor activity rhythm, while retaining apparently normal phase shifting responses to light (45). Deletion of both *mCry1* and *mCry2* resulted in mice with arrhythmic locomotor activity patterns (45).

These data supported the roles of cryptochromes in mammals as components of the oscillator itself, in addition to a putative role in the light entrainment pathway. But how? The answer came with analysis of Clock/Bmal1 transcriptional repression by the Per and cryptochrome proteins—cryptochromes were found to be very potent transcriptional repressors even in the absence of the Per proteins (46, 47). They were also shown to cycle in the SCN in the same phase as the mPer proteins, and were down-regulated in *Clock/Clock* mutant mice, implying that they were also direct targets of the Clock/Bmal1 complex (46, 47). Unlike the fly cryptochrome, mCry activity was light independent. In short, cryptochromes had become repressors, while the Per proteins played an as yet unknown role.

Their role was partially revealed based on analysis of *Bmal1* levels in mammalian circadian mutants. In wild type animals, *Bmal1* levels displayed a consistent circadian rhythm with peak levels opposite to those of the *mPer* genes (48). In addition, levels of Bmal1 were altered in *Clock*, *Per2*, and *cryptochrome* deficient mice - suggesting that regulation of *Bmal1* levels was critical in the normal functioning of the oscillator (49). Levels were lower in *Clock* mutant mice - consistent with the hypothesis that Clock and Bmal1 indirectly activated *Bmal1* transcription (50). Paradoxically, the levels of *Bmal1* were also lower in *cryptochrome* mutant mice, suggesting genetic activation by the cryptochromes as well (51). Finally, levels of *Bmal1* (and the *per* and *cry* genes) were higher in $Per2^{Brdm}$ mice - suggesting that Per2 was functioning as an activator, rather than repressor, of these target genes (51). More importantly, the observation of a second feedback loop in the regulation of mammalian rhythms emphasized the complexity of mammalian circadian regulation.

A more complete understanding on the functioning of the cryptochrome and Per proteins in the mammalian clock came from biochemical analyses. First, many components of the mammalian core clock undergo daily rhythms in phosphorylation, localization, and activity (52). The Per proteins appear to play a critical role in modulating the localization of the cryptochromes, thereby gating the activity of the repressor complex on the positive activators Clock/Bmal1. They also may be required to maintain appropriate phosphorylation of the cryptochromes by casein kinases (53, 54). In the nucleus, the repressor complex consisting of cryptochromes, per proteins, and casein kinases, interacts with Clock/Bmal1 on DNA to quench the normal transcriptional output of the complex (52). This may occur through modulation of the p300/Clock/Bmal1 complex, resulting in changes in histone H3 acetylation, RNA polymerase II recruitment, and concomitant activation (55, 56). The exact mechanism by which this cryptochrome-containing complex inhibits p300 dependent aceytltransferase activity is as yet undetermined.

3.2   Beyond the PAS

Accessory factors have also been shown to play important roles in circadian clock function in flies and mammals. In flies, three protein kinases, Doubletime (Discs overgrown; mammalian casein kinase 1 epsilon) and Casein Kinase 2α phosphorylate Per, while Shaggy (mammalian glycogen synthase kinase 3), promotes Tim phosphorylation (19, 57, 58). While Dbt results in destabilization of Per, Shaggy regulates nuclear entry of the Per/Tim complex.   Two additional *Drosophila* transcription factors, Vrille and Pdp1, also function in the fly clock by directly activating fly *Clock* and constituting yet another feedback loop (59-61).   Collectively, these proteins function to refine and modulate circadian clock function in the fly by regulating aspects such as nuclear entry (and therefore function) of Per/Tim complexes as well as their protein levels (stability), or by transcriptionally regulating key components (*Vrille* and *Pdp1*).

A number of accessory factors have also been identified in mammals including the mammalian ortholog of Doubletime, casein kinase 1ε (62).   A naturally occurring mutation in hamsters resulting in short locomotor activity (*tau*) has been identified as casein kinase 1ε, confirming its role in the function of the mammalian clockworks (62, 63).   Furthermore, characterization of a human advanced phase sleep syndrome revealed a mutation in *Per2* causing less efficacious phosphorylation by casein kinase 1ε in vitro (64).   A highly homologous ortholog, casein kinase 1δ, may be involved as well, as both proteins have been shown to progressively phosphorylate both the Per and cryptochrome proteins in mammals (52). More recently, the involvement of several nuclear hormone receptor family members has been investigated in mammals.   RARα and RXRα have been shown to interact with Clock and negatively regulate Clock/Bmal1 transcriptional activity (65).   Importantly, these studies showed retinoic acid was capable of phase shifting *Per2* mRNA rhythms in the periphery (65). Earlier work had shown that glucocorticoids were capable of phase-shifting the molecular rhythms of *Per* and *cry* messages in fibroblast cells (discussed later) (66).   More recently, Schibler and colleagues have shown that REV-ERBα functions as a component of the mammalian clock (67).   As a trans-repressor, it is a central determinant of *Bmal1's* circadian expression (antiphasic to the *per* and *cry* genes).   In addition, *Rev-erbα* expression is regulated by Clock/Bmal1 and is circadian.   Finally, mice deficient in *Rev-erbα* have a short period length of locomotor activity rhythms.   These data definitively demonstrate its critical role in clock function.   A schematic of the functioning of the core clock is depicted in Figure 3.

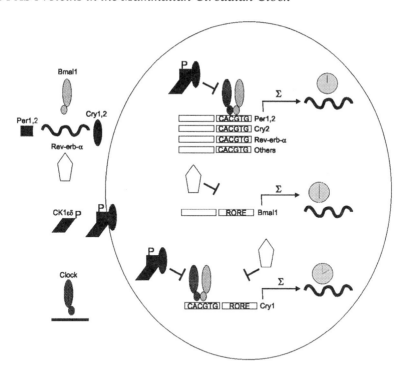

Figure 3. Current understanding of mammalian core (SCN) clock mechanism. Clock and Bmal1 activate transcription through CACGTG containing E-box response elements present in target structural genes such as the circadian-regulated repressors *Cry1,2*, *Per1,2*, and *REV-ERBα*. As their transcript levels rise, Cry1,2 and Per1,2 proteins are synthesized, and complex with Casein kinase 1δ and ε. This "repressor-some" complex translocates to the nucleus, where it inhibits the transcriptional capacity of Clock/Bmal1 heterodimers, resulting in decreasing levels of their own transcripts. Subsequently, as repressor protein levels fall, the complex degrades, and Clock/Bmal1 transcriptional activity is stimulated. A second loop involving the nuclear hormone receptor family member REV-ERBα has recently been characterized (67). REV-ERBα is a trans-repressor, which binds to ROR elements (RORE) present in the *Bmal1* and *Cry1* structural genes. This results in cycling levels of *Bmal1* that are higher when REV-ERBα levels are low, due to the regulation of *Rev-erbα* by the Clock/Bmal1/repressor-some loop. More recently, *Cry1* has been found to have both RORE and E-box regulatory elements, providing a novel mechanism to regulate phase control (56). Clock/Bmal1 E-box regulated promoters generally peak at CT6 (SCN), while the REV-ERBα/RORE regulated *Bmal1* peaks 12-hours out of phase. A sigma (Σ) denotes that phase control of circadian-regulated target genes is a result of complex interactions of enhancer sequences. The phase of expression of cycling genes is indicated by a clock-dial, with CT0 defined as 6 am.

4.   PHOTIC INPUT TO THE CLOCK

As mentioned previously, light input to circadian oscillators is critical to the daily resetting of circadian clocks to a precise 24 hour period. Light is perceived by the eye in mammals, travels down the retino-hypothalamic tract to the suprachiasmatic nucleus where depending on the luminance levels and time of day it can either have no effect, phase advance, or phase delay the clock (68-70).   There, glutaminergic and PACAP neurotransmission appears to be critical for the resetting of the circadian oscillator (71-74).   In fact, PACAP signaling through the VPAC2 receptor appears to be required for maintenance of normal circadian rhythmicity, in addition to entrainment (75, 76).   Much recent attention has focused on the photoreceptors involved in clock function in mammals.   Whereas Cry functions as both a photoreceptor and clock component in flies, deletion of either of the mammalian *cry* genes had only a very modest effect on resetting of the oscillator (if any at all) (45).   Surprisingly, visual photoreceptors, conventional rods and cones, also appear to have a redundant role in circadian photoentrainment as certain visually blind people and mice have apparently normal responses to light (77, 78).   Recently, physiological and genetic support has been generated supporting a role for melanopsin in non-imaging forming photoperception (78-81).   Melanopsin deficient animals have an attenuated phase shifting response to light (79, 80).   These studies indicate that melanopsin may be the primary circadian photoreceptor in mammals, however it is also likely that other photoreceptors play a role as well - for example, rods and cones.   A preliminary conclusion is that mammalian circadian photoperception, like that of plants and flies, utilizes multiple photoperceptive systems. This may be a result of the importance of light as an environmental signal, or a reflection of the different photoreceptors required to efficiently capture the varying light qualities and intensities in the natural environment.

A definitive mechanism by which light information transmitted to the SCN is communicated to the clock is as yet unclear.   Early investigation focused on the involvement of immediate early genes, such as *Fos*, *Jun*, and *Zif268* (82-84).   These transcription factors were excellent candidates because they responded differentially to very brief pulses of light, largely ignoring it during the subjective day, and responding more vigorously in the early or late subjective night. These observations were further supported by the use of antisense technology. By injecting antisense (and appropriate control oligos) against *c-Fos* and *JunB*, phase shifting in response to light was abrogated (85). However, reverse genetic experiments in mice lacking *Fos* failed to confirm its required role as a mediator of light information to the clock (86). Recently, genetic support for the role of *CREB* has emerged, as a mouse phosphorylation mutant of *CREB* fails to phase shift in response to light (87). The circuit, then, would be as follows: PACAP and glutamate neurotransmission at the cell surface stimulates the cAMP pathway, which in turn promotes protein kinase A (PKA) activation and CREB phosphorylation, resulting in an increase in CREB-dependent transcription. The observation that the known clock components, mPer1 and mPer2, responded acutely to light at appropriate times during the subjective night has suggested that they may function as the endpoint in the light circuit to the clock. Although genetic support for this hypothesis exists, it does not rule out as yet unidentified factors also playing a role in communicating light information to components of the core clock.

5.   PERIPHERAL CLOCKS AND OUTPUT

Having established a transcriptional/translational feedback loop utilizing bHLH-PAS proteins as the driving force of circadian rhythms, the important question of how these circadian changes in gene expression translated into observed daily variations in physiology remained. This problem was compounded by the complexity of higher eukaryotic physiology—circadian regulation of liver function was certain to be different from that in SCN. A key observation explaining this phenomenon came once again from the study of the fly. Using a luciferase-based reporter assay, transgenic flies were found to have functioning, independent oscillators in not just their heads, but in their bodies as well (88). Shortly afterwards, it was noted that many mammalian clock genes, such as the *per* genes, displayed robust rhythms in peripheral organs of mammals as well (26, 89). Notably, the timing of the oscillations were several hours delayed with respect to the SCN for the *per* and *cry* genes. These observations suggested the presence of peripheral clocks, and argued for a role of the SCN as their master conductor.

This hypothesis was greatly strengthened by the observations of Schibler and colleagues. By placing rat fibroblasts into culture medium and subjecting them to serum shock, they could "start" the endogenous oscillator present in fibroblasts resulting in robust rhythms in *per* gene transcription (90). Not only would these clocks start, but the rhythms were near 24-hour, and persisted for several days post-shock, and could be phase shifted by the administration of certain glucocorticoids (91). Because these studies were done entirely ex-vivo, this result implied that fibroblasts harbored their own independent oscillator, which drove circadian transcriptional rhythms. Extending these observations, transgenic mice were fashioned where reporter genes were driven by circadian promoter sequences. By placing tissues derived from these mice in culture and monitoring reporter gene activity, robust circadian rhythms were observed in heart, lung, liver, and other peripheral tissues that persisted for several days in culture (89). The kinetics were slightly different, and various tissues would "regress to the median" faster than others, but collectively these results implied that many (if not most) tissues and cells harbored their own autonomous circadian clocks. If these oscillators were truly independent from the core SCN clock, would it be possible to decouple them with the appropriate stimuli? Strangely, yes, by restricting feeding to the subjective day period, it was possible to phase shift the liver clock, while the SCN clock maintained normal nocturnal locomotor activity consolidation (92).

Two *Clock* and *Cycle* orthologs, *Npas2/Mop4* and *Bmal2/Mop9*, respectively, may play an important role in peripheral clock function (33, 93, 94). These molecules share many biochemical properties with their central clock siblings, including heterodimerization, DNA binding, and transcriptional activation with their cognate partners. While *Clock* and *Bmal1* are clearly required for maintenance of normal circadian function, deletion of *Npas2* did not lead to an observable circadian phenotype in locomotor activity (95). Furthermore, the absolute requirement of *Bmal1* for maintenance of circadian locomotor activity suggests that *Bmal2/Mop9* is not sufficient for this critical central clock output. However, rhythmic expression of *Per2* message was abolished in *Npas2* knockout animals, suggesting that *Npas2* functions in the clockworks of the forebrain oscillator. In addition, the DNA binding activity of Npas2, Clock, and Bmal1 is apparently altered by a change in redox state of NAD cofactors, suggesting that change in redox state (by simple neuronal activity) may reset both the core and peripheral oscillators (96). More recently, Npas2 has been shown to harbor a heme prosthetic group, which may confer carbon monoxide-responsive DNA binding capabilities to this protein (97). Further work will be required to cement these provocative hypotheses, however modulation of clock function by cellular metabolism (including redox state) seems likely.

The discovery and characterization of peripheral clocks set the stage for the important link between chronobiology and transcriptional output. Because the circadian clock is a coupled transcriptional/translational feedback loop, and because peripheral organs maintained robust circadian oscillations in clock genes, other circadian output genes could be discovered that could determine the observed tissue-specific chronobiology. The first large-scale screens for circadian-expressed genes in mammals utilized differential display technologies, and identified tens of known and unknown genes with circadian patterns of steady state mRNA levels (98, 99). The advent and use of DNA arrays greatly facilitated the discovery process, allowing for truly genome scale analysis of circadian transcription. Several of these studies were published concomitantly, and several basic rules emerged (reviewed in (100)). First, depending on the experimental design and execution, anywhere from 3-10% of genes in a given tissue were found to have circadian expression patterns, a surprisingly large number. Second, the vast majority of all circadian transcriptional output was tissue-specific; this implied that relatively few genes cycled everywhere, notably including the known clock components. Analysis of circadian deficient mouse mutants suggested that relatively few of the newly found genes were direct targets of known components of the clock, *Clock* and *Bmal1*, consistent with the observation that relatively few genes cycled everywhere. Instead, it seems likely that circadian transcription is initiated, and certain targets are themselves transcription factors, which subsequently regulate their distinct target genes. Frequently, key and rate limiting genes were identified with circadian patterns of mRNA expression. For example, Alas1, the rate-limiting enzyme in heme biosynthesis, and HMG-CoA-reductase, the rate-limiting enzyme in cholesterol biosynthesis, were found to have circadian patterns in their steady state mRNA levels. Many genes of unknown function were also found with robust circadian expression patterns, opening the door to the discovery of as yet undescribed circadian biology.

6. SUMMARY

The past ten years has seen an explosion in the understanding of mammalian circadian biology. Key to the understanding of the molecular clockworks, has been the role of PAS transcription factors. The discovery of Per as the charter member of the PAS family, and its key role in *Drosophila* and mammalian clocks, touched off a odyssey of biochemical, genetic, and behavioral studies revealing the clock as a transcriptional/translational feedback loop. Other bHLH-PAS members functioned prominently in this pathway as well, as the subsequent discovery of Clock and Bmal1, revealed them as the positive activators of core clock transcription and provided the important clue to the functioning of the core feedback loop. This basic feedback loop was completed when the Per1 and Per2 proteins were found to interact with the Cry proteins and contribute to the repressor axis of the clock.

But, the story is not yet complete. Quantitative trait analyses (QTL) indicate that more than a dozen other components of the oscillator may yet be discovered (101). While it is improbable that these factors will be PAS members, it is all but certain that they will function in concert with existing clock components to refine the circadian clock. Not only is the current understanding of the core oscillator incomplete, many issues surrounding the input to and output from the oscillator and the roles of PAS members in these processes are unclear as well. Subsequent biochemical, genetic, and behavioral analyses should increase our understanding of these processes, and lead to an increasingly rich model for the control of the sleep-wake cycle, and the many other physiologies under clock control.

REFERENCES

1. Dunlap, J. C. 1999. Molecular bases for circadian clocks. *Cell* 96:271-90.
2. Reppert, S. M., and D. R. Weaver. 2002. Coordination of circadian timing in mammals. *Nature* 418:935-41.
3. Young, M. W., and S. A. Kay. 2001. Time zones: a comparative genetics of circadian clocks. *Nat. Rev. Genet.* 2:702-15.
4. Pittendrigh, C. S. 1967. Circadian systems. I. The driving oscillation and its assay in Drosophila pseudoobscura. *Proc. Natl. Acad. Sci. USA* 58:1762-7.
5. Konopka, R. J., and S. Benzer. 1971. Clock mutants of Drosophila melanogaster. *Proc. Natl. Acad. Sci. USA* 68:2112-6.
6. Reddy, P., W. A. Zehring, D. A. Wheeler, V. Pirrotta, C. Hadfield, J. C. Hall, and M. Rosbash. 1984. Molecular analysis of the period locus in Drosophila melanogaster and identification of a transcript involved in biological rhythms. *Cell* 38:701-10.
7. Bargiello, T. A., F. R. Jackson, and M. W. Young. 1984. Restoration of circadian behavioural rhythms by gene transfer in Drosophila. *Nature* 312:752-4.
8. Huang, Z. J., I. Edery, and M. Rosbash. 1993. PAS is a dimerization domain common to Drosophila period and several transcription factors. *Nature* 364:259-62.
9. Crosthwaite, S. K., J. C. Dunlap, and J. J. Loros. 1997. Neurospora wc-1 and wc-2: transcription, photoresponses, and the origins of circadian rhythmicity. *Science* 276:763-9.
10. Gu, Y. Z., J. B. Hogenesch, and C. A. Bradfield. 2000. The PAS superfamily: sensors of environmental and developmental signals. *Annu. Rev. Pharmacol. Toxicol.* 40:519-61.
11. Burbach, K. M., A. Poland, and C. A. Bradfield. 1992. Cloning of the Ah-receptor cDNA reveals a distinctive ligand-activated transcription factor. *Proc. Natl. Acad. Sci. USA* 89:8185-9.
12. Dolwick, K. M., H. I. Swanson, and C. A. Bradfield. 1993. In vitro analysis of Ah receptor domains involved in ligand-activated DNA recognition. *Proc. Natl. Acad. Sci. USA* 90:8566-70.
13. Hardin, P. E., J. C. Hall, and M. Rosbash. 1990. Feedback of the Drosophila period gene product on circadian cycling of its messenger RNA levels. *Nature* 343:536-40.
14. Zwiebel, L. J., P. E. Hardin, J. C. Hall, and M. Rosbash. 1991. Circadian oscillations in protein and mRNA levels of the period gene of Drosophila melanogaster. *Biochem. Soc. Trans.* 19:533-7.
15. Baylies, M. K., T. A. Bargiello, F. R. Jackson, and M. W. Young. 1987. Changes in abundance or structure of the per gene product can alter periodicity of the Drosophila clock. *Nature* 326:390-2.
16. Gekakis, N., L. Saez, A. M. Delahaye-Brown, M. P. Myers, A. Sehgal, M. W. Young, and C. J. Weitz. 1995. Isolation of timeless by PER protein interaction: defective

interaction between timeless protein and long-period mutant PERL. *Science* 270:811-5.

17. Myers, M. P., K. Wager-Smith, C. S. Wesley, M. W. Young, and A. Sehgal. 1995. Positional cloning and sequence analysis of the Drosophila clock gene, timeless. *Science* 270:805-8.

18. Zeng, H., Z. Qian, M. P. Myers, and M. Rosbash. 1996. A light-entrainment mechanism for the Drosophila circadian clock. *Nature* 380:129-35.

19. Price, J. L., J. Blau, A. Rothenfluh, M. Abodeely, B. Kloss, and M. W. Young. 1998. double-time is a novel Drosophila clock gene that regulates PERIOD protein accumulation. *Cell* 94:83-95.

20. Hao, H., D. L. Allen, and P. E. Hardin. 1997. A circadian enhancer mediates PER-dependent mRNA cycling in Drosophila melanogaster. *Mol. Cell. Biol.* 17:3687-93.

21. Onate, S. A., S. Y. Tsai, M. J. Tsai, and B. W. O'Malley. 1995. Sequence and characterization of a coactivator for the steroid hormone receptor superfamily. *Science* 270:1354-7.

22. Sun, Z. S., U. Albrecht, O. Zhuchenko, J. Bailey, G. Eichele, and C. C. Lee. 1997. RIGUI, a putative mammalian ortholog of the Drosophila period gene. *Cell* 90:1003-11.

23. Tei, H., H. Okamura, Y. Shigeyoshi, C. Fukuhara, R. Ozawa, M. Hirose, and Y. Sakaki. 1997. Circadian oscillation of a mammalian homologue of the Drosophila period gene. *Nature* 389:512-6.

24. Albrecht, U., Z. S. Sun, G. Eichele, and C. C. Lee. 1997. A differential response of two putative mammalian circadian regulators, mper1 and mper2, to light. *Cell* 91:1055-64.

25. Shearman, L. P., M. J. Zylka, D. R. Weaver, L. F. Kolakowski, Jr., and S. M. Reppert. 1997. Two period homologs: circadian expression and photic regulation in the suprachiasmatic nuclei. *Neuron* 19:1261-9.

26. Zylka, M. J., L. P. Shearman, D. R. Weaver, and S. M. Reppert. 1998. Three period homologs in mammals: differential light responses in the suprachiasmatic circadian clock and oscillating transcripts outside of brain. *Neuron* 20:1103-10.

27. Vitaterna, M. H., D. P. King, A. M. Chang, J. M. Kornhauser, P. L. Lowrey, J. D. McDonald, W. F. Dove, L. H. Pinto, F. W. Turek, and J. S. Takahashi. 1994. Mutagenesis and mapping of a mouse gene, Clock, essential for circadian behavior. *Science* 264:719-25.

28. Antoch, M. P., E. J. Song, A. M. Chang, M. H. Vitaterna, Y. Zhao, L. D. Wilsbacher, A. M. Sangoram, D. P. King, L. H. Pinto, and J. S. Takahashi. 1997. Functional identification of the mouse circadian Clock gene by transgenic BAC rescue. *Cell* 89:655-67.

29. King, D. P., Y. Zhao, A. M. Sangoram, L. D. Wilsbacher, M. Tanaka, M. P. Antoch, T. D. Steeves, M. H. Vitaterna, J. M. Kornhauser, P. L. Lowrey, et al. 1997. Positional cloning of the mouse circadian clock gene. *Cell* 89:641-53.

30. King, D. P., M. H. Vitaterna, A. M. Chang, W. F. Dove, L. H. Pinto, F. W. Turek, and J. S. Takahashi. 1997. The mouse Clock mutation behaves as an antimorph and maps within the W19H deletion, distal of Kit. *Genetics* 146:1049-60.

31. Gekakis, N., D. Staknis, H. B. Nguyen, F. C. Davis, L. D. Wilsbacher, D. P. King, J. S. Takahashi, and C. J. Weitz. 1998. Role of the CLOCK protein in the mammalian circadian mechanism. *Science* 280:1564-9.

32. Hogenesch, J. B., Y. Z. Gu, S. Jain, and C. A. Bradfield. 1998. The basic-helix-loop-helix-PAS orphan MOP3 forms transcriptionally active complexes with circadian and hypoxia factors. *Proc. Natl. Acad. Sci. USA* 95:5474-9.

33. Hogenesch, J. B., W. K. Chan, V. H. Jackiw, R. C. Brown, Y. Z. Gu, M. Pray-Grant, G. H. Perdew, and C. A. Bradfield. 1997. Characterization of a subset of the basic-helix-loop-helix-PAS superfamily that interacts with components of the dioxin signaling pathway. *J. Biol. Chem.* 272:8581-93.

34. Ikeda, M., and M. Nomura. 1997. cDNA cloning and tissue-specific expression of a novel basic helix-loop- helix/PAS protein (BMAL1) and identification of alternatively spliced variants with alternative translation initiation site usage. *Biochem. Biophys. Res. Commun.* 233:258-64.

35. Bunger, M. K., L. D. Wilsbacher, S. M. Moran, C. Clendenin, L. A. Radcliffe, J. B. Hogenesch, M. C. Simon, J. S. Takahashi, and C. A. Bradfield. 2000. Mop3 is an essential component of the master circadian pacemaker in mammals. *Cell* 103:1009-17.

36. Rutila, J. E., V. Suri, M. Le, W. V. So, M. Rosbash, and J. C. Hall. 1998. CYCLE is a second bHLH-PAS clock protein essential for circadian rhythmicity and transcription of Drosophila period and timeless. *Cell* 93:805-14.

37. Allada, R., N. E. White, W. V. So, J. C. Hall, and M. Rosbash. 1998. A mutant Drosophila homolog of mammalian Clock disrupts circadian rhythms and transcription of period and timeless. *Cell* 93:791-804.

38. Darlington, T. K., K. Wager-Smith, M. F. Ceriani, D. Staknis, N. Gekakis, T. D. Steeves, C. J. Weitz, J. S. Takahashi, and S. A. Kay. 1998. Closing the circadian loop: CLOCK-induced transcription of its own inhibitors per and tim. *Science* 280:1599-603.

39. Zheng, B., U. Albrecht, K. Kaasik, M. Sage, W. Lu, S. Vaishnav, Q. Li, Z. S. Sun, G. Eichele, A. Bradley, et al. 2001. Nonredundant roles of the mPer1 and mPer2 genes in the mammalian circadian clock. *Cell* 105:683-94.

40. Miyamoto, Y., and A. Sancar. 1998. Vitamin B2-based blue-light photoreceptors in the retinohypothalamic tract as the photoactive pigments for setting the circadian clock in mammals. *Proc. Natl. Acad. Sci. USA* 95:6097-102.

41. Zhong, H. H., A. S. Resnick, M. Straume, and C. Robertson McClung. 1997. Effects of synergistic signaling by phytochrome A and cryptochrome1 on circadian clock-regulated catalase expression. *Plant Cell* 9:947-55.

42. Somers, D. E., P. F. Devlin, and S. A. Kay. 1998. Phytochromes and cryptochromes in the entrainment of the Arabidopsis circadian clock. *Science* 282:1488-90.

43. Stanewsky, R., M. Kaneko, P. Emery, B. Beretta, K. Wager-Smith, S. A. Kay, M. Rosbash, and J. C. Hall. 1998. The cryb mutation identifies cryptochrome as a circadian photoreceptor in Drosophila. *Cell* 95:681-92.

44. Ceriani, M. F., T. K. Darlington, D. Staknis, P. Mas, A. A. Petti, C. J. Weitz, and S. A. Kay. 1999. Light-dependent sequestration of TIMELESS by CRYPTOCHROME. Science 285:553-6.

45. van der Horst, G. T., M. Muijtjens, K. Kobayashi, R. Takano, S. Kanno, M. Takao, J. de Wit, A. Verkerk, A. P. Eker, D. van Leenen, et al. 1999. Mammalian Cry1 and Cry2 are essential for maintenance of circadian rhythms. *Nature* 398:627-30.

46. Kume, K., M. J. Zylka, S. Sriram, L. P. Shearman, D. R. Weaver, X. Jin, E. S. Maywood, M. H. Hastings, and S. M. Reppert. 1999. mCRY1 and mCRY2 are essential components of the negative limb of the circadian clock feedback loop. *Cell* 98:193-205.

47. Griffin, E. A., Jr., D. Staknis, and C. J. Weitz. 1999. Light-independent role of CRY1 and CRY2 in the mammalian circadian clock. *Science* 286:768-71.

48. Honma, S., M. Ikeda, H. Abe, Y. Tanahashi, M. Namihira, K. Honma, and M. Nomura. 1998. Circadian oscillation of BMAL1, a partner of a mammalian clock gene Clock, in rat suprachiasmatic nucleus. *Biochem. Biophys. Res. Commun.* 250:83-7.

49. Shearman, L. P., S. Sriram, D. R. Weaver, E. S. Maywood, I. Chaves, B. Zheng, K. Kume, C. C. Lee, G. T. van der Horst, M. H. Hastings, et al. 2000. Interacting molecular loops in the mammalian circadian clock. *Science* 288:1013-9.

50. Oishi, K., H. Fukui, and N. Ishida. 2000. Rhythmic expression of BMAL1 mRNA is altered in Clock mutant mice: differential regulation in the suprachiasmatic nucleus and peripheral tissues. *Biochem. Biophys. Res. Commun.* 268:164-71.

51. Yu, W., M. Nomura, and M. Ikeda. 2002. Interactivating feedback loops within the mammalian clock: BMAL1 is negatively autoregulated and upregulated by CRY1, CRY2, and PER2. *Biochem. Biophys. Res. Commun.* 290:933-41.

52. Lee, C., J. P. Etchegaray, F. R. Cagampang, A. S. Loudon, and S. M. Reppert. 2001. Posttranslational mechanisms regulate the mammalian circadian clock. *Cell* 107:855-67.

53. Akashi, M., Y. Tsuchiya, T. Yoshino, and E. Nishida. 2002. Control of intracellular dynamics of mammalian period proteins by casein kinase I epsilon (CKIepsilon) and CKIdelta in cultured cells. *Mol. Cell. Biol.* 22:1693-703.

54. Eide, E. J., E. L. Vielhaber, W. A. Hinz, and D. M. Virshup. 2002. The circadian regulatory proteins BMAL1 and cryptochromes are substrates of casein kinase Iepsilon. *J. Biol. Chem.* 277:17248-54.

55. Takahata, S., T. Ozaki, J. Mimura, Y. Kikuchi, K. Sogawa, and Y. Fujii-Kuriyama. 2000. Transactivation mechanisms of mouse clock transcription factors, mClock and mArnt3. *Genes Cells* 5:739-47.

56. Etchegaray, J. P., C. Lee, P. A. Wade, and S. M. Reppert. 2003. Rhythmic histone acetylation underlies transcription in the mammalian circadian clock. *Nature* 421:177-82.

57. Martinek, S., S. Inonog, A. S. Manoukian, and M. W. Young. 2001. A role for the segment polarity gene shaggy/GSK-3 in the Drosophila circadian clock. *Cell* 105:769-79.

58. Lin, J. M., V. L. Kilman, K. Keegan, B. Paddock, M. Emery-Le, M. Rosbash, and R. Allada. 2002. A role for casein kinase 2alpha in the Drosophila circadian clock. *Nature* 420:816-20.

59. Blau, J., and M. W. Young. 1999. Cycling vrille expression is required for a functional Drosophila clock. *Cell* 99:661-71.

60. Glossop, N. R., J. H. Houl, H. Zheng, F. S. Ng, S. M. Dudek, and P. E. Hardin. 2003. VRILLE feeds back to control circadian transcription of Clock in the Drosophila circadian oscillator. *Neuron* 37:249-61.

61. Cyran, S. A., A. M. Buchsbaum, K. L. Reddy, M. C. Lin, N. R. Glossop, P. E. Hardin, M. W. Young, R. V. Storti, and J. Blau. 2003. vrille, Pdp1, and dClock form a second feedback loop in the Drosophila circadian clock. *Cell* 112:329-41.

62. Lowrey, P. L., K. Shimomura, M. P. Antoch, S. Yamazaki, P. D. Zemenides, M. R. Ralph, M. Menaker, and J. S. Takahashi. 2000. Positional syntenic cloning and functional characterization of the mammalian circadian mutation tau. *Science* 288:483-92.

63. Ralph, M. R., and M. Menaker. 1988. A mutation of the circadian system in golden hamsters. *Science* 241:1225-7.

64. Toh, K. L., C. R. Jones, Y. He, E. J. Eide, W. A. Hinz, D. M. Virshup, L. J. Ptacek, and Y. H. Fu. 2001. An hPer2 phosphorylation site mutation in familial advanced sleep phase syndrome. *Science* 291:1040-3.

65. McNamara, P., S. P. Seo, R. D. Rudic, A. Sehgal, D. Chakravarti, and G. A. FitzGerald. 2001. Regulation of CLOCK and MOP4 by nuclear hormone receptors in the vasculature: a humoral mechanism to reset a peripheral clock. *Cell* 105:877-89.

66. Le Minh, N., F. Damiola, F. Tronche, G. Schutz, and U. Schibler. 2001. Glucocorticoid hormones inhibit food-induced phase-shifting of peripheral circadian oscillators. *EMBO J.* 20:7128-36.

67. Preitner, N., F. Damiola, L. Lopez-Molina, J. Zakany, D. Duboule, U. Albrecht, and U. Schibler. 2002. The orphan nuclear receptor REV-ERBalpha controls circadian transcription within the positive limb of the mammalian circadian oscillator. *Cell* 110:251-60.

68. Lucas, R. J., M. S. Freedman, D. Lupi, M. Munoz, Z. K. David-Gray, and R. G. Foster. 2001. Identifying the photoreceptive inputs to the mammalian circadian system using transgenic and retinally degenerate mice. *Behav. Brain Res.* 125:97-102.

69. Rusak, B. 1979. Neural mechanisms for entrainment and generation of mammalian circadian rhythms. *Fed. Proc.* 38:2589-95.

70. McGuire, R. A., W. M. Rand, and R. J. Wurtman. 1973. Entrainment of the body temperature rhythm in rats: effect of color and intensity of environmental light. *Science* 181:956-7.

71. Meijer, J. H., E. A. van der Zee, and M. Dietz. 1988. Glutamate phase shifts circadian activity rhythms in hamsters. *Neurosci. Lett.* 86:177-83.

72. Hannibal, J., J. M. Ding, D. Chen, J. Fahrenkrug, P. J. Larsen, M. U. Gillette, and J. D. Mikkelsen. 1997. Pituitary adenylate cyclase-activating peptide (PACAP) in the retinohypothalamic tract: a potential daytime regulator of the biological clock. *J. Neurosci.* 17:2637-44.

73. Chen, D., G. F. Buchanan, J. M. Ding, J. Hannibal, and M. U. Gillette. 1999. Pituitary adenylyl cyclase-activating peptide: a pivotal modulator of glutamatergic regulation of the suprachiasmatic circadian clock. *Proc. Natl. Acad. Sci. USA* 96:13468-73.

74. Hannibal, J., F. Jamen, H. S. Nielsen, L. Journot, P. Brabet, and J. Fahrenkrug. 2001. Dissociation between light-induced phase shift of the circadian rhythm and clock gene expression in mice lacking the pituitary adenylate cyclase activating polypeptide type 1 receptor. *J. Neurosci.* 21:4883-90.

75. Shen, S., C. Spratt, W. J. Sheward, I. Kallo, K. West, C. F. Morrison, C. W. Coen, H. M. Marston, and A. J. Harmar. 2000. Overexpression of the human VPAC2 receptor in the suprachiasmatic nucleus alters the circadian phenotype of mice. *Proc. Natl. Acad. Sci. USA* 97:11575-80.

76. Harmar, A. J., H. M. Marston, S. Shen, C. Spratt, K. M. West, W. J. Sheward, C. F. Morrison, J. R. Dorin, H. D. Piggins, J. C. Reubi, et al. 2002. The VPAC(2) receptor is essential for circadian function in the mouse suprachiasmatic nuclei. *Cell* 109:497-508.

77. Freedman, M. S., R. J. Lucas, B. Soni, M. von Schantz, M. Munoz, Z. David-Gray, and R. Foster. 1999. Regulation of mammalian circadian behavior by non-rod, non-cone, ocular photoreceptors. *Science* 284:502-4.

78. Menaker, M. 2003. Circadian rhythms. Circadian photoreception. *Science* 299:213-4.

79. Ruby, N. F., T. J. Brennan, X. Xie, V. Cao, P. Franken, H. C. Heller, and B. F. O'Hara. 2002. Role of melanopsin in circadian responses to light. *Science* 298:2211-3.

80. Panda, S., T. K. Sato, A. M. Castrucci, M. D. Rollag, W. J. DeGrip, J. B. Hogenesch, I. Provencio, and S. A. Kay. 2002. Melanopsin (Opn4) requirement for normal light-induced circadian phase shifting. *Science* 298:2213-6.

81. Lucas, R. J., S. Hattar, M. Takao, D. M. Berson, R. G. Foster, and K. W. Yau. 2003. Diminished pupillary light reflex at high irradiances in melanopsin- knockout mice. *Science* 299:245-7.

82. Rusak, B., H. A. Robertson, W. Wisden, and S. P. Hunt. 1990. Light pulses that shift rhythms induce gene expression in the suprachiasmatic nucleus. *Science* 248:1237-40.

83. Kornhauser, J. M., D. E. Nelson, K. E. Mayo, and J. S. Takahashi. 1990. Photic and circadian regulation of c-fos gene expression in the hamster suprachiasmatic nucleus. *Neuron* 5:127-34.

84. Lin, J. T., J. M. Kornhauser, N. P. Singh, K. E. Mayo, and J. S. Takahashi. 1997. Visual sensitivities of nur77 (NGFI-B) and zif268 (NGFI-A) induction in the suprachiasmatic nucleus are dissociated from c-fos induction and behavioral phase-shifting responses. *Brain Res. Mol. Brain Res.* 46:303-10.

85. Wollnik, F., W. Brysch, E. Uhlmann, F. Gillardon, R. Bravo, M. Zimmermann, K. H. Schlingensiepen, and T. Herdegen. 1995. Block of c-Fos and JunB expression by antisense oligonucleotides inhibits light-induced phase shifts of the mammalian circadian clock. *Eur. J. Neurosci.* 7:388-93.

86. Honrado, G. I., R. S. Johnson, D. A. Golombek, B. M. Spiegelman, V. E. Papaioannou, and M. R. Ralph. 1996. The circadian system of c-fos deficient mice. *J. Comp. Physiol. [A]* 178:563-70.

87. Gau, D., T. Lemberger, C. von Gall, O. Kretz, N. Le Minh, P. Gass, W. Schmid, U. Schibler, H. W. Korf, and G. Schutz. 2002. Phosphorylation of CREB Ser142 regulates light-induced phase shifts of the circadian clock. *Neuron* 34:245-53.

88. Plautz, J. D., M. Kaneko, J. C. Hall, and S. A. Kay. 1997. Independent photoreceptive circadian clocks throughout Drosophila. *Science* 278:1632-5.

89. Yamazaki, S., R. Numano, M. Abe, A. Hida, R. Takahashi, M. Ueda, G. D. Block, Y. Sakaki, M. Menaker, and H. Tei. 2000. Resetting central and peripheral circadian oscillators in transgenic rats. *Science* 288:682-5.

90. Balsalobre, A., F. Damiola, and U. Schibler. 1998. A serum shock induces circadian gene expression in mammalian tissue culture cells. *Cell* 93:929-37.

91. Balsalobre, A., S. A. Brown, L. Marcacci, F. Tronche, C. Kellendonk, H. M. Reichardt, G. Schutz, and U. Schibler. 2000. Resetting of circadian time in peripheral tissues by glucocorticoid signaling. *Science* 289:2344-7.

92. Stokkan, K. A., S. Yamazaki, H. Tei, Y. Sakaki, and M. Menaker. 2001. Entrainment of the circadian clock in the liver by feeding. *Science* 291:490-3.

93. Zhou, Y. D., M. Barnard, H. Tian, X. Li, H. Z. Ring, U. Francke, J. Shelton, J. Richardson, D. W. Russell, and S. L. McKnight. 1997. Molecular characterization of two mammalian bHLH-PAS domain proteins selectively expressed in the central nervous system. *Proc. Natl. Acad. Sci. USA* 94:713-8.

94. Hogenesch, J. B., Y. Z. Gu, S. M. Moran, K. Shimomura, L. A. Radcliffe, J. S. Takahashi, and C. A. Bradfield. 2000. The basic helix-loop-helix-PAS protein MOP9 is a brain-specific heterodimeric partner of circadian and hypoxia factors. *J. Neurosci.* 20:RC83.

95. Reick, M., J. A. Garcia, C. Dudley, and S. L. McKnight. 2001. NPAS2: an analog of clock operative in the mammalian forebrain. *Science* 293:506-9.

96. Rutter, J., M. Reick, L. C. Wu, and S. L. McKnight. 2001. Regulation of clock and NPAS2 DNA binding by the redox state of NAD cofactors. Science 293:510-4.

97.   Dioum, E. M., J. Rutter, J. R. Tuckerman, G. Gonzalez, M. A. Gilles-Gonzalez, and S.
       L. McKnight. 2002. NPAS2: a gas-responsive transcription factor. *Science* 298:2385-
       7.

98.   Morris, M. E., N. Viswanathan, S. Kuhlman, F. C. Davis, and C. J. Weitz. 1998. A
       screen for genes induced in the suprachiasmatic nucleus by light. *Science* 279:1544-7.

99.   Kornmann, B., N. Preitner, D. Rifat, F. Fleury-Olela, and U. Schibler. 2001. Analysis
       of circadian liver gene expression by ADDER, a highly sensitive method for the
       display of differentially expressed mRNAs. *Nucleic Acids Res.* 29:E51-1.

100.   Sato, T. K., S. Panda, S. A. Kay, and J. B. Hogenesch. 2003. DNA arrays:
       applications and implications for circadian biology. *J. Biol. Rhythms* 18:96-105.

101.   Shimomura, K., S. S. Low-Zeddies, D. P. King, T. D. Steeves, A. Whiteley, J. Kushla,
       P. D. Zemenides, A. Lin, M. H. Vitaterna, G. A. Churchill, et al. 2001. Genome-wide
       epistatic interaction analysis reveals complex genetic determinants of circadian
       behavior in mice. *Genome Res.* 11:959-80.

Index